This is the first book to provide a thorough and systematic description of the entire subject of particle field holography. The use of holography to study very small objects in a dynamic volume is a technique of importance for scientists and engineers across a variety of disciplines for obtaining information about the size, shape and velocity of small objects such as dust particles, fuel droplets, raindrops, pollen, bubbles, etc. Dr Vikram has made major contributions to the field, and here provides a coherent, comprehensive and self-contained treatment of the theory, practice and applications.

After a brief historical background, the fundamental holographic expressions and notations that will be used throughout the book are introduced. The author then covers the general theory and mathematical formulations of in-line Fraunhofer holography, before moving on to examine system design and practical considerations, image analysis and reconstruction, aberrations, hologram fringe contrast and its enhancement, non-image-plane analysis, velocimetry and high speed holography, and finally the off-axis approach.

The volume is written to satisfy the needs of researchers in the technique, practising engineers dealing with applications, and advanced students in science or engineering departments. All the necessary mathematical formulations, figures and photographs, experimental procedures and results, and literature citations are therefore included.

T0276095

CAMBRIDGE STUDIES IN MODERN OPTICS: 11

Particle field holography

TITLES IN THIS SERIES

1 Interferometry (Second Edition)
W. H. Steel

2 Optical Holography – Principles, Techniques and Applications
P. Hariharan

3 Fabry-Perot Interferometers
G. Hernandez

4 Optical Instabilities
edited by R. W. Boyd, M. G. Raymer and L. M. Narducci

5 Coherence, Cooperation and Fluctuations
edited by F. Haake, L. M. Narducci and D. Walls

6 Holographic and Speckle Interferometry (Second Edition)
R. Jones and C. Wykes

7 Laser Chemical Processing for Microelectronics
edited by K. G. Ibbs and R. M. Osgood

8 Multiphoton Processes
edited by S. J. Smith and P. L. Knight

9 The Elements of Nonlinear Optics
P. N. Butcher and D. Cotter

10 Optical Solitons – Theory and Experiment
edited by J. R. Taylor

11 Particle Field Holography
C. S. Vikram

12 Ultrafast Fiber Switching Devices and Systems
M. N. Islam

Particle field holography

CHANDRA S. VIKRAM
Center for Applied Optics
The University of Alabama in Huntsville

CAMBRIDGE UNIVERSITY PRESS
Cambridge, New York, Melbourne, Madrid, Cape Town, Singapore, São Paulo

Cambridge University Press
The Edinburgh Building, Cambridge CB2 2RU, UK

Published in the United States of America by Cambridge University Press, New York

www.cambridge.org
Information on this title: www.cambridge.org/9780521411271

First published 1992
This digitally printed first paperback version 2005

A catalogue record for this publication is available from the British Library

Library of Congress Cataloguing in Publication data

Vikram, Chandra S., 1950–
Particle field holography / Chandra S. Vikram.
p. cm. – (Cambridge studies in modern optics ; 11)
Includes bibliographical references and index.
ISBN 0-521-41127-0
1. Particles. 2. Holography. I. Title. II. Series.
TA418.78.V55 1992
620.43–dc20 91-30087 CIP

ISBN-13 978-0-521-41127-1 hardback
ISBN-10 0-521-41127-0 hardback

ISBN-13 978-0-521-01830-2 paperback
ISBN-10 0-521-01830-7 paperback

In the memory of my father Dr Pratap Singh and to my mother Mrs Kailash Vati Singh

Chandra S. Vikram

Contents

Foreword

It was with some considerable pleasure that I accepted Chandra Vikram's invitation to prepare a foreword for his book on particle field holography. He has made major contributions to this field – both to its science and its technology. This book is a fine summary of the current state-of-the-art and goes hand-in-hand with the recently published SPIE Milestone Series volume on Selected Papers on Holographic Particle Diagnostics (MS21).[1]

Certainly it is always flattering to be asked to write a foreword to a volume such as this. In this particular case, however, I happen to have a special relationship to the subject matter of particle field holography. I had the good fortune to be one of the co-founders of this particular area of endeavor and have continued to be involved in it from time to time with some modest contributions of my own, but more importantly with some significant work jointly with professional colleagues and with first-rate graduate students as part of their individual Ph.D. theses.

It is hard to believe that this particular application of holography started nearly 30 years ago and in a rather interesting way. As I chronicled in a recent review paper.[2]

> The first suggestions about using lasers in particle and droplet size measurement come from Bernard Silverman and his co-workers at the Air Force Cambridge Research Laboratories in Bedford, Massachusetts in 1962–63. They attempted to use laser shadowgraphs to measure the particle size of a sample of objects. For large objects (large angular subtense at the recording plane, i.e. small distance between the object and the recording plane) the results were not very good but could be of some value, but for smaller objects (small angular subtense, i.e. large distance between the object and the

> recording plane) it worked very badly indeed. . . . The problem was, of course, diffraction since the better the collimation of the incident beam the better the coherence and the more prominent the diffraction effects, and with a laser beam the effects of diffraction are almost perfectly present. These were the early days of laser, and hence, we should not be too critical of these experiments today.
> Silverman and his colleagues turned to a group of us (Bouché, Parrent and myself) at Technical Operations in Burlington, Massachusetts to see if we could make some sense out of these shadowgraphs. Since we spent a good deal of our time worrying about diffraction and partially coherent light it was a natural problem for us to look at. It was, of course, second nature to us to recognize these patterns as Fresnel diffraction patterns associated with the field containing the object – but are they of any value in determining the particle cross-sectional size? The answer is yes if the resulting Fresnel pattern is thought of as an interference pattern formed between the light diffracted by the particle and the uninterrupted light that propagates to the receiving plane.

This understanding then led to the recognition that these Fresnel patterns were, in fact, in-line holograms. An important special condition exists however; if the pattern is recorded in the far-field of the individual particle then the recorded interference pattern that is the hologram is produced by the interference of the background illumination with the far-field (i.e. Fraunhofer) diffraction pattern of the cross-section of the particle. And so this application of holography was born.

Vikram has put together a comprehensive volume on the subject of particle field holography as we understand it today. The general theory is first developed and then system design considerations are explored. Recording conditions and parameters are clearly defined as well as the properties of the reconstructed image. It must be remembered that this method is a two-step imaging method that captures in a single holographic record the information about the cross-section shape of all particles in the volume together with their relative positions in the volume. A stationary image of all these particles can then be produced from the hologram for study and measurement. Other chapters in the book deal with aberrations and their control, hologram fringe-contrast and its enhancement. Also important is the discussion of non-image plane analysis and the use of the off-axis approach.

Finally, in this foreword I would like to pay tribute to all those who have contributed to this field including especially, Chandra Vikram. It was a pleasure for me to work and publish in this field with a variety of outstanding researchers including George Parrent, Edmund Bouché, Bernard Silverman, John Ward, John DeVelis, the late George Reynolds, William Zinky, Glenn Tyler, Paul Dunn, Stephen Cartwright, Ed Boettner, Joe Crane, Julius Knapp, John Zeiss and Phillip Malyak.

Enjoy the world of particle field holography that is explored by this book.

Rochester, New York Brian J. Thompson

1. C. S. Vikram (Ed), *Selected Papers on Holographic Particle Diagnostics*, SPIE Milestone Series Volume MS21, The International Society for Optical Engineering, Bellingham, Washington, 1990.
2. B. J. Thompson, Holographic methods for particle size and velocity measurement—Recent Advances, in *Holographic Optics II: Principles and Applications, Proc. SPIE*, **1136**, 308–26 (1989).

Preface

The study of very small objects in a dynamic volume is of significant importance in modern science and technology. Particles, bubbles, aerosols, droplets, etc. play key roles in processes dealing with nozzles, jets, combustion, turbines, rocket engines, cavitation, fog, raindrops, pollution, and so on. The quantities to be studied are size, shape, velocity, phase, etc. Holography provides an excellent imaging tool with aperture-limited resolution in a volume rather than just in a plane. The purpose of this book is to describe this subject matter under one cover for the first time.

After a brief historical background in Chapter 1, Chapter 2 introduces the fundamental expressions and notations of holography to be used throughout the book. Chapter 3 covers general theory and mathematical formulations of in-line far-field holography. The recording and reconstruction processes are discussed in detail and the relationships obtained are used throughout the book. Chapters 4 and 5 deal with a large number of the design and practical problems often encountered. Aspects like film size and resolution requirements, source coherence requirements, allowable object velocity, hologram recording and processing techniques, image enhancement, ultimate resolution limits and uncertainties are detailed in these chapters. Chapter 6 provides detailed analysis to relate object and image spaces. The aspects covered are location and magnifications for quantitative analysis of the image. Chapter 7 covers third-order aberrations dealing with holography of small objects. Limits due to these aberrations on the ultimate resolution and some ways to control the aberrations are described. First-order aberrations caused by cylindrical test-sections are also covered in this chapter. These concern the experimental situations of tunnels and medical ampules. Chapter 8 describes the role of the hologram fringe contrast on the resolution. Relationships between the object shape and the resolution and methods to improve the contrast are also included. Certain non-image plane analysis techniques are described in Chapter 9. The analysis of the hologram itself, in a misfocused image plane, and in Fourier transform plane are considered along with their role in accurate measurements and automated analysis. Chapter 10 covers developments in the subject for velocimetry and transient phenomena analysis. The in-line Fraunhofer approach is very common in particle field holography due to its simplicity. Nevertheless there are applications and advantages for the off-axis method. Chapter 11 is devoted to the off-axis method dealing with holography of particle fields.

I wrote this book for researchers in the technique, engineers dealing with applications, and as a possible graduate or senior level text. Thus it covers necessary mathematical formulations, wide and up-to-date cited literature, and necessary experimental procedures and results.

Many researchers contributed very valuably to this book through their excellent published work. Particularly, the work of Professor B. J. Thompson and his coworkers has contributed very significantly to my own understanding of the subject. Professor Thompson's review articles have been extremely informative and useful to me.

My special thanks go to Professor H. J. Caulfield for providing several valuable criticisms and inspiration throughout the process, and the opportunity and atmosphere for things such as writing this book.

Mrs Sharron A. Barefoot eased the entire process by typing the manuscript with remarkable efficiency and cheerfulness. Thanks go to my wife Bina, our daughter Preeti and our son Tushar for their patience and encouragement. Special thanks to Bina for her prompt proofreadings during various stages of this project.

Huntsville, Alabama Chandra S. Vikram

Acknowledgements

The author would like to thank the publishers listed below, and the authors, for permission to reproduce figures: *Japanese Journal of Applied Physics* (Figure 5.12); *Journal of Modern Optics* (Figures 5.27 and 6.4); Optical Society of America (Figures 5.14, 5.20, 7.13, 7.14 and 9.9); Society of Photo-Optical Instrumentation Engineers (Figures 4.10, 5.2, 5.21, 5.24, 5.25, 8.4, 8.5, 8.6, 10.1, 10.2, 10.3 and 10.7); Springer-Verlag (Figure 9.6); Wissenschaftliche Verlagsgesellschaft mbH (Figures 4.8, 4.9, 4.11, 4.12, 4.13 and 4.14).

1

Historical background

There are a large number of situations in modern science and technology requiring the study of very small objects in a volume. These small objects, for example, could be fog, sprays, dust particles, burning coal particles, cavitation and other bubbles in water, fuel droplets in a combustion chamber, etc. In practical situations these objects are small and micrometer order resolution is often required. This resolution itself is not a problem with conventional microscopic techniques. However, a serious problem arises due to the need to study a dynamic volume and not just a plane. An imaging system that can resolve a diameter d has a depth of field of only about d^2/λ, where λ is the wavelength of the light used. For $d = 10\ \mu\text{m}$, $\lambda = 0.5\ \mu\text{m}$, the depth of field is only 0.2 mm! Clearly this is not satisfactory.

Light scattering and diffraction methods[1–8] depend on models. These models are often based on assumptions of the object shape, pre-knowledge of the refractive index and other physical parameters. These assumptions are good enough when rapid and mean particle size distribution rather than exact shape and other parameters are needed. An example is size distribution measurements in the ceramic industry.[9] These methods being real-time in nature, are very valuable in particular situations. Some of these methods are described in Chapter 9.

These non-imaging methods are not adequate in a large number of situations of practical interest. In cavitation studies, bubbles in water should be distinguished from dirt and other particles. In marine biology, organisms are identifiable from their shapes. The required inverse processing for the desired shape parameter from the scattering or diffraction data is very difficult if not impossible. In the studies of boundary layers, droplet and bubble formation, solid propellant combustion, spraying nozzles, etc. both the size and shape of the field of interest are required. Obviously some kind of imaging is needed with good resolution capability which is useful in a volume rather than just in a plane.

The historical development in this connection started with the work of Thompson[10,11] dealing with the Fraunhofer diffraction patterns of small objects. In this, collimated quasimonochromatic light is used to illuminate small objects. At sufficient distance, the Fraunhofer diffraction pattern of the object can be observed or recorded. Due to the coherent background, the pattern contains the Fraunhofer diffraction by the object multiplied by a sine function. The sine function depends only upon the position vector in the observation plane, the recording wavelength, and the object-recording plane separation. The coherent backgound can be eliminated by spatial filtering to observe the pure diffraction pattern.[11] In the coherent background method, first the sine term can be used to determine the object distance and then the broad diffraction pattern to determine the object size. In the pure diffraction pattern approach using the spatial filtering, the size can be determined if the distance is already known. Critical study of the diffraction technique was performed by Parrent and Thompson.[12] Silverman, Thompson and Ward[13] applied the technique to study fog droplets using a Q-switched pulsed ruby laser. Thompson[14] provided further insight into the subject. These diffraction techniques can be used for distance and size analysis if the shape is known. Also, the distinguished diffraction pattern must be observed and therefore a small sample volume with very clean optical surfaces is needed. A more detailed discussion and application of the technique is provided in Section 9.1. This approach, being indirect analysis from a pattern, lacks the analysis of the shape and exact size in a distribution.

The recorded photographic distribution of the complex diffraction pattern is basically a Gabor[15] or in-line hologram. Thompson,[14] DeVelis, Parrent and Thompson,[16] Thompson, Parrent, Ward and Justh,[17] Thompson and Ward[18] and Knox[19] demonstrated the image reconstruction and also the negligible effect of the twin image problem of Gabor holography. This special form of Gabor holography, i.e. in-line Fraunhofer (or far-field) holography assumes the object to be at least one far-field away from the recording plane. This assumption means that the misfocused image in the background is practically insignificant. Thus, a simple but very powerful tool for studying small objects was developed. A pulsed laser beam, such as one from a Q-switched ruby laser can be passed through the test

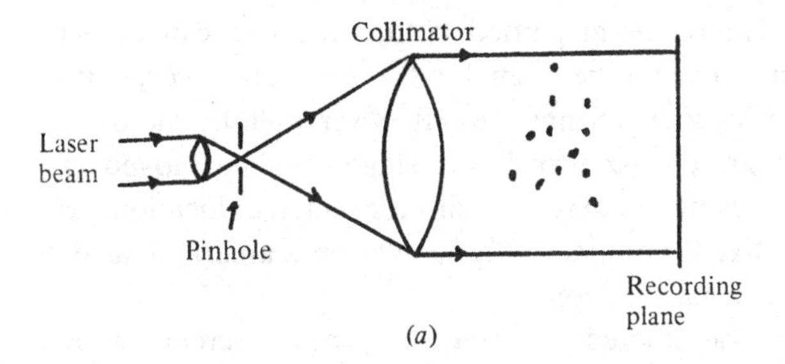

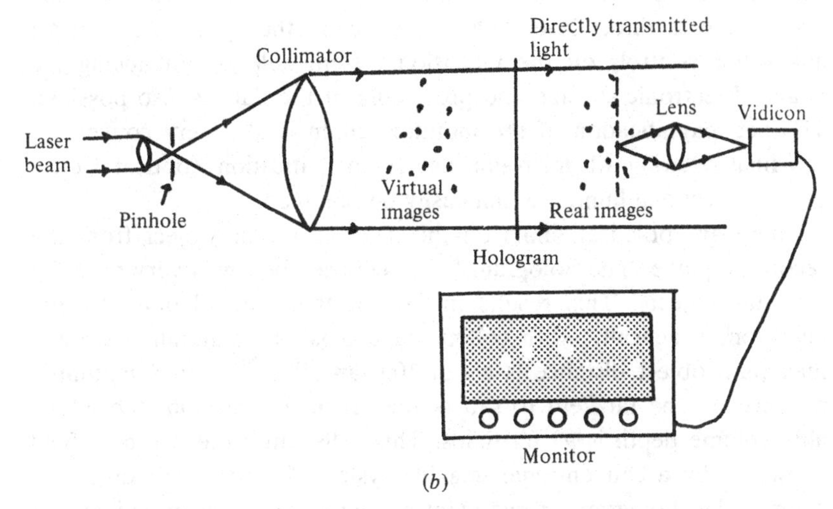

Fig. 1.1 In-line Fraunhofer holography with collimated light. (*a*) Recording. (*b*) Reconstruction with original recording arrangement.

volume freezing the scene information in a hologram. The reconstructed image can be studied in detail later using a continuous wave laser. The early development of the technique has been discussed in detail by DeVelis and Reynolds.[20] Theoretical analysis of the process has also been discussed on different occasions.[20–4]

In the simplest form, the in-line Fraunhofer holographic technique can be described by Figure 1.1. A collimated laser beam pulse illuminates the volume containing the small objects. The hologram between the directly transmitted light and the field diffracted by the objects is recorded on a suitably high resolution photographic emulsion. Upon reconstruction, real, virtual and directly transmitted light is observed. The images are generally observed using a closed circuit television system.[17] The real images are commonly used. A screen can be put on the real image plane for lensless viewing. However, very small images have to be magnified before the eye can detect them. A 10 μm image cannot be seen by the eye. However, with 500× magnification, it becomes 5 mm – easily observable by the eye.

The hologram, on an $x-y-z$ translation stage, can be moved to focus and frame by frame analysis of different image locations is performed. With a fixed hologram the television camera can be moved to perform the same function.

The image can also be studied by projection onto a screen or by using a travelling microscope. However, the closed circuit television system avoids direct laser light exposure on the eyes. The contrast and other controls on the television system help by enhancing the image. Electronic storage and processing of the data is also possible. The net magnification of the monitor screen is also very convenient for final viewing of microobjects. A magnification (optical × electronic) of several hundreds can easily be obtained.

There are practical limits on the distance of the object from the recording plane. The hologram fringe contrast becomes very poor for very far objects. This results in loss of their recordability by the emulsion. Practically, with collimated beams, the generally accepted maximum object distance is about 100 far-fields.[20] With a minimum distance of one far-field to satisfy the far-field condition, the allowable volume depth is 99 far-fields. This is 99 times the depth of field allowable by a conventional imaging system. In fact, with divergent beams, the hologram fringe contrast enhances and the allowable object space depth can be well over 2000 far-fields.[25,26] The techniques available to enhance the hologram fringe contrast are discussed in Chapter 8.

In the presence of both large and small objects, the resolution and the far-field condition may not be met. Also, with increase in the density of the objects, the uninterrupted cross-section of the beam acting as the reference reduces. In these, as well as some other specialized situations (see Chapter 10), a separate reference beam is required. This off-axis or Leith–Upatnieks[27–9] approach was introduced with pulsed lasers by Brooks, Heflinger, Wuerker and Briones.[30] The technique with special reference to particle fields is considered in detail in Chapter 11.

So, with a suitably pulsed laser, a transient or dynamic volume containing very small objects (often called microobjects due to size range generally encountered) can be holographically recorded and reconstructed later. Velocimetry and high speed holography are also possible using suitable multiplexing. Good resolution power and the large depth of field simultaneously resulted in enormous interest in the field of particle field holography. The activity has centered on the application of the method in several physical conditions as well as on the critical study, refinement, special recording and analyzing, etc. of the technique itself. This is evident from a significant number of review articles and extended discussions.[31–47]

The technique has been applied to the study of aerosols, cloud particles, agricultural sprays, cavitation and laser induced breakdown in water, multiphase flow visualization, combustion of solid particles and liquid droplets, bubble chambers, processes in steam turbines and rocket engines, marine plankton behavior, contaminating particles in sealed antibiotic ampules, etc. New applications are periodically being reported. The technique itself is being continuously refined for better resolution, ease of operation, accuracy of measurements, etc.

2

Introduction to holography

Different aspects of imaging by holography have been discussed in detail for example by Collier, Burckhardt, and Lin,[1] Smith,[2] Caulfield[3] and Hariharan[4]. The introduction in this chapter should familiarize the reader with the basic concepts. The chapter also acts as a general reference for the notations and the relationships to be used throughout the book. Readers with knowledge of holography may omit this chapter and use it only for special relationships.

Particular details such as recording emulsion resolution requirements, source coherence requirements, object stability and laser pulse duration, etc. are not considered in this chapter. These are described as the need arises throughout the book.

2.1 Recording and reconstruction with plane holograms

Let us assume that the origin of the Cartesian $(\bar{x}, \bar{y}, z)$ coordinate system is centered in the recording medium. When dealing with a thin plane recording medium such as a thin holographic emulsion, the recording plane can be described by the $(\bar{x}, \bar{y})$ plane as shown in Figure 2.1. The object is defined by the points (x_o, y_o, z_o) in the object space. Similarly, a reference beam, if derived by a point source, can be characterized by the coordinates (x_R, y_R, z_R) of the point. A plane reference wave can be described by $z_R = -\infty$, $x_R/z_R = \sin \phi_{x,R}$ and $y_R/z_R = \sin \phi_{y,R}$ where $\phi_{x,R}$ and $\phi_{y,R}$ are the angles that the reference beam makes with the positive direction of the z-axis in $(\bar{x}, z)$ and $(\bar{y}, z)$ planes respectively. Notice that the object and the reference points are on the LHS of the recording plane in Figure 2.1(a). With our coordinate system z_o is a negative quantity.

The reconstruction arrangement is shown in Figure 2.1(b). The

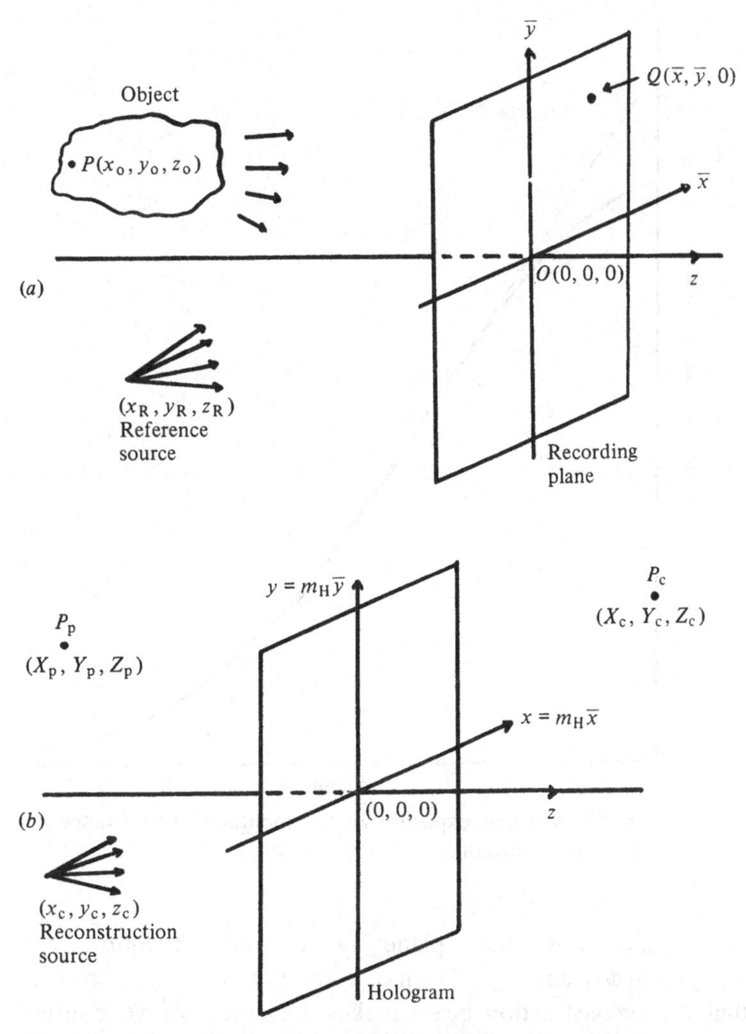

Fig. 2.1 The coordinate system for holography with a thin plane recording medium. An object point $P(x_o, y_o, z_o)$ and a general hologram point $Q(\bar{x}, \bar{y}, 0)$ are shown, (*a*) Recording with source wavelength λ_o. (*b*) Reconstruction arrangement with source wavelength $\lambda_c = n\lambda_o$ and hologram scaling m_H. P_p and P_c represent primary and conjugate image points respectively of the point P.

original hologram is sometimes magnified (or demagnified) by m_H. The coordinate system during the reconstruction (x, y, z) is thus $(m_H\bar{x}, m_H\bar{y}, z)$. In common applications without the hologram scaling, m_H is unity and both coordinate systems become identical. The reconstruction can be provided by the point source situated at

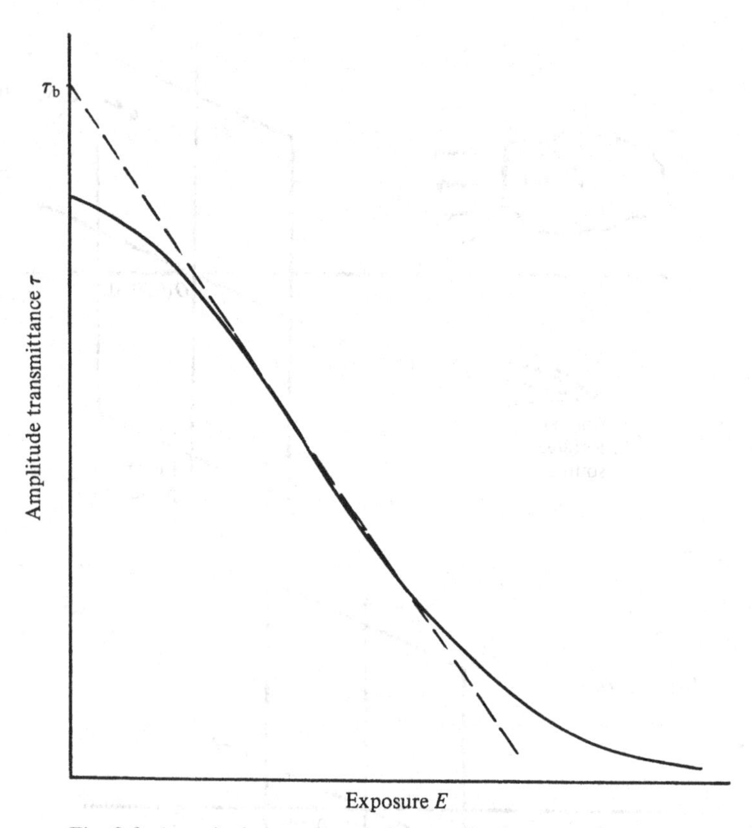

Fig. 2.2 A typical exposure vs the amplitude transmittance curve of a negative emulsion.

(x_c, y_c, z_c). For the plane wave reconstruction, $z_c = -\infty$, $x_c/z_c = \sin\phi_{x,c}$ and $y_c/z_c = \sin\phi_{y,c}$ where $\phi_{x,c}$ and $\phi_{y,c}$ are the angles that the reconstruction beam makes with the positive z-direction in (x, z) and (y, z) planes respectively. Upon reconstruction, two images (called primary and conjugate) are formed. For the object point at P, the primary and the conjugate images are situated at the points P_p and P_c respectively.

The complex light amplitude at the recording plane due to the object and the reference beams can be denoted by $O(\bar{x}, \bar{y})$ and $R(\bar{x}, \bar{y})$ respectively. The irradiance distribution is therefore

$$I(\bar{x}, \bar{y}) = |O(\bar{x}, \bar{y}) + R(\bar{x}, \bar{y})|^2 \tag{2.1}$$

Suppose the path difference between the object and the reference beams all over the recording plane is well within the coherence limits of the source. Also suppose the recording emulsion can resolve the fine fringe pattern formed by the complex interference. The quantities $O(\bar{x}, \bar{y})$ and $R(\bar{x}, \bar{y})$ are assumed to be scalar considering both waves mutually polarized parallel. Lasers with plane polarized light are commonly used. The preferred laser orientation is such that the direction of polarization is perpendicular to the plane containing the propagation directions of object and reference beams.[1]

Under these circumstances, the exposure variation is proportional to the irradiance variation $I(\bar{x}, \bar{y})$. The light amplitude transmittance of the processed negative hologram is

$$T(\bar{x}, \bar{y}) = \tau_b - KI(\bar{x}, \bar{y}) \tag{2.2}$$

where K is a positive film constant. For a positive emulsion, K is negative. τ_b is the intercept on the linear portion of $\tau-E$ (amplitude transmittance–exposure) curve with τ-axis. This linear model is generally accepted for common applications. Equation (2.2) can be verified using Figure 2.2. With thin phase holograms, such as bleached holograms, the phase variation rather than the amplitude transmittance variation is a linear function of $I(\bar{x}, \bar{y})$.[2]

Equations (2.1) and (2.2) yield

$$\begin{aligned} T(\bar{x}, \bar{y}) = \tau_b &- K[|O(\bar{x}, \bar{y})|^2 + |R(\bar{x}, \bar{y})|^2] \\ &- KO(\bar{x}, \bar{y})R^*(\bar{x}, \bar{y}) - KO^*(\bar{x}, \bar{y})R(\bar{x}, \bar{y}). \end{aligned} \tag{2.3}$$

Now, if the reconstruction field at the processed and scaled hologram (x, y) plane is $C(x, y)$, then the amplitude transmitted by the hologram is

$$\begin{aligned} \psi(x, y) &= C(x, y)T(x, y) \\ &= \{\tau_b - K[|O(x/m_H, y/m_H)|^2 \\ &\quad + |R(x/m_H, y/m_H)|^2]\}C(x, y) \\ &\quad - KO(x/m_H, y/m_H)C(x, y)R^*(x/m_H, y/m_H) \\ &\quad - KO^*(x/m_H, y/m_H)C(x, y)R(x/m_H, y/m_H). \end{aligned} \tag{2.4}$$

The reconstructed amplitude of Equation (2.4) basically contains three terms. To describe them briefly in the beginning, let us consider the hologram without scaling ($m_H = 1$) and the same recording and reconstruction beams, i.e. the wavelength ratio $n = 1$ and $C(x, y) = R(x, y)$. Also, let us assume that the absolute value of $R(x, y)$ or $C(x, y)$ does not vary much over the recording plane. This is the practical situation for commonly used spherical waves under

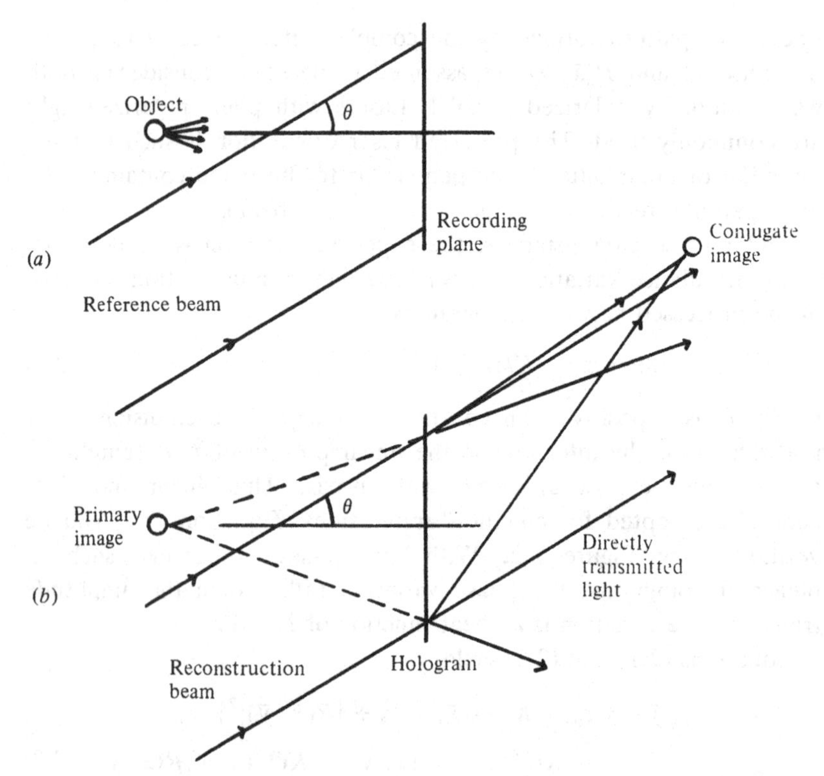

Fig. 2.3 Holography with a collimated reference and reconstruction beams. (*a*) Recording. (*b*) Reconstruction with a beam the same as the original reference beam.

paraxial approximation or plane waves. If the reference beam intensity is kept sufficiently large as compared to the object beam intensity, then

$$|R(x, y)|^2 \gg |O(x, y)|^2. \tag{2.5}$$

In this situation, the first term in the RHS of Equation (2.4) is basically the directly transmitted reconstruction beam described by the amplitude $C(x, y)$ or $R(x, y)$. The second term, which is proportional to $O(x, y)$, is the exact replica of the original object field $O(x, y)$ modulated by some constant. The image is termed 'primary' or 'orthoscopic' and with current assumptions 'virtual'. The third term is proportional to $R^2(x, y)O^*(x, y)$. This image is the complex conjugate of the original field $O(x, y)$. The phase of $O^*(x, y)$ is subtracted from twice the phase of $R(x, y)$. The wave is therefore called phase 'conjugate' to the original $O(x, y)$. This means an inverted wavefront and the image is formed on the opposite side of

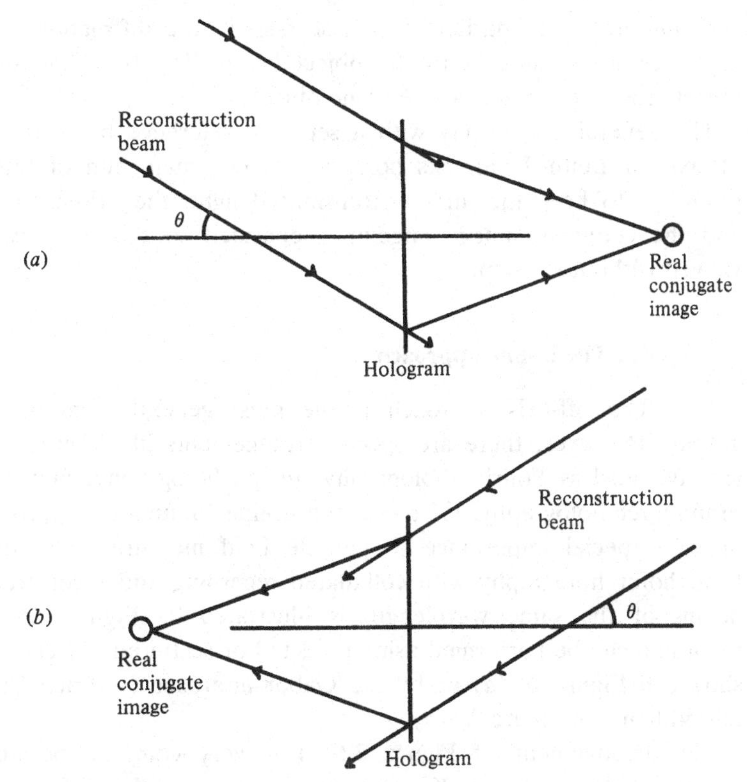

Fig. 2.4 Methods for obtaining distortion-free real conjugate images. (*a*) Reconstruction using a beam which is the mirror image of the reference beam with respect to the hologram. (*b*) Reconstruction with a beam conjugate to the original reference beam.

the hologram from the illuminating source. This image is highly aberrated. A typical recording and reconstruction with plane waves is shown in Figure 2.3. The exposed recording medium, after processing, is replaced in its original position for the image reconstruction.

To obtain a conjugate real image which is aberration-free, two reconstruction geometries are possible. These geometries are shown in Figure 2.4. The first method, as shown in Figure 2.4(*a*), involves the reconstruction beam orientation appearing symmetrical to the reference beam with respect to the optical axis. The second method, as shown in Figure 2.4(*b*), requires the reconstruction beam to be conjugate or antiparallel to the original reference beam. The image coincides with the original subject location.

The conjugate image suffers a depth inversion effect and that is why it is called 'pseudoscopic'. For plane objects, this effect is not very important. In particle field holography, the diffraction by the object cross-section acts as the object beam. The boundary or the object cross-section acts as the plane object.

The general holography with a separate reference beam is called off-axis or Leith–Upatnieks holography. The main aim of this approach is to keep the directly transmitted light, the primary image, and the conjugate image angularly separated so that they can be viewed without overlap.

2.2 The in-line approach

The off-axis approach is the most general form of holography. However, there are special arrangements like Fourier holography, lensless Fourier holography, image holography, and in-line Fraunhofer holography. Of these, the in-line Fraunhofer approach is of very special importance to particle field measurements. In-line Fraunhofer holography with collimated recording and reconstruction beams of the same wavelength is illustrated by Figure 2.5. The recording can be performed using reflected or scattered object light as shown in Figure 2.5(*a*) or by the Gabor approach (diffracted/transmitted light) of Figure 2.5(*b*).

The arrangement of Figure 2.5(*b*) is very common because it involves only one beam. For opaque objects, only the diffraction by the object cross-section acts as the object beam. Consequently, only the object cross-section is reconstructed. This information is sufficient in most particle field holography applications. The reconstruction arrangement is shown in Figure 2.5(*c*). The directly transmitted light and both the images are along the same line. An observer, focusing on one image also sees the out-of-focus twin image and the strong background. Early methods to eliminate the twin image[1] were not very effective and convenient. This limitation of the Gabor holography was ultimately solved by off-axis or Leith–Uptanieks holography.

A special case of in-line approach is when the object is very small satisfying the far-field condition. The effect of the twin image on the reconstruction then becomes very small and the image boundaries are distinct. This aspect resulted in the development of in-line far-field or Fraunhofer holography. Because of its simplicity (just a beam has to be passed through the test volume), the arrangement is very common and is thus the one discussed in most of this book.

The general in-line Fraunhofer holographic arrangement with a

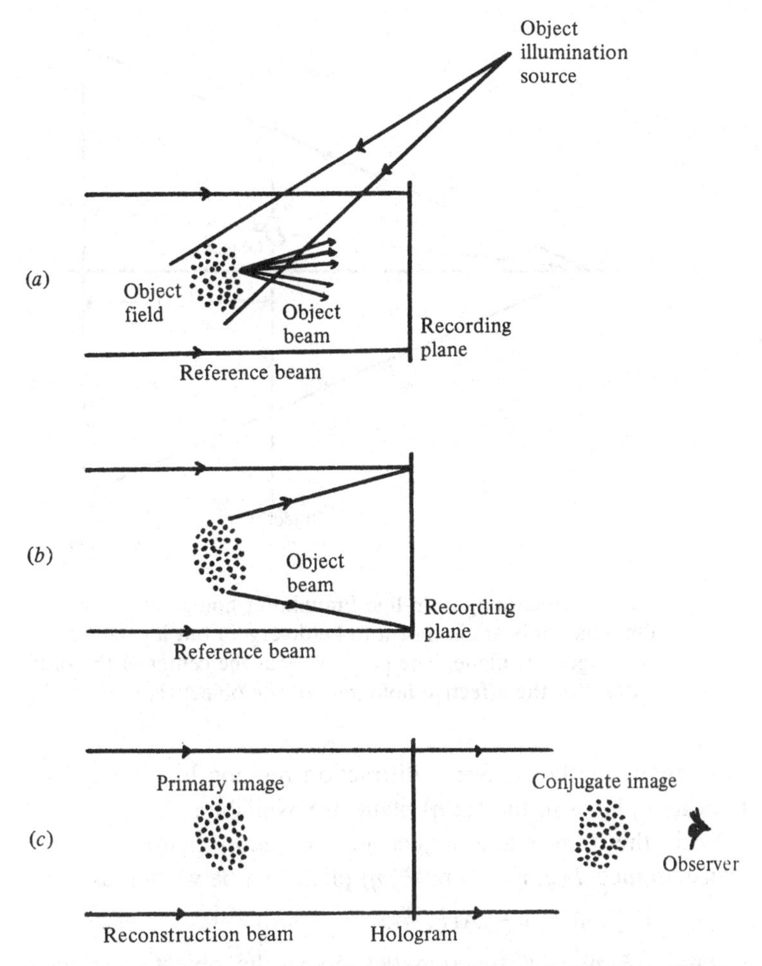

Fig. 2.5 In-line holography with a collimated beam. (*a*) Recording with a scattered or reflected object field. (*b*) Recording with a diffracted/transmitted object field. (*c*) Reconstruction.

back illuminated beam can be described by Figure 2.6. The source at S on the optical axis illuminates the micro-object centered at P. The illuminating beam can be plane or convergent. Suppose the beam diffracts at the object cross-section in the (ξ, η) plane which is parallel to the recording (x, y) plane. This assumption restricts the effective object in the (ξ, η) plane (i.e. thin planar object). The assumption is not necessary from the general holographic point of view. The hologram is still formed and the image is still reconstructed even without this assumption. However, the assumption results in a simpler explanation of several aspects of in-line Fraunhofer holography

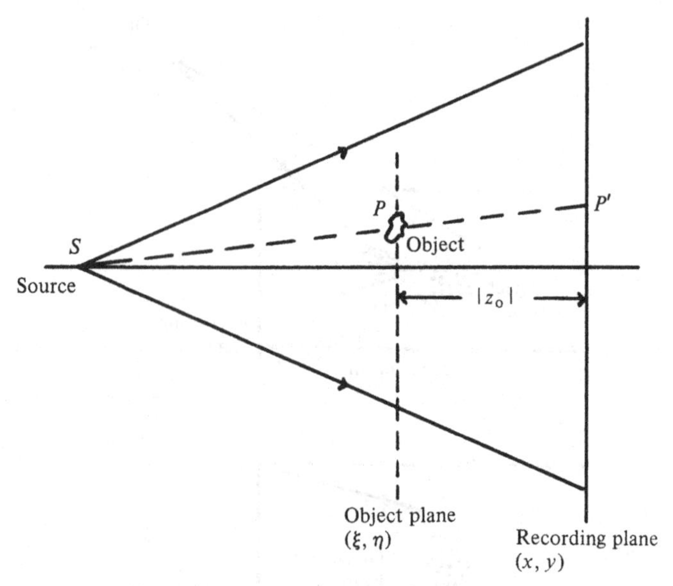

Fig. 2.6 Recording of in-line Fraunhofer holograms. The illumination beam is shown to be divergent but it can also be convergent or plane. The point P' is at the center of the diffraction pattern or the effective hologram of the object centered at P.

throughout the book. Some diffraction relationships for objects other than thin planes in the (ξ, η) plane are available.[5,6]

With the thin plane object in the (ξ, η) plane, the amplitude transmittance $T(\xi, \eta)$ in the (ξ, η) plane can be written as

$$T(\xi, \eta) = 1 - A(\xi, \eta), \tag{2.6}$$

where $A(\xi, \eta)$ is a function describing the object. For an opaque object

$$A(\xi, \eta) = \begin{cases} 1 & \text{within the object cross-section} \\ 0 & \text{otherwise} \end{cases} \tag{2.7}$$

The advantage of this type of representation of the transmittance in the object plane is that the reconstructed field can be characterized by the function $A(\xi, \eta)$ (see Chapter 3). Some of the forms of $A(\xi, \eta)$ for non-opaque objects are described in Section 5.9.

2.3 Image–object position relationships

An extended object can be assumed to be made of a large number of point objects. The image–object point relationships derived by Meier[7] are well accepted. The object, the reference and the reconstruction beams at the hologram plane (x, y) can be expressed as real and imaginary parts:

$$O(x, y) = |O(x, y)| \exp[-i\phi_o(x, y)],$$
$$R(x, y) = |R(x, y)| \exp[-i\phi_R(x, y)],$$
$$C(x, y) = |C(x, y)| \exp[-i\phi_c(x, y)], \tag{2.8}$$

where $\phi_o(x, y)$, $\phi_R(x, y)$ and $\phi_c(x, y)$ are the respective phases. The phase Φ_p of the primary image and the phase Φ_c of the conjugate image can be determined from the second and the third terms respectively of the RHS of Equation (2.4). Therefore

$$\Phi_p = \phi_c(x, y) + \phi_o(x/m_H, y/m_H) - \phi_R(x/m_H, y/m_H) \tag{2.9}$$

and

$$\Phi_c = \phi_c(x, y) - \phi_o(x/m_H, y/m_H) + \phi_R(x/m_H, y/m_H). \tag{2.10}$$

At the general point $Q(\bar{x}, \bar{y})$ during the recording (see Figure 2.1), the phase of the object beam relative to the phase at the origin $O(0, 0, 0)$ is

$$\begin{aligned}\phi_o(\bar{x}, \bar{y}) &= \frac{2\pi}{\lambda_o}(\overline{PQ} - \overline{PO})\\ &= \frac{2\pi}{\lambda_o}\{[(\bar{x} - x_o)^2 + (\bar{y} - y_o)^2 + z_o^2]^{1/2}\\ &\quad - (x_o^2 + y_o^2 + z_o^2)^{1/2}\}.\end{aligned} \tag{2.11}$$

Under paraxial approximation $z_o^2 > x_o^2 + y_o^2$, we obtain:

$$\begin{aligned}\phi_o(\bar{x}, \bar{y}) = \frac{2\pi}{\lambda_o}\Bigg[&\frac{1}{2z_o}(\bar{x}^2 + \bar{y}^2 - 2\bar{x}x_o - 2\bar{y}y_o)\\ &- \frac{1}{8z_o^3}(\bar{x}^4 + \bar{y}^4 + 2\bar{x}^2\bar{y}^2 - 4\bar{x}^3x_o\\ &- 4\bar{y}^3y_o - 4\bar{x}^2\bar{y}y_o - 4\bar{x}\bar{y}^2x_o + 6\bar{x}^2x_o^2\\ &+ 6\bar{y}^2y_o^2 + 2\bar{x}^2y_o^2 + 2\bar{y}^2x_o^2 + 8\bar{x}\bar{y}x_oy_o - 4\bar{x}x_o^3\\ &- 4\bar{y}y_o^3 - 4\bar{x}x_oy_o^2 - 4\bar{x}x_o^2y_o)\\ &+ \text{higher order terms}\Bigg].\end{aligned} \tag{2.12}$$

A similar relationship can be written for the reference beam by changing coordinates x_o, y_o, z_o, into x_R, y_R, z_R respectively. For the reconstruction beam, the phase $\phi_c(x, y)$ is similar to one in Equation (2.12) but with $\bar{x}$, $\bar{y}$ replaced by x, y; x_o, y_o, z_o replaced by x_c, y_c,

z_c; and λ_o replaced by λ_c. The first-order terms of the primary wave phase then become

$$\Phi_p^{(1)} = \frac{2\pi}{\lambda_c}\frac{1}{2}\left[\frac{x^2 + y^2 - 2xX_p - 2yY_p}{Z_p}\right]. \quad (2.13)$$

The third-order terms can similarly be determined. They are discussed in Chapter 7 dealing with primary aberrations.

The wave represents the primary image point $P_p(X_p, Y_p, Z_p)$ under the paraxial approximation as:

$$X_p = \frac{m_H^2 x_o z_o z_R + n m_H x_o z_c z_R - n m_H x_R z_c z_o}{m_H^2 z_o z_R + n z_c z_R - n z_c z_o}. \quad (2.14)$$

$$Y_p = \frac{m_H^2 y_o z_o z_R + n m_H y_o z_c z_R - n m_H y_R z_c z_o}{m_H^2 z_o z_R + n z_c z_R - n z_c z_o}. \quad (2.15)$$

$$Z_p = \frac{m_H^2 z_c z_o z_R}{m_H^2 z_o z_R - n z_c z_R + n z_c z_o}. \quad (2.16)$$

Similarly, for the conjugate image point

$$X_c = \frac{m_H^2 x_c z_o z_R - n m_H x_o z_c z_R + n m_H x_R z_c z_o}{m_H^2 z_o z_R - n z_c z_R + n z_c z_o}. \quad (2.17)$$

$$Y_c = \frac{m_H^2 y_c z_o z_R - n m_H y_o z_c z_R + n m_H y_R z_c z_o}{m_H^2 z_o z_R - n z_c z_R + n z_c z_o}. \quad (2.18)$$

$$Z_c = \frac{m_H^2 z_c z_o z_R}{m_H^2 z_o z_R - n z_c z_R + n z_c z_o}. \quad (2.19)$$

Both these images, in general, can be real or imaginary depending on the sign of the longitudinal coordinate of the image with respect to the reconstruction arrangement. For a collimated reference and/or reconstruction beam, the coordinate ratios can be replaced by angles as mentioned in the beginning of Section 2.1.

2.3.1 *Magnifications*

The transverse or lateral magnification M_p of the primary image can be obtained by differentiation of X_p or Y_p. We obtain

$$M_p = \frac{dX_p}{dx_o} = \frac{dY_p}{dy_o} = \frac{m_H}{1 + m_H^2 z_o/n z_c - z_o/z_R}. \quad (2.20)$$

Similarly, the magnification for the conjugate image is

$$M_c = \frac{dX_c}{dx_o} = \frac{dY_c}{dy_o} = \frac{m_H}{1 - m_H^2 z_o/n z_c - z_o/z_R} \quad (2.21)$$

Longitudinal magnification, i.e. dZ_p/dz_o for the primary image or dZ_c/dz_o for the conjugate image becomes

$$M_{long} = -\frac{1}{n} M^2_{transverse} \tag{2.22}$$

where M_{long} and $M_{transverse}$ are the corresponding longitudinal and transverse magnifications respectively.

3

General theory of in-line Fraunhofer holography

In this chapter we introduce the basic theory of in-line Fraunhofer holography. Starting with recording in-line holograms, the image in the reconstruction will be described under the far-field condition. Besides basic understanding of the process, the chapter will provide analytical guidelines for further discussion on various aspects throughout the book.

3.1 Recording process

Tyler and Thompson[1] have discussed the recording and reconstruction processes in detail when collimated beams of the same wavelength are used. Generally the recording and reconstruction are performed with different wavelengths. This is due to the practical need of a pulsed laser (such as ruby) to record a high speed event. The reconstruction requires a continuous wave (CW) laser (such as HeNe) for detailed analysis.

The general arrangement of recording in-line holograms is described in Figure 3.1. The recording source of coherent light of wavelength λ_o is situated at the point S on the optical axis. The microobject is situated in the (ξ, η) plane at $P(\xi_o, \eta_o)$ and the recording emulsion at the (x, y) plane. With respect to the recording plane, the longitudinal coordinates of the object and source planes are z_o and z_R respectively. Thus, the collimated beam recording means $z_R = -\infty$.

The complex light amplitude due to the source at the object plane can be written as

$$E_o = B \exp\left[-\frac{ik_o(\xi^2 + \eta^2)}{2(z_R - z_o)}\right], \tag{3.1}$$

where B is a constant and $k_o = 2\pi/\lambda_o$. If $A(\xi, \eta)$ represents the object distribution (see Section 2.2 for details), then Huygens–Fresnel Principle[2] can be used to determine the field distribution at the recording plane as

$$\psi(x, y) = \frac{iB}{\lambda_o z_o} \exp(-ik_o z_o) \iint_{-\infty}^{+\infty} [1 - A(\xi, \eta)]$$
$$\times \exp\left\{-\frac{ik_o}{2z_o}[(x-\xi)^2 + (y-\eta)^2]\right\}$$
$$\times \exp\left[-\frac{ik_o(\xi^2 + \eta^2)}{2(z_R - z_o)}\right] d\xi\, d\eta. \quad (3.2)$$

Without the object at $P(\xi, \eta)$, the amplitude transmittance of the object plane is unity. With the object, it becomes $1 - A(\xi, \eta)$. It is also noticeable that z_o is a negative quantity. The part of the integral without $A(\xi, \eta)$ can be solved resulting in:

$$\psi(x, y) = -\left(\frac{B}{m_o}\right) \exp(-ik_o z_o) \exp\left[-\frac{ik_o(x^2 + y^2)}{2z_R}\right]$$
$$\times \left(1 - \frac{im_o}{\lambda_o z_o} \iint_{-\infty}^{+\infty} A(\xi, \eta) \exp\left\{-\frac{ik_o}{2z_o}[(x-\xi)^2\right.\right.$$
$$\left. + (y-\eta)^2]\right\} \exp\left[\frac{ik_o(x^2 + y^2)}{2z_R}\right]$$
$$\left. \times \exp\left[-\frac{ik_o(\xi^2 + \eta^2)}{2(z_R - z_o)}\right] d\xi\, d\eta\right), \quad (3.3)$$

where m_o is defined as

$$m_o = (1 - z_o/z_R)^{-1}. \quad (3.4)$$

The quantity m_o is often called the magnification from the recording configuration. In Figure 3.1, the line from the source S through the object point $P(\xi_o, \eta_o, z_o)$ in the object plane meets the recording plane at a point $P'(x_o, y_o, 0)$. It can easily be determined that the in-plane distances of the points P and P' from the optical axis are related as

$$(x_o^2 + y_o^2)^{1/2} = m_o(\xi_o^2 + \eta_o^2)^{1/2}. \quad (3.5)$$

Suppose the irradiance distribution at the (x, y) plane is recorded on a negative emulsion. Also let us assume that the linear portion of the exposure vs amplitude transmittance curve is used. Then the light amplitude transmittance of the processed transparency or the hologram is (see Figure 2.2).

$$T(x, y) = \tau_b - K|\psi(x, y)|^2, \quad (3.6)$$

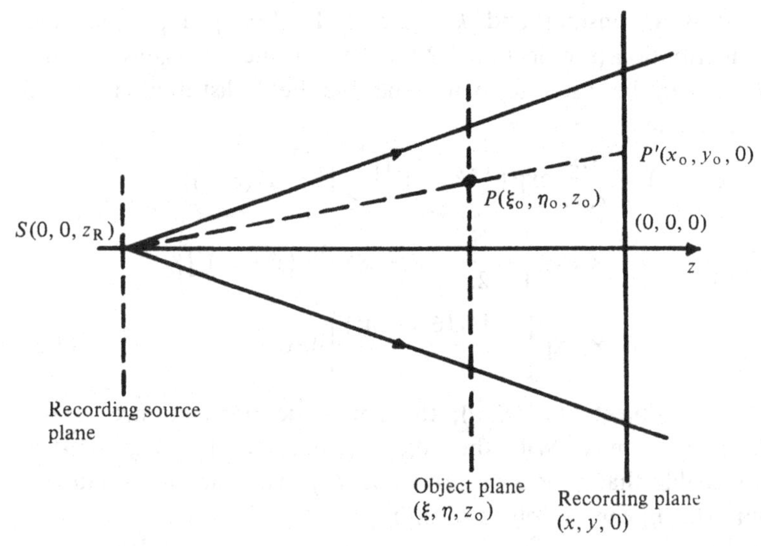

Fig. 3.1 Schematic diagram of the recording arrangement of in-line holograms.

where K is a constant and τ_b is the intercept on the linear portion of the $\tau - E$ curve with the τ-axis. The linear model represented by Equation (3.6) is well accepted in optics and adopted in particle field holography applications.[1,3,4]

Equations (3.3) and (3.6) can be combined to obtain the amplitude transmittance of the hologram. Here we assume that the recording emulsion is capable of resolving the spatial frequencies present in the irradiance distribution. The source, usually a laser light, is supposed to be spatially and temporally coherent over the recording aperture. These, as well as other related parameters are discussed in more detail in Chapter 4. For the time being we consider the idealized hologram to study the reconstruction.

3.2 Reconstruction

For the reconstruction, let us consider that the hologram is illuminated by the source of wavelength λ_c. As shown in Figure 3.2, the longitudinal coordinate of the source plane with respect to the hologram is z_c. The complex amplitude of the reconstruction beam at the hologram plane is therefore

$$E_c = C \exp\left[-\frac{ik_c(x^2 + y^2)}{2z_c}\right], \tag{3.7}$$

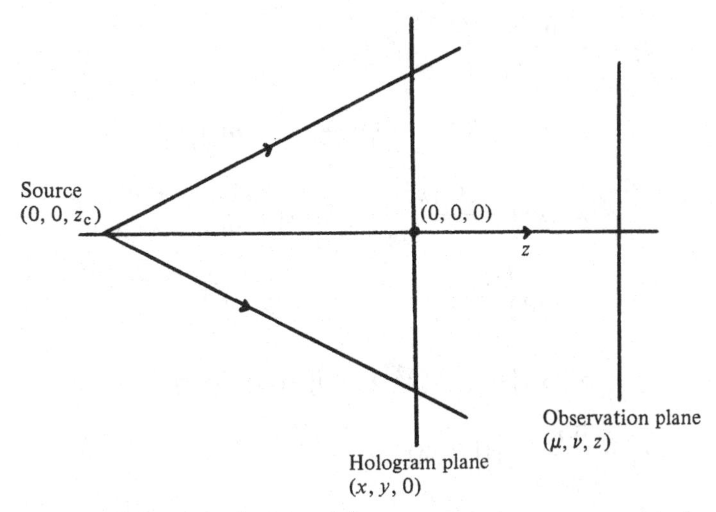

Fig. 3.2. Schematic diagram of the reconstruction arrangement of in-line holography.

where $k_c = 2\pi/\lambda_c$. Huygens–Fresnel Principle can again be used to determine the light amplitude as a 'forward' distance z (positive) from the hologram in the transverse plane (μ, ν) as

$$\psi(\mu, \nu) = -\frac{\mathrm{i}}{\lambda_c z}\exp(\mathrm{i}k_c z)\iint_{-\infty}^{+\infty} E_c T(x, y) \times \exp\left\{\frac{\mathrm{i}k_c}{2z}[(\mu - x)^2 + (\nu - y)^2]\right\}\mathrm{d}x\,\mathrm{d}y. \quad (3.8)$$

Equations (3.3), (3.6) and (3.7) can be substituted in Equation (3.8) to obtain

$$\psi(\mu, \nu) = -\frac{\mathrm{i}CB'}{\lambda_c z}\exp(\mathrm{i}k_c z)(I_1 + I_2 + I_3 + I_4). \quad (3.9)$$

where

$$B' = \tau_b - K(B/m_o)^2 \quad (3.10)$$

and after denoting

$$\Gamma = K(B/m_o)^2/B', \quad (3.11)$$

we have defined

$$I_1 = \iint_{-\infty}^{+\infty} \exp\left[-\frac{\mathrm{i}k_c(x^2 + y^2)}{2z_c}\right] \times \exp\left\{\frac{\mathrm{i}k_c}{2z}[(\mu - x)^2 + (\nu - y)^2]\right\}\mathrm{d}x\,\mathrm{d}y \quad (3.12)$$

$$I_2 = -\frac{\mathrm{i}\Gamma m_o}{\lambda_o z_o}\iiiint_{-\infty}^{+\infty} A^*(\xi,\eta)$$
$$\times \exp\left\{\frac{\mathrm{i}k_o}{2z_o}[(x-\xi)^2+(y-\eta)^2]\right\}$$
$$\times \exp\left[-\frac{\mathrm{i}k_o(x^2+y^2)}{2z_R}\right]\exp\left[\frac{\mathrm{i}k_o(\xi^2+\eta^2)}{2(z_R-z_o)}\right]$$
$$\times \exp\left\{\frac{\mathrm{i}k_c}{2z}[(\mu-x)^2+(\nu-y)^2]\right\}$$
$$\times \exp\left[-\frac{\mathrm{i}k_c}{2z_c}(x^2+y^2)\right]\mathrm{d}x\,\mathrm{d}y\,\mathrm{d}\xi\,\mathrm{d}\eta, \tag{3.13}$$

$$I_3 = \frac{\mathrm{i}\Gamma m_o}{\lambda_o z_o}\iiiint_{-\infty}^{+\infty} A(\xi,\eta)$$
$$\times \exp\left\{-\frac{\mathrm{i}k_o}{2z_o}[(x-\xi)^2+(y-\eta)^2]\right\}$$
$$\times \exp\left[\frac{\mathrm{i}k_o(x^2+y^2)}{2z_R}\right]\exp\left[-\frac{\mathrm{i}k_o(\xi^2+\eta^2)}{2(z_R-z_o)}\right]$$
$$\times \exp\left\{\frac{\mathrm{i}k_c}{2z}[(\mu-x)^2+(\nu-y)^2]\right\}$$
$$\times \exp\left[-\frac{\mathrm{i}k_c(x^2+y^2)}{2z_c}\right]\mathrm{d}x\,\mathrm{d}y\,\mathrm{d}\xi\,\mathrm{d}\eta \tag{3.14}$$

and

$$I_4 = -\frac{\Gamma m_o^2}{(\lambda_o z_o)^2}\iint_{-\infty}^{+\infty}\exp\left\{\frac{\mathrm{i}k_c}{2z}[(\mu-x)^2+(\nu-y)^2]\right\}$$
$$\times \exp\left[-\frac{\mathrm{i}k_c(x^2+y^2)}{2z_c}\right]\left(\iint_{-\infty}^{+\infty} A(\xi,\eta)\right.$$
$$\times \exp\left\{-\frac{\mathrm{i}k_o}{2z_o}[(x-\xi)^2+(y-\eta)^2]\right\}$$
$$\times \exp\left[-\frac{\mathrm{i}k_o(\xi^2+\eta^2)}{2(z_R-z_o)}\right]\mathrm{d}\xi\,\mathrm{d}\eta\iint_{-\infty}^{+\infty} A^*(\xi,\eta)$$
$$\times \exp\left\{\frac{\mathrm{i}k_o}{2z_o}[(x-\xi)^2+(y-\eta)^2]\right\}$$
$$\times \exp\left[\frac{\mathrm{i}k_o(\xi^2+\eta^2)}{2(z_R-z_o)}\right]\left.\mathrm{d}\xi\,\mathrm{d}\eta\right)\mathrm{d}x\,\mathrm{d}y. \tag{3.15}$$

At this stage, Equation (3.9) gives the general light amplitude at a forward distance z from the hologram. The case of particular interest here is to see the distribution at the plane of the reconstructed image. This is the conjugate image plane defined at the longitudinal distance of $z = Z_c$ (see Chapter 2) as

$$Z_c = -z_o M_c/n, \tag{3.16}$$

where n is the illuminating to the recording wavelength ratio given by

$$n = \lambda_c/\lambda_o. \tag{3.17}$$

M_c is the transverse magnification of the conjugate image given as

$$M_c = (1 - z_o/nz_c - z_o/z_R)^{-1} \tag{3.18}$$

Equations (3.4), (3.16) and (3.18) can be combined to get another useful form:

$$M_c/m_o = 1 - Z_c/z_c. \tag{3.19}$$

Solution of I_1 is straightforward resulting in, at $z = Z_c$,

$$I_1 = -\frac{\lambda_c z_c Z_c \mathrm{i}}{Z_c - z_c} \exp\left[\frac{\mathrm{i}k_c R^2}{2(Z_c - z_c)}\right] \tag{3.20}$$

where

$$R^2 = \mu^2 + \nu^2. \tag{3.21}$$

For solving the integral I_2 given by Equation (3.13), the x^2 and y^2 terms in the exponential completely disappear at $z = Z_c$. The remaining integral can be solved by performing Fourier transformation twice. Thus, we obtain

$$I_2 = -\frac{\mathrm{i}\Gamma(\lambda_c Z_c)^2 m_o}{\lambda_o z_o M_c^2} \exp\left[\frac{\mathrm{i}k_c(1 - 1/M_c)R^2}{2Z_c}\right] \times \exp\left(\frac{\mathrm{i}k_o m_o R^2}{2z_R M_c^2}\right) A^*\left(\frac{\mu}{M_c}, \frac{\nu}{M_c}\right). \tag{3.22}$$

The integral I_3 can be simplified using the position of the primary image plane (see Chapter 2):

$$Z_p = z_o M_p/n \tag{3.23}$$

and

$$M_p = (1 + z_o/nz_c - z_o/z_R)^{-1}. \tag{3.24}$$

By doing so, the integral given by Equation (3.14) becomes, at the conjugate image point:

$$I_3 = \frac{\mathrm{i}\Gamma m_o}{\lambda_o z_o} \iiiint_{-\infty}^{+\infty} A(\xi, \eta)$$
$$\times \exp\left\{\frac{\mathrm{i}nk_c}{2z_o M_p}[(x - \xi M_p)^2 + (y - \eta M_p)]\right\}$$
$$\times \exp\left[-\frac{\mathrm{i}k_c n(1 - M_p)(\xi^2 + \eta^2)}{2z_o}\right]$$
$$\times \exp\left\{\frac{\mathrm{i}k_c}{2Z_c}[(\mu - x)^2 + (v - y)^2]\right\}$$
$$\times \exp\left[-\frac{\mathrm{i}nk_c(\xi^2 + \eta^2)}{2(z_R - z_o)}\right]\mathrm{d}x\,\mathrm{d}y\,\mathrm{d}\xi\,\mathrm{d}\eta. \tag{3.25}$$

The solution of $x-y$ integral in Equation (3.25) is known (see for example equation (6) of Ref. 1 to yield:

$$I_3 = -\frac{\Gamma m_o}{\lambda_o z_o}\frac{\lambda_c Z_c Z_p}{Z_p - Z_c} \iint_{-\infty}^{+\infty} A(\xi, \eta)$$
$$\times \exp\left\{-\frac{\mathrm{i}k_c[(\mu - \xi M_p)^2 + (v - \eta M_p)^2]}{2(Z_p - Z_c)}\right\}$$
$$\times \exp\left[-\frac{\mathrm{i}k_c n(1 - M_p)(\xi^2 + \eta^2)}{2z_o}\right]$$
$$\times \exp\left[-\frac{\mathrm{i}nk_c(\xi^2 + \eta^2)}{2(z_R - z_o)}\right]\mathrm{d}\xi\,\mathrm{d}\eta, \tag{3.26}$$

which, by a coordinate transform, becomes

$$I_3 = -\frac{\Gamma m_o}{\lambda_o z_o}\frac{\lambda_c Z_c Z_p}{Z_p - Z_c}\frac{1}{M_p^2} \iint_{-\infty}^{+\infty} A\left(\frac{\xi'}{M_p}, \frac{\eta'}{M_p}\right)$$
$$\times \exp\left\{-\frac{\mathrm{i}k_c[(\mu - \xi')^2 + (v - \eta')^2]}{2(Z_p - Z_c)}\right\}$$
$$\times \exp\left[-\frac{\mathrm{i}k_c n}{2M_p^2}\left(\frac{1 - M_p}{z_o} - \frac{m_o}{z_R}\right)(\xi'^2 + \eta'^2)\right]\mathrm{d}\xi'\,\mathrm{d}\eta'. \tag{3.27}$$

Comparing with Equation (3.3), I_3 given by Equation (3.27) represents a hologram of a magnified (M_p times) object at a distance $Z_p - Z_c$, recorded with the wavelength λ_c. The coefficient of the term $(\xi'^2 + \eta'^2)$ gives the object–source plane distance.

Integral I_4 can be solved by the method of stationary phase.[5] The integration over x and y can be solved after obtaining the critical points $[\mu/(1 - Z/z_c), v/(1 - Z/z_c)]$. I_4 given by Equation (3.15) thus becomes, at $z = Z_c$:

$$I_4 = -\frac{\lambda_c i\Gamma m_o^2}{(\lambda_o z_o)^2}\left(\frac{1}{Z_c} - \frac{1}{z_c}\right)^{-1} \exp\left[\frac{ik_c R^2}{2(Z - z_c)}\right] \iint_{-\infty}^{+\infty} A(\xi, \eta)$$
$$\times \exp\left\{-\frac{ik_o}{2z_o}\left[\left(\frac{\mu}{1 - Z_c/z_c} - \xi\right)^2 + \left(\frac{\nu}{1 - Z_c/z_c} - \eta\right)^2\right]\right\}$$
$$\times \exp\left[-\frac{ik_o(\xi^2 + \eta^2)}{2(z_R - z_o)}\right] d\xi\, d\eta \iint_{-\infty}^{+\infty} A^*(\xi, \eta)$$
$$\times \exp\left\{\frac{ik_o}{2z_o}\left[\left(\frac{\mu}{1 - Z_c/z_c} - \xi\right)^2 + \left(\frac{\nu}{1 - Z_c/z_c} - \eta\right)^2\right]\right\}$$
$$\times \exp\left[\frac{ik_o(\xi^2 + \eta^2)}{2(z_R - z_o)}\right] d\xi\, d\eta. \tag{3.28}$$

3.3 Image form in the far-field case

In Section 3.2 we have seen the general image form in in-line holography when divergent beams not necessarily of the same wavelength are used for recording and reconstruction. Equation (3.9) gives the general reconstructed light amplitude and the integrals I_1, I_2, I_3 and I_4 are given by Equations (3.20), (3.22), (3.27) and (3.28) respectively for the conjugate image plane. Besides the phase term, I_1 in Equation (3.20) represents a directly transmitted light of uniform absolute amplitude. The modulus of I_2 given by Equation (3.22) represents an image of the object magnified by M_c. However, the simultaneously present I_3 and I_4 terms will generally spoil the image distribution. In fact, due to these terms the Gabor method[6–8] is generally not suitable, leading to the need for off-axis or Leith–Upatnieks method.[9,10]

Now consider the case of a very small object such that the $\xi^2 + \eta^2$ (this implies to a two-dimensional object) variation over the object cross-section is very small as compared to wavelength times the longitudinal distances involved in RHS of the integral of Equation (3.27). In that situation, I_3 given by Equation (3.27) becomes a Fourier transform and its contribution can be generally neglected. Practically, it has been found that objects as close as one far-field away give a reasonably good image without other terms having serious effect.[11]

Similarly, under the far-field approximation, the term I_4 given by Equation (3.28) becomes the product of two Fourier transforms and its contribution becomes too small to have a practical effect as compared to the term I_2. Thus, under the far-field approximation, the reconstructed field at $z = Z_c$ becomes

$$\psi(\mu, \nu) = \frac{iCB'}{Z_c\lambda_c}\exp(ik_cZ_c)\left\{\frac{i\lambda_c z_c Z_c}{Z_c - z_c}\right.$$
$$\times \exp\left[\frac{ik_cR^2}{2(Z_c - z_c)}\right] + \frac{i\Gamma m_o(\lambda_c Z_c)^2}{\lambda_o z_o M_c^2}$$
$$\times \exp\left[\frac{ik_c(1 - 1/M_c)R^2}{2Z_c}\right]$$
$$\left.\times \exp\left(\frac{ik_o m_o R^2}{2z_R M_c^2}\right)A^*\left(\frac{\mu}{M_c}, \frac{\nu}{M_c}\right)\right\}. \quad (3.29)$$

Equations (3.16), (3.17) and (3.19) can be used to prove that

$$\frac{\lambda_c z_c Z_c}{Z_c - z_c} = \frac{m_o(\lambda_c Z_c)^2}{\lambda_o z_o M_c^2}. \quad (3.30)$$

Also Equations (3.4), (3.16), (3.17) and (3.19) yield:

$$\frac{k_c(1 - 1/M_c)}{2Z_c} + \frac{k_o m_o}{2z_R M_c^2} = \frac{k_c}{2(Z_c - z_c)}. \quad (3.31)$$

Substituting Equations (3.30) and (3.31) in Equation (3.29) we obtain:

$$\psi(\mu, \nu) = -\frac{CB'z_c}{Z_c - z_c}\exp(ik_cZ_c)\exp\left[\frac{ik_cR^2}{2(Z_c - z_c)}\right]$$
$$\times\left[1 + \Gamma A^*\left(\frac{\mu}{M_c}, \frac{\nu}{M_c}\right)\right]. \quad (3.32)$$

The irradiance distribution is therefore

$$I(\mu, \nu) = |\psi(\mu, \nu)|^2 = I_0\left|1 + \Gamma A^*\left(\frac{\mu}{M_c}, \frac{\nu}{M_c}\right)\right|^2 \quad (3.33)$$

where I_0 is the background irradiance. Thus, we see that the image is erect with an uniform coherent background. For simplicity, if we consider an opaque object, then the object distribution is real. In that situation, Equation (3.33) gives the irradiance within the image (magnified M_c times) cross-section as

$$I_{\text{image}} = I_0(1 + \Gamma)^2 \quad (3.34)$$

The contrast of the image thus depends on the value of Γ and should be as large as possible. For a negative hologram, Γ is positive. This means the image is bright in the darker background. This model of opaque objects provides insight into the imaging process in many practical situations. However, phase objects such as bubbles, droplets, etc. are also studied using the technique. A detailed discussion on this aspect is presented elsewhere in the text (see Section 5.9).

So far, we have considered the general recording process of two-dimensional objects described by Equations (3.3) and (3.6). These expressions are generally valid for one-dimensional objects (such as long thin wire, fibers, etc.) also but, the far-field condition that $\xi^2 + \eta^2$ variation is small cannot be met. However, the coordinate system can be rotated so that $A(\xi)$ can describe the one-dimensional object distribution. The far-field condition will only apply in ξ-direction. A careful look at integrals I_1 and I_2 given by Equations (3.12) and (3.13) reveals that their analytical forms given by Equations (3.20) and (3.22) respectively are derived without the far-field assumption. Thus these results are true for one as well as two-dimensional objects so long as the integrals I_3 and I_4 can be neglected. This is generally true so long as the far-field condition is satisfied. However, one-dimensional objects give better high frequency interference fringe contrast at the recording stage than similar diameter two-dimensional objects. This aspect results in better recordability of the hologram, as discussed in detail in Chapter 8.

3.4 Collimated beams of the same wavelength

For better physical understanding of the imaging process particularly regarding neglecting the integrals I_3 and I_4, let us consider the collimated recording and reconstruction beams of the same wavelength λ. Also let us consider the object to be on the optical axis for simplicity as this will not change any physical meaning of the process for the collimated beam case incident normally to the recording plane. Under these circumstances, we obtain $z_c = z_R = -\infty$, $Z_c = -Z_p = -z_o$, $M_c = M_p = 1$ and $m_o = 1$. Then I_1, I_2, I_3 and I_4 given by Equations (3.20), (3.22), (3.26) and (3.28) respectively can be written as

$$I_1 = -i\lambda z_o, \tag{3.35}$$

$$I_2 = -i\Gamma\lambda z_o A^*(\mu, \nu), \tag{3.36}$$

$$I_3 = \frac{\Gamma}{2}\exp\left(-\frac{\pi i R^2}{2\lambda z_o}\right) \times \iint_{-\infty}^{+\infty} A^*(\xi, \eta)\exp\left[\frac{\pi i}{\lambda z_o}(\mu\xi + \nu\eta)\right]d\xi\, d\eta, \tag{3.37}$$

and

$$I_4 = \frac{i\Gamma}{\lambda z_o}\left|\iint_{-\infty}^{+\infty} A(\xi, \eta)\exp\left[\frac{2\pi i}{\lambda z_o}(\mu\xi + \nu\eta)\right]d\xi\, d\eta\right|^2, \tag{3.38}$$

where a two-dimensional object is considered such that

$$2\pi(\xi^2 + \eta^2)_{\max} \ll \lambda|z_o| \tag{3.39}$$

Here $(\xi^2 + \eta^2)^{1/2}$ represents the maximum spatial extent of the object. The condition given by Equation (3.39) is known as the far-field condition.

For an opaque object of radius a, the object has circular symmetry:

$$A(\xi, \eta) = \text{circ}(r/a) = \begin{cases} 1 & \text{for } r \leq a \\ 0 & \text{for } r > a \end{cases} \tag{3.40}$$

where

$$r = (\xi^2 + \eta^2)^{1/2}. \tag{3.41}$$

The two-dimensional Fourier transform of this function is given by

$$\iint_{-\infty}^{+\infty} A(\xi, \eta) \exp\left[-\frac{2\pi i}{\lambda|z_o|}(\mu\xi + \nu\eta)\right] d\xi\, d\eta = \pi a^2 \left[\frac{2J_1(2\pi ar/\lambda|z_o|)}{2\pi ar/\lambda|z_o|}\right], \tag{3.42}$$

where J_1 denotes the first-order Bessel function. Substituting the value from Equation (3.42) into Equations (3.37) and (3.38), Equations (3.35)–(3.38) give the reconstructed irradiance $I(R)$ as

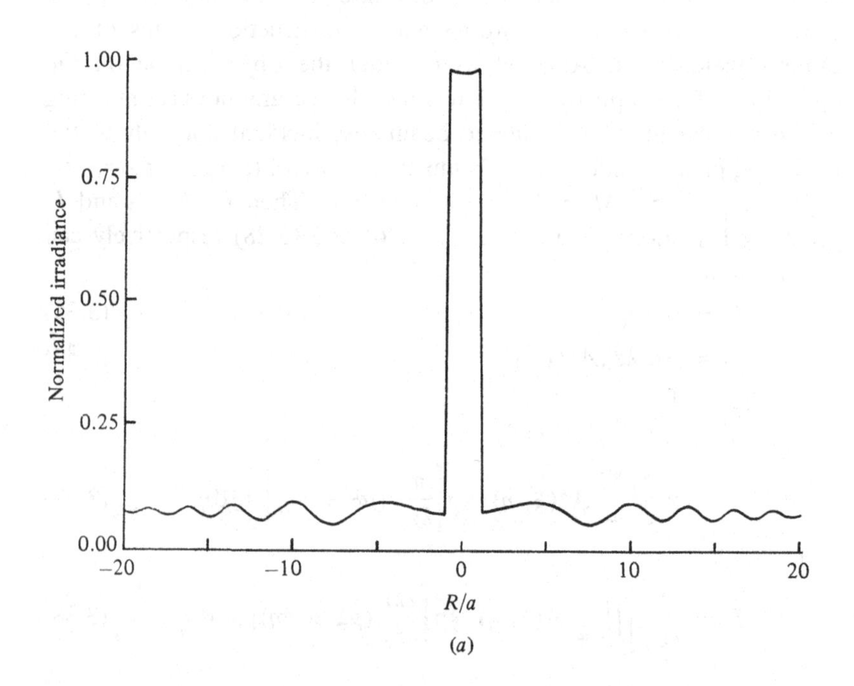

(*a*)

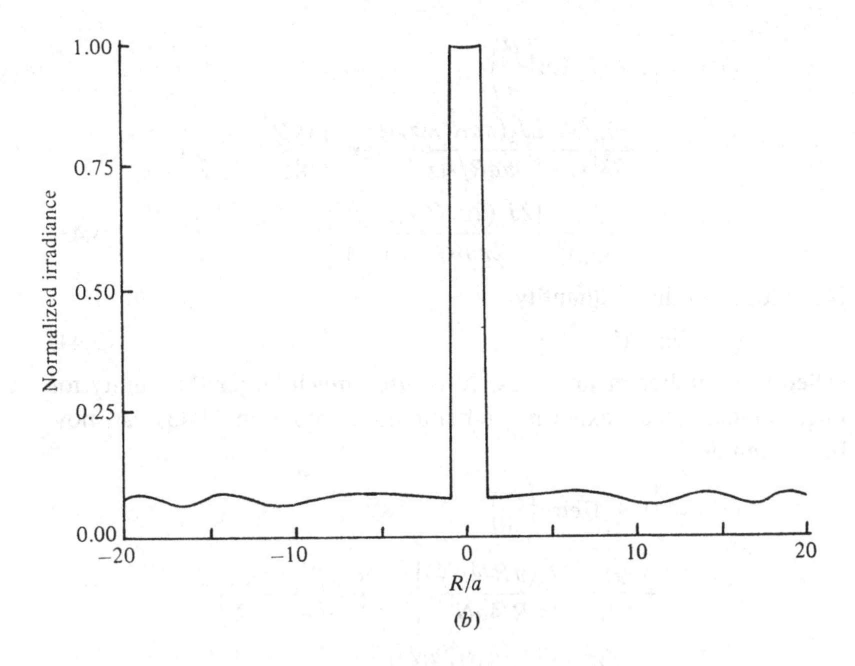

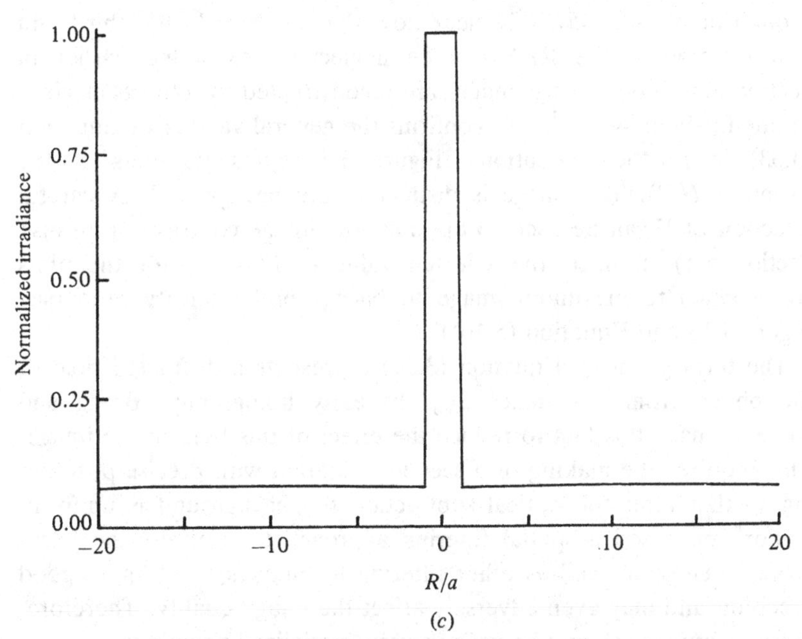

Fig. 3.3 Image irradiance distribution for a circular cross-section object. Plots are from Equation (3.45) for $\Gamma = 2.8$. (*a*), (*b*) and (*c*) are for $N = 5$, 10 and 50 respectively.

$$I(R) \propto \left| 1 + \Gamma \operatorname{circ}\left(\frac{R}{a}\right) + \frac{\pi a^2 \Gamma}{2\lambda|z_o|} \frac{2J_1(\pi a R/\lambda|z_o|)}{\pi a R/\lambda|z_o|} \exp\left(\frac{ikR^2}{4|z_o|} - \frac{\pi}{2}\right) - \frac{\Gamma \pi^2 a^4}{(\lambda z_o)^2} \left[\frac{2J_1(2\pi a R/\lambda|z_o|)}{2\pi a R/\lambda|z_o|}\right]^2 \right|^2. \tag{3.43}$$

Now, let us define a quantity

$$N = \lambda|z_o|/(2a)^2, \tag{3.44}$$

called the number of far-fields. N is often much larger than unity for micro-objects in an experimental situation. Equation (3.43) can now be written as

$$I(R) \propto \left| 1 + \Gamma \operatorname{circ}\left(\frac{R}{a}\right) + \frac{\pi \Gamma}{8N} \left[\frac{2J_1(\pi R/4aN)}{\pi R/4aN}\right] \exp\left(\frac{\pi i R^2}{8a^2 N} - \frac{\pi i}{2}\right) - \frac{\Gamma \pi^2}{16N^2} \left[\frac{2J_1(\pi R/2aN)}{\pi R/2aN}\right]^2 \right|^2. \tag{3.45}$$

From Equation (3.45) it is clear now that as $N \gg 1$, the third and fourth terms of the RHS can be neglected. As stated earlier in Section 3.3, good quality images are reconstructed for objects as close as one far-field away.[11] This confirms the general validity of Equation (3.33) in practical situations. Figure 3.3 represents plots of the variation $I(R)$. The image is distinct in the background. A careful selection of Γ can be used to optimize the image contrast[12] (see also Section 5.5). In fact, the selected value of $\Gamma(= 2.8)$ for the plots corresponds to maximum image to background intensity ratio (see Figure 5.13 and Equation (5.16)).

The term given by Equation (3.37) represents a diffracted field of the object from a distance $2z_o$. In early holography, Bragg and Rogers[13] used this fact to reduce the effect of this field in the image. This requires the making of a second hologram with precise positioning so that after the optical subtraction the background is uniform. Gabor[8] proposed a spatial filtering approach to suppress the background. These as well as other filtering techniques[14–17] require good precision and may even adversely affect the image quality. Therefore, these techniques should be used in very specialized situations.

In the in-line Fraunhofer holography, the virtual image and the intermodulation (described by the integrals I_3 and I_4 respectively)

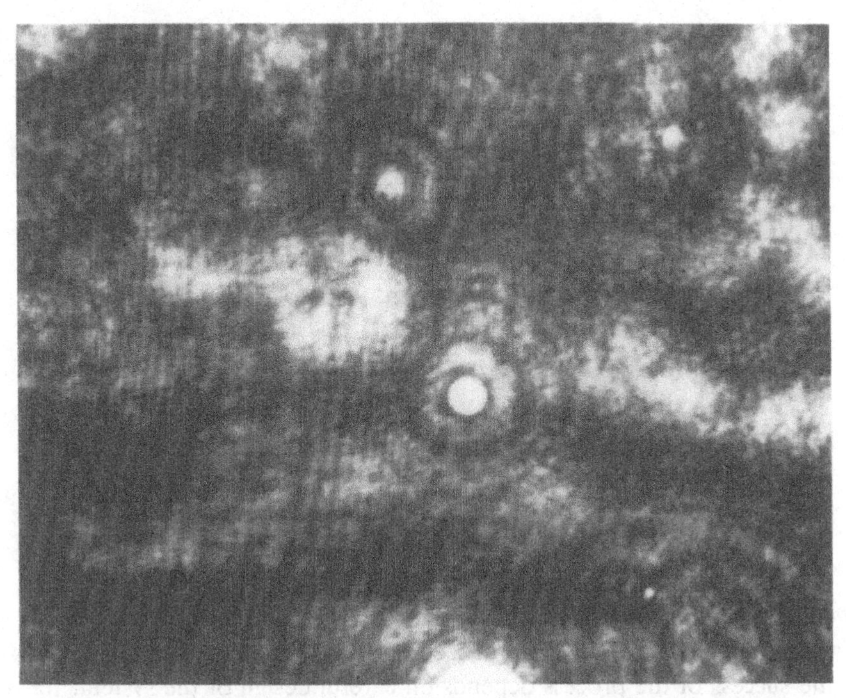

Fig. 3.4 Reconstructed image of a 100 μm air bubble in water as seen on the video monitor. Diffraction patterns due to out-of-focus images are also present.

play very little part. The images are generally viewed using a closed circuit television system which may employ electronic filtering devices to enhance entire image qualities like image contrast, brightness, magnification, etc.[18-19] Even contrast reversal using a positive hologram, i.e. using negative K in equation (3.6) which shows dark image in bright background[20] is electronically possible. Research, is continuing to decode and/or suppress undesired components in the image electronically.[21-6]

For most applications dealing with practical particle field holography, an ordinary closed circuit television system is sufficient and in common use. Figure 3.4 shows such an image on the monitor screen.

4

System design considerations

In Chapter 3 we saw that the image is formed from in-line holograms. Also present is a uniform coherent background and negligible effects of the real image and intermodulation in the Fraunhofer case. The basic assumption was that the recording was perfect, i.e. all spatial frequencies present were recorded properly, the laser had sufficient coherence, the object was static or the laser pulse width was very short, and so on. Since in practice, all these parameters have limits, the success of the process depends on careful design of the system. In this chapter we shall deal with these considerations. Starting with the irradiance distribution at the recording stage, numerous physical parameters will be discussed for system design considerations.

4.1 Recording irradiance distribution

Recalling Equation (3.1) suppose the object in the (ξ, η) plane is illuminated by the generally divergent beam and the hologram is recorded at the (x, y) plane. Object–hologram and source–hologram distances are z_o and z_R respectively. Equation (3.3) gives the light amplitude distribution at the recording plane as

$$\psi(x, y) = -\left(\frac{B}{m_o}\right)\exp(-ikz_o)\exp\left[-\frac{ik_o(x^2+y^2)}{2z_R}\right]$$
$$\times\left(1 - \frac{im_o}{\lambda_o z_o}\iint_{-\infty}^{+\infty} A(\xi, \eta)\right.$$
$$\times\exp\left\{-\frac{ik_o}{2z_o}[(x-\xi)^2 + (y-\eta)^2]\right\}$$
$$\left.\times\exp\left[\frac{ik_o}{2z_R}(x^2+y^2)\right]\exp\left[-\frac{ik_o(\xi^2+\eta^2)}{2(z_R - z_o)}\right]\right). \tag{4.1}$$

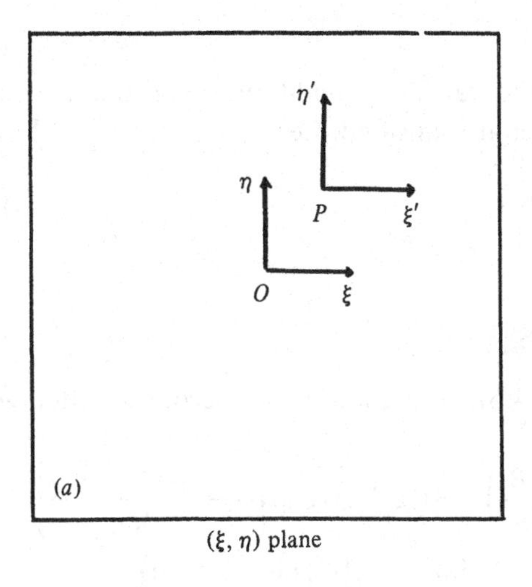

(ξ, η) plane

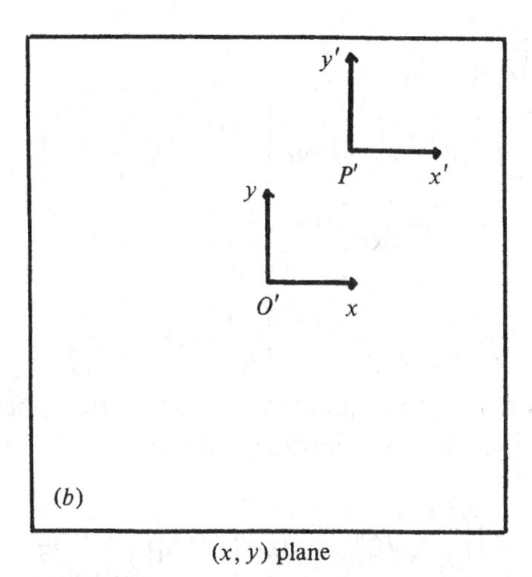

(x, y) plane

Fig. 4.1 Schematic diagram representing the coordinate transformations from the origins O and O' to the points P and P' respectively. With respect to O and O', the coordinates of P and P' are (ξ_o, η_o) and (x_o, y_o) respectively.

As shown in Figure 4.1, coordinate centers at P and P' are related by the transformations:

$$x' = x - x_o, \tag{4.2}$$

$$y' = y - y_o, \tag{4.3}$$

$$\xi' = \xi - \xi_o, \tag{4.4}$$

and

$$\eta' = \eta - \eta_o. \tag{4.5}$$

Let us also assume the far-field conditions over the maximum spatial extent of the object are satisified, i.e.,

$$\frac{(\xi'^2 + \eta'^2)_{\max}}{\lambda_o|z_o|} \ll 1 \tag{4.6}$$

and

$$\frac{(\xi'^2 + \eta'^2)_{\max}}{\lambda_o|z_R - z_o|} \ll 1. \tag{4.7}$$

Using equations (4.2)–(4.6), Equation (4.1) becomes, after some algebra:

$$\begin{aligned}
\psi(x', y') &= -\left(\frac{B}{m_o}\right)\exp(ik_o|z_o|)\exp\left[-\frac{ik_o(x^2+y^2)}{2z_R}\right] \\
&\quad\times\left\{1+\frac{im_o}{\lambda_o|z_o|}\exp\left[\frac{ik_o(x'^2+y'^2)}{2m_o|z_o|}\right]\right. \\
&\quad\times\left.\tilde{A}\left(\frac{x'}{\lambda_o|z_o|}, \frac{y'}{\lambda_o|z_o|}\right)\right\} \\
&= -\left(\frac{B}{m_o}\right)\exp(ik_o|z_o|)\exp\left[-\frac{ik_o(x^2+y^2)}{2z_R}\right] \\
&\quad\times\left\{1+\frac{im_o}{\lambda_o|z_o|}\exp\left[\frac{ik_o(x'^2+y'^2)}{2m_o|z_o|}\right.\right. \\
&\quad\left.\left.+\,i\phi\left(\frac{x'}{\lambda_o|z_o|}, \frac{y'}{\lambda_o|z_o|}\right)\right]\tilde{A}_o\left(\frac{x'}{\lambda_o|z_o|}, \frac{y'}{\lambda_o|z_o|}\right)\right\},
\end{aligned} \tag{4.8}$$

where the fact that z_o is a negative quantity has been incorporated. $\tilde{A}(x'/\lambda_o|z_o|, y'/\lambda_o|z_o|)$ is the Fourier transform of the object distribution given by

$$\begin{aligned}
\tilde{A}\left(\frac{x'}{\lambda_o|z_o|}, \frac{y'}{\lambda_o|z_o|}\right) &= \iint_{-\infty}^{+\infty} A(\xi', \eta')\exp\left\{-2\pi i\left[\left(\frac{x'}{\lambda_o|z_o|}\right)\xi'\right.\right. \\
&\quad\left.\left.+\left(\frac{y'}{\lambda_o|z_o|}\right)\eta'\right]\right\}d\xi' d\eta' \\
&= \tilde{A}_o\left(\frac{x'}{\lambda_o|z_o|}, \frac{y'}{\lambda_o|z_o|}\right) \\
&\quad\times\exp\left[i\phi\left(\frac{x'}{\lambda_o|z_o|}, \frac{y'}{\lambda_o|z_o|}\right)\right].
\end{aligned} \tag{4.9}$$

$\tilde{A}_o(\ )$ is the real amplitude and $\phi(\ )$ is the phase of the Fourier transform.

Equation (4.8) is for two-dimensional object cross-sections governed by the far-field conditions described by Equations (4.6) and (4.7). For one-dimensional object cross-sections (such as a long thin wire), the coordinate system can be rotated such that the object distribution is described by $A(\xi)$. The new far-field conditions also apply only in the ξ-direction:

$$\xi'^2_{\max}/\lambda_o|z_o| \ll 1 \tag{4.10}$$

and

$$\xi'^2_{\max}/\lambda_o|z_R - z_o| \ll 1 \tag{4.11}$$

The integrals in Equation (4.1) become separable in ξ' and η'. After the necessary algebra we obtain:

$$\begin{aligned}\psi(x', y') &= -\left(\frac{B}{m_o}\right)\exp(\mathrm{i}k_o|z_o|)\exp\left[-\frac{\mathrm{i}k_o(x^2+y^2)}{2z_R}\right] \\ &\quad\times\left\{1-\left(\frac{m_o}{\lambda_o|z_o|}\right)^{1/2}\exp\left[\mathrm{i}\left(\frac{k_o x'^2}{2m_o|z_o|}-\frac{\pi}{4}\right)\right]\tilde{A}\left(\frac{x'}{\lambda_o|z_o|}\right)\right\} \\ &= -\left(\frac{B}{m_o}\right)\exp(\mathrm{i}k_o|z_o|)\exp\left[-\frac{\mathrm{i}k_o(x^2+y^2)}{2z_R}\right] \\ &\quad\times\left\{1-\left(\frac{m_o}{\lambda_o|z_o|}\right)^{1/2}\right. \\ &\quad\left.\times\exp\left[\frac{\mathrm{i}k_o x'^2}{2m_o|z_o|}-\frac{\pi\mathrm{i}}{4}+\mathrm{i}\phi\left(\frac{x'}{\lambda_o|z_o|}\right)\right]\tilde{A}_o\left(\frac{x'}{\lambda_o|z_o|}\right)\right\},\end{aligned} \tag{4.12}$$

where $\tilde{A}(x'/\lambda_o|z_o|)$ is a one-dimensional Fourier transform given by

$$\begin{aligned}\tilde{A}\left(\frac{x'}{\lambda_o|z_o|}\right) &= \int_{-\infty}^{+\infty} A(\xi')\exp\left\{-2\pi\mathrm{i}\left[\left(\frac{x'}{\lambda_o|z_o|}\right)\xi'\right]\right\}\mathrm{d}\xi' \\ &= \tilde{A}_o\left(\frac{x'}{\lambda_o|z_o|}\right)\exp\left[\mathrm{i}\phi\left(\frac{x'}{\lambda_o|z_o|}\right)\right].\end{aligned} \tag{4.13}$$

Equation (4.8) gives the irradiance distribution for two-dimensional objects as

$$\begin{aligned}I(x', y') &= |\psi(x', y')|^2 \\ &= \left(\frac{B}{m_o}\right)^2\left\{1-\frac{2m_o}{\lambda_o|z_o|}\left[\sin\left(\frac{\pi r'^2}{\lambda_o m_o|z_o|}\right)\right.\right. \\ &\quad\times\operatorname{Re}\tilde{A}\left(\frac{x'}{\lambda_o|z_o|},\frac{y'}{\lambda_o|z_o|}\right)+\cos\left(\frac{\pi r'^2}{\lambda_o m_o|z_o|}\right) \\ &\quad\left.\times\operatorname{Im}\tilde{A}\left(\frac{x'}{\lambda_o|z_o|},\frac{y'}{\lambda_o|z_o|}\right)\right]\end{aligned}$$

$$+\left(\frac{m_o}{\lambda_o|z_o|}\right)^2\left|\widetilde{A}\left(\frac{x'}{\lambda_o|z_o|}, \frac{y'}{\lambda_o|z_o|}\right)\right|^2\Bigg\}$$

$$=\left(\frac{B}{m_o}\right)^2\Bigg\{1-\left(\frac{2m_o}{\lambda_o|z_o|}\right)\sin\Bigg[\frac{\pi r'^2}{\lambda_o m_o|z_o|}$$

$$+\phi\left(\frac{x'}{\lambda_o|z_o|}, \frac{y'}{\lambda_o|z_o|}\right)\Bigg]\widetilde{A}_o\left(\frac{x'}{\lambda_o|z_o|}, \frac{y'}{\lambda_o|z_o|}\right)$$

$$+\left(\frac{m_o}{\lambda_o|z_o|}\right)^2\widetilde{A}_o^2\left(\frac{x'}{\lambda_o|z_o|}, \frac{y'}{\lambda_o|z_o|}\right)\Bigg\}, \tag{4.14}$$

where

$$r'^2 = x'^2 + y'^2 \tag{4.15}$$

and Re and Im denote real and imaginary parts respectively.

Similarly, Equation (4.12) gives the irradiance distribution in the one-dimensional case as

$$I(x') = \left(\frac{B}{m_o}\right)^2\Bigg\{1-2\left(\frac{m_o}{\lambda_o|z_o|}\right)^{1/2}\Bigg[\cos\left(\frac{\pi x'^2}{m_o\lambda_o|z_o|}-\frac{\pi}{4}\right)$$

$$\times \operatorname{Re}\widetilde{A}\left(\frac{x'}{\lambda_o|z_o|}\right) - \sin\left(\frac{\pi x'^2}{m_o\lambda_o|z_o|}-\frac{\pi}{4}\right)$$

$$\times \operatorname{Im}\widetilde{A}\left(\frac{x'}{\lambda_o|z_o|}\right)\Bigg] + \left(\frac{m_o}{\lambda_o|z_o|}\right)\left|\widetilde{A}\left(\frac{x'}{\lambda_o|z_o|}\right)\right|^2\Bigg\}$$

$$=\left(\frac{B}{m_o}\right)^2\Bigg\{1-2\left(\frac{m_o}{\lambda_o|z_o|}\right)^{1/2}\cos\Bigg[\frac{\pi x'^2}{m_o\lambda_o|z_o|}$$

$$+\phi\left(\frac{x'}{\lambda_o|z_o|}\right)-\frac{\pi}{4}\Bigg]\widetilde{A}_o\left(\frac{x'}{\lambda_o|z_o|}\right)$$

$$+\left(\frac{m_o}{\lambda_o|z_o|}\right)\widetilde{A}_o^2\left(\frac{x'}{\lambda_o|z_o|}\right)\Bigg\}. \tag{4.16}$$

For the particular case of the collimated recording beam ($m_o = 1$) the relationships given by Equations (4.14) and (4.16) become those derived by Tyler and Thompson.[1]

For an opaque object with circular cross-section of diameter $2a$, the two-dimensional Fourier transform is (see Equation (3.42)) given by

$$\widetilde{A}\left(\frac{x'}{\lambda_o|z_o|}, \frac{y'}{\lambda_o|z_o|}\right) = \pi a^2\left[\frac{2J_1(2\pi a r'/\lambda_o|z_o|)}{2\pi a r'/\lambda_o|z_o|}\right] \tag{4.17}$$

Substituting Equation (4.17) into Equation (4.14) we obtain

$$I(x', y') = \left(\frac{B}{m_o}\right)^2 \left\{1 - \frac{2m_o\pi a^2}{\lambda_o|z_o|} \times \sin\left(\frac{\pi r'^2}{m_o\lambda_o|z_o|}\right)\left[\frac{2J_1(2\pi ar'/\lambda_o|z_o|)}{2\pi ar'/\lambda_o|z_o|}\right] + \left(\frac{m_o\pi a^2}{\lambda_o|z_o|}\right)^2 \left[\frac{2J_1(2\pi ar'/\lambda_o|z_o|)}{2\pi ar'/\lambda_o|z_o|}\right]^2\right\}. \tag{4.18}$$

Similarly, for a long thin wire of diameter $2a$, the one-dimensional Fourier transform is given by

$$\tilde{A}\left(\frac{x'}{\lambda_o|z_o|}\right) = 2a\left[\frac{\sin(2\pi ax'/\lambda_o|z_o|)}{2\pi ax'/\lambda_o|z_o|}\right]. \tag{4.19}$$

Substituting this in Equation (4.16) gives the irradiance distribution for a one-dimensional opaque object of diameter $2a$ as

$$I(x', y') = \left(\frac{B}{m_o}\right)^2 \left\{1 - 4a\left(\frac{m_o}{\lambda_o|z_o|}\right)^{1/2} \cos\left(\frac{\pi x'^2}{m_o\lambda_o|z_o|} - \frac{\pi}{4}\right) \times \left[\frac{\sin(2\pi ax'/\lambda_o|z_o|)}{2\pi ax'/\lambda_o|z_o|}\right] + \frac{4m_o a^2}{\lambda_o|z_o|} \times \left[\frac{\sin(2\pi ax'/\lambda_o|z_o|)}{2\pi ax'/\lambda_o|z_o|}\right]^2\right\}. \tag{4.20}$$

Plots of distributions given by Equations (4.18) and (4.20) in different numerical situations are available in the literature.[1–4] The validity of the far-field or Fraunhofer approximation in the irradiance distribution has been experimentally demonstrated.[1,2] Slimani, Grehan, Gouesbet and Allano[5] used Lorenz–Mie Theory and established the approximation. In Figure 4.2 we have plotted the irradiance distributions given by Equation (4.18) in a particular numerical situation. Fine interference fringes given by sine and cosine distributions are modulated by relatively broad diffraction patterns – containing central maxima and side lobes. Figure 4.3 shows a positive made from a hologram (negative) of air bubbles at different distances and of different sizes in water. In theory, as many side lobes as possible should be recorded. However, in practice the resolution and dynamic range of available recording materials set limits on the effective aperture and hence the resolution. Fortunately even under these limits, the technique is useful in a large number of situations of practical interest. Another interesting aspect of the holographic irradiance distribution is that the fine interference term does not contain information about the object size (a) but it does depend on its distance (z_o). Once z_o is known, it is possible to determine the

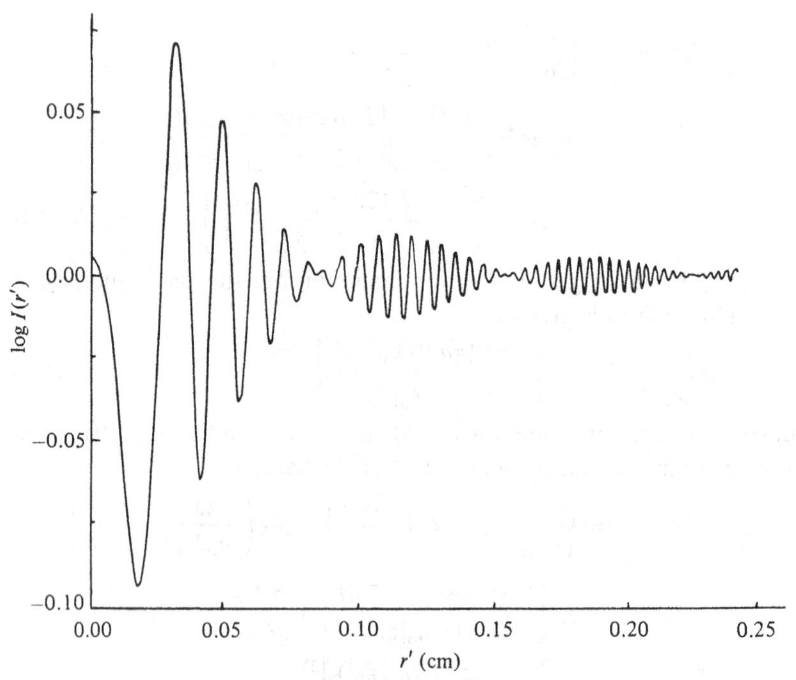

Fig. 4.2 Variation of log $I(r')$ against r' for $m_o = 1$, $\lambda_o = 0.6943\ \mu$m, $a = 50\ \mu$m and $z_o = 10$ cm. $I(r')$ is the normalized intensity $I(x', y')/(B/m_o)^2$ for a circular cross-section object.

object size (a) from the envelope or the diffraction part of the distribution. The possibility of this type of direct position and size analysis from the hologram is discussed in Chapter 9 in more detail.

4.2 Film resolution requirements

The resolution of the image upon reconstruction depends on the successful recording of the irradiance distribution at the recording plane. Starting with the image resolution requirements, we will deal here with the film resolution requirements.

At the time of reconstruction, if r' is the radius of the hologram aperture and the real image at distance Z_c is observed as shown in Figure 4.4 then according to the Rayleigh resolution limit (Section 4.8), the resolved image diameter $2a$ is:

$$2a|_{\text{image}} = 1.22\lambda_c Z_c/r', \tag{4.21}$$

which, with the help of Equations (3.16) and (3.17) becomes

$$2a|_{\text{image}} = 1.22\lambda_o|z_o|M_c/r'. \tag{4.22}$$

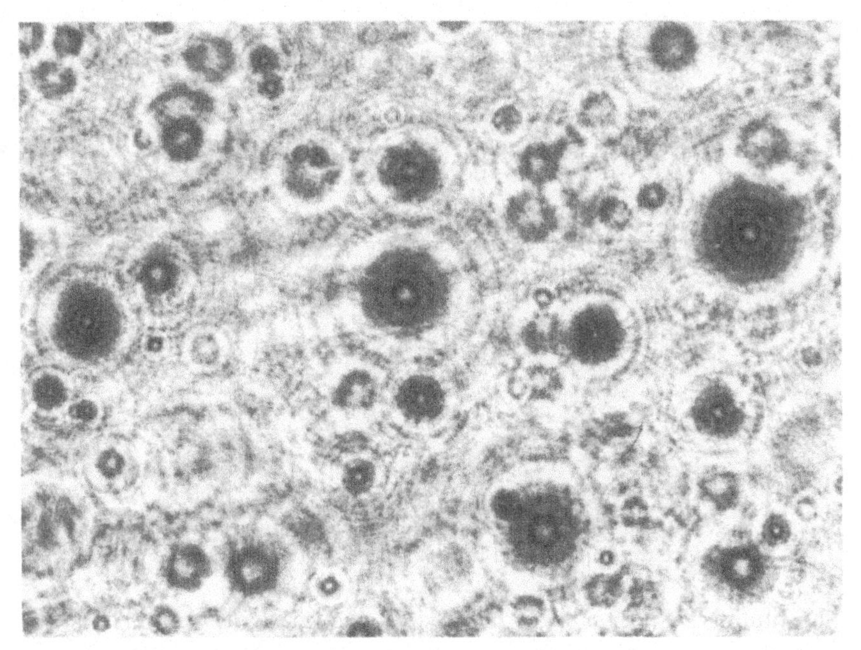

Fig. 4.3 An enlarged portion of a hologram of air bubbles in water.

In the object space, this diameter has to be divided by M_c giving the resoluble particle diameter $2a$ as

$$2a = 1.22\lambda_o|z_o|/r'. \tag{4.23}$$

Thus, for a particle of diameter $2a$ to be resolved, the minimum hologram radius should be

$$r'_{\min} = 1.22\lambda_o|z_o|/2a. \tag{4.24}$$

Substituting this value in the argument of the Bessel function in Equation (4.18), the argument comes out to be about 3.83. This, as shown in Figure 4.5, corresponds to the first zero of the Bessel function. This means that the central maximum must be recorded in order to resolve the particle. From experience this central maximum is sufficient.[6,7] However, there is no harm in a more conservative approach of recording a few side lobes.[3,8] From a very crude approximation for non-zero values of g, we have

$$J_1(g) \sim \left(\frac{2}{\pi g}\right)^{1/2} \cos\left(g - \frac{3\pi}{4}\right), \tag{4.25}$$

which gives the mth side lobe corresponding to

$$g - \frac{3\pi}{4} = (m + \tfrac{1}{4})\pi. \tag{4.26}$$

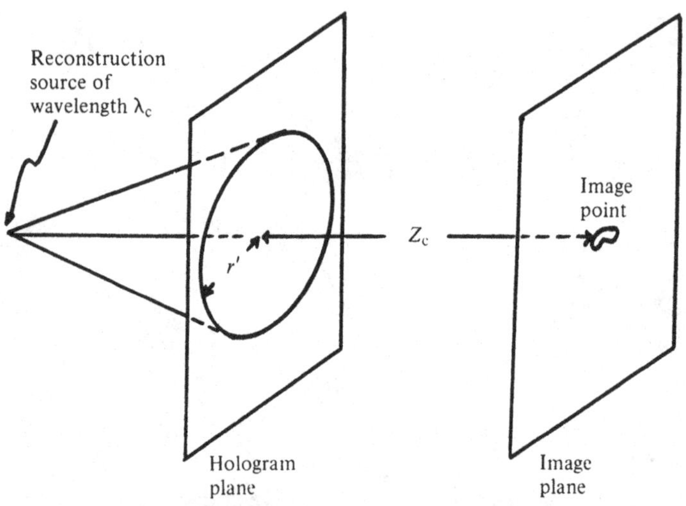

Fig. 4.4 Diagram for the resolution calculation due to a hologram aperture of radius r' and reconstruction wavelength λ_c.

Substituting g for $2\pi ar'/\lambda_o z_o$ in the diffraction pattern, we obtain

$$r'_{\min} = (1 + m)\lambda_o|z_o|/2a, \tag{4.27}$$

if at least m side lobes are to be recorded. For determining the frequency of fine fringes, the argument of the sine term in RHS of Equation (4.18) can be used. The difference between the nth and the $(n + 1)$th fringe being 2π, we obtain

$$r'^2_{n+1} - r'^2_n \approx 2r'\Delta r' = 2\lambda_o m_o|z_o|. \tag{4.28}$$

From Equations (4.27) and (4.28), the fringe spacing is

$$\Delta r' = 2am_o/(1 + m), \tag{4.29}$$

which is the minimum fringe spacing at the outer side of the diffraction pattern or the edge of the mth side lobe. Using the criterion that three side lobes are sufficient for the information storage,[3]

$$\Delta r' = am_o/2. \tag{4.30}$$

Thus, the film must resolve the frequency $2/am_o$. Accounting for the sampling theorem and recording at least half the cut-off frequency let us allow for a factor of 4 making the film resolution requirements $8/am_o$. For $m_o = 1$ and a 5 μm diameter object, this requirement comes to be 3200 lines per millimeter. This is not a problem with modern holographic recording materials, particularly since our calculation has been conservative at every stage. For example, considering only the central maximum ($m = 0$) yields the requirement as only 800

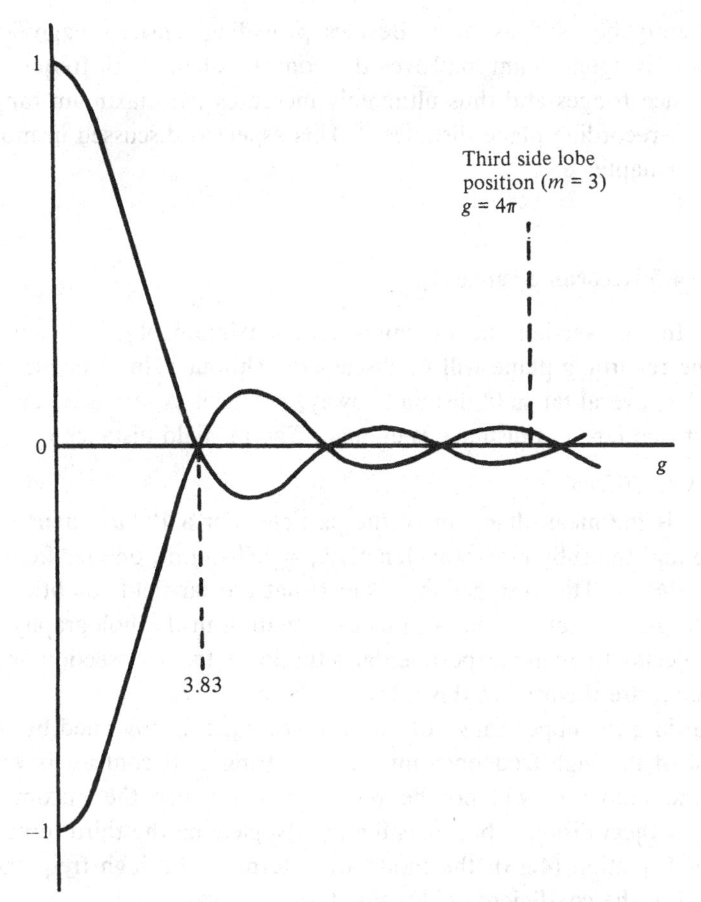

Fig. 4.5 Distribution of the function $2J_1(g)/g$ to determine r' in the hologram for a given number of side lobes of the diffraction pattern to be recorded.

lines per millimeter. Although the particle cross-section was assumed to be circular, there is very little change in the results for other shapes of the same nominal size. Here, it is very interesting to note that when divergent beams are used for the recording ($m_o > 1$), the film resolution requirements are reduced by a factor of m_o. This means that at significant magnification, ordinary films can be used for the purpose of in-line Fraunhofer holography. In fact, when very small objects, say 1–10 μm in diameter, are to be recorded, use of lenses is required to ease the film resolution requirements.[3,8,9] The divergent beam provides a lensless method in this connection. In fact, the divergent beam coming directly from the spatial filter can very conveniently be used as such. Besides providing lensless magnification, the divergent beam improves the contrast of the high frequency interference fringes and thus ultimately increases the maximum range of object–recording plane distance.[10] This aspect is discussed in more detail in Chapter 8.

4.3 Recording range

In this section the minimum and maximum object distances from the recording plane will be discussed. Although, in principle, z_o should be several far-field distances away, practical experience[9] shows that just one far-field distance is enough. The far-field distance is

$$\delta = d^2/\lambda_o, \tag{4.31}$$

where d is the mean diameter of the particle. For a 100 μm diameter particle and the ruby laser wavelength $\lambda_o = 0.6943$ μm, one far-field is about 1.44 cm. This distance is so short that the far-field condition is generally readily satisfied in a practical situation in the holography of microobjects. In many experimental situations, the test-section windows, etc., are themselves this thick.

Regarding the upper range of the distance z_o, it is governed by the contrast of the high frequency interference fringes. If contrast is very poor, the hologram will not be recorded and hence the maximum allowed subject distance becomes limited. Neglecting the third term in RHS of Equation (4.18), the modulation term of the high frequency fringes, i.e. the coefficient of this sine term, is

$$M = \frac{2\pi m_o a^2}{\lambda_o |z_o|} \cdot \frac{|2J_1(2\pi a r'/\lambda_o |z_o|)|}{2\pi a r'/\lambda_o |z_o|}. \tag{4.32}$$

Using Equation (4.27) and considering that at least three side lobes of the diffraction patterns are to be recorded, Equation (4.32) becomes

$$M = \frac{m_o |J_1(4\pi)|}{4N} \approx \frac{0.04 m_o}{N}, \tag{4.33}$$

where N is the number of far-fields given by Equation (4.31) representing the object–recording plane separation. Obviously, depending on the minimum allowable modulation M, there is a maximum allowable value of the distance defined by the number of far-fields, N. System noise, non-uniformity of the emulsion, etc., will further degrade the contrast of the fringes to be recorded. Consequently, the actual allowable maximum value of N should generally be lower than the theoretically predicted value[11] depending on the situation[12]. Thus in a particular experimental situation, the maximum allowable N can at best be determined by calibration. However, the accepted upper value of N in the collimated beam ($m_o = 1$) is generally 100.[12,13] Thus for the collimated beam case, one can generally write

$$1 < N \leqslant 100. \tag{4.34}$$

The recording distance range is very limited particularly for very small particles. For a ruby laser $\lambda_o = 0.6943$ μm and a 5 μm diameter object, 100 far-fields is only about 4 mm. In these situations, the recording plane can be relayed for proper film placement. Pre-magnification of the subject volume using lenses also eases the film resolution requirements.[8,9] According to Equation (4.34) the largest particle in the subject volume should be at least one far-field away from the recording plane and the smallest one at the most 100 far-fields away. Equation (4.31) thus shows that a size range of the order of 10 is allowable in the simplest case. It is interesting to note that using divergent beams ($m_o > 1$) also increases the modulation as seen in Equation (4.33) and hence the range of N.[10,13] In fact, the reported farthest far-field with divergent beams is as high as 2058.[13]

It is also interesting to note that for a similar diameter one-dimensional object, the corresponding modulation derived using Equation (4.20) at $2\pi a x'/\lambda_o z_o = 3.75\pi$ comes out to be

$$M \approx 0.12 m_o^{1/2}/N^{1/2}. \tag{4.35}$$

Comparing Equations (4.33) and (4.35) we observe that the modulation is generally higher in the case of a one-dimensional object. This will allow better recordability in the case of one-dimensional objects as well as longer range of values of N. This kind of object shape dependent change in modulation[4] as well as methods to improve the contrast of the high frequency fringes is considered in Chapter 8.

4.4 Film format

The proper size of the recording emulsion can be determined from the discussions of Sections 4.2 and 4.4. From Figure 4.6, let us consider a microobject in the object space which gives the cone of the diffraction pattern whose minimum radius r'_{min} at the recording plane is to be stored. Thus, the film size must at least take care of the minimum number of side lobes (3 as decided in Section 4.2) and maximum number of far-fields (100 for $m_o = 1$ as decided in Section 4.3).

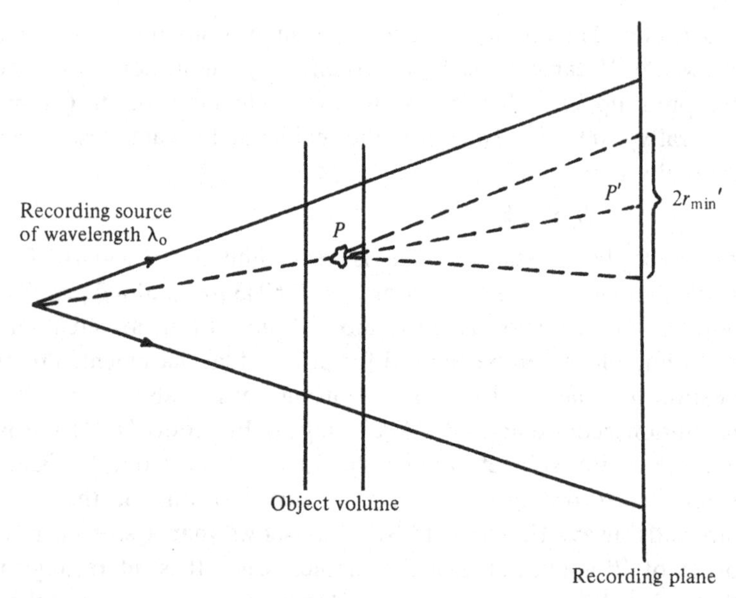

Fig. 4.6 Schematic diagram for determining the minimum film size for recording an in-line hologram of the microobject at the point P.

For two-dimensional objects Equations (4.27) and (4.31) give

$$\begin{aligned} r'_{\min} &= (1 + m)Nd \\ &= 4Nd \quad \text{for} \quad m = 3 \end{aligned} \tag{4.36}$$

where d is the object diameter. The minimum film half width should correspond to the larger of:

(1) The largest number of far-fields for the smaller objects: say 0.20 cm, for $N = 100$, $d = 5\ \mu$m. This consideration is in view of the fact that the far-field for smaller particles is physically very small. Or:

(2) The smallest number of far-fields for the larger particles, say $N = 1$, so that object–recording plane distance is not very large. For $N = 1$, $d = 500\ \mu$m, $r'_{\min}$ comes out to be again 0.2 cm.

These conditions are considered so that the recording distance is not unrealistically large or small and still covers a range of microobject sizes. However, in the unlikely event of there being large particles at many far-fields, Equation (4.36) is available. Thus, a very small film size is actually required for one particle. For the volume to be studied, the actual film size required will therefore be slightly larger than the cross-section of the volume so that edge cutting of the diffraction pattern for the outer particles is not a problem. Also, if the sample volume is pre-magnified using lenses or the divergent recording beam ($m_o > 1$), the film size should be properly scaled up. If a very large number of far-fields is allowed by special techniques,[13] then a proper increase in the film size to satisfy Equation (4.36) will also be necessary. For example, for $N = 2058$ and $d = 5\ \mu$m allowed from Ref. 13, r'_{min} from Equation (4.36) is about 4 cm, which is now a considerable film size even for a single particle.

Another interesting aspect of the divergent beam is that it improves the contrast and hence a greater number of side bands in the diffraction patterns can be recorded than in the collimated beam case.[10] To get full advantage of this capability the film size should be properly increased according to the equation[10]

$$r'/r'_{coll} = m_o^{2/3}, \tag{4.37}$$

where r'_{coll} corresponds to the collimated beam case ($m_o = 1$). Equation (4.37) is obtained by equating the general modulation given by Equation (4.32) to that for the $m_o = 1$ case.

To conclude the common in-line Fraunhofer holography with collimated beams, the film size should be about the subject volume cross-sectional area.[3] For special techniques where a large number of far-fields is allowed, or a lens is used to pre-magnify the subject volume, increased film size will be accordingly required.

4.5 Recording source coherence requirements

The general theory of in-line Fraunhofer holography discussed so far assumes a completely coherent light source. The commonly used output from a laser source is spatially coherent and the temporal coherence length is generally more than sufficient. Still, in less common applications where thermal, electron or x-ray sources are to be used, the coherence aspect becomes important. A laser operating in transverse multimode operation can also be spatially partially coherent.

There are a few studies available on the effects of spatial and temporal coherence in in-line Fraunhofer holography.[14–16] Without going into the rigorous theory on partial coherence, we are describing the important requirements useful in successful recording of the hologram. At the recording stage the contrast of the fine interference fringes is reduced by the degree of coherence. This means a less effective size of hologram and hence reduced resolution capability.

The degree of coherence can be determined from knowledge of the source and even measured if needed by simple experiments. For our purpose in in-line Fraunhofer holography, we are discussing here the requirements from a practical point of view. For spatial coherence, the size of a non-laser source gives enough information to fulfill the needs. Temporal coherence requirements can be satisfied by keeping the maximum path difference well within the coherence length.

4.5.1 *Spatial coherence*

The radiation from a laser oscillating in one transverse mode, say the commonly used TEM_{00} mode, is spatially coherent. However when thermal sources, such as a mercury arc lamp, are used for recording holograms, spatial coherence properties for a given source size need to be carefully investigated. The well-known van Cittert–Zernike theorem[17,18] can be used to determine the degree of spatial coherence or in simple words, the maximum allowable pinhole diameter of the source.

There are two ways of obtaining a beam to illuminate the subject volume in in-line holography. One, illustrated in Figure 4.7(*a*), is when the light from the aperture is used as such (divergent beam) and the other is when a collimating lens is ultimately used as seen in Figure 4.7(*b*). For an infinitesimally small aperture, the source can be considered spatially coherent. However, in practice the aperture has a certain size and each point within the pinhole plane acts as an independent source. Even if the source wavelength is ideal with zero spectral width, these infinite source points will provide infinite superimposed holograms at the recording plane. Consequently, the hologram fringe visibility reduces. Ultimately at certain conditions, the fringes will be completely smeared or lost, resulting in no hologram formation. According to van Cittert–Zernike Theorem,[17,18] a circular pinhole of radius r_o gives the loss of the contrast or the degree of spatial coherence as

$$|\mu_s| = \left| \frac{J_1(2\pi r_o \theta/\lambda_o)}{\pi r_o \theta/\lambda_o} \right|. \tag{4.38}$$

The distribution given by Equation (4.38) is basically the Fourier transform of the circular aperture and is represented in Figure 4.5. As seen in Section 4.4, common in-line Fraunhofer holography of a microobject requires a maximum beam cross-section of less than 1 cm. If the aperture containing the pinhole is 200 cm away from the

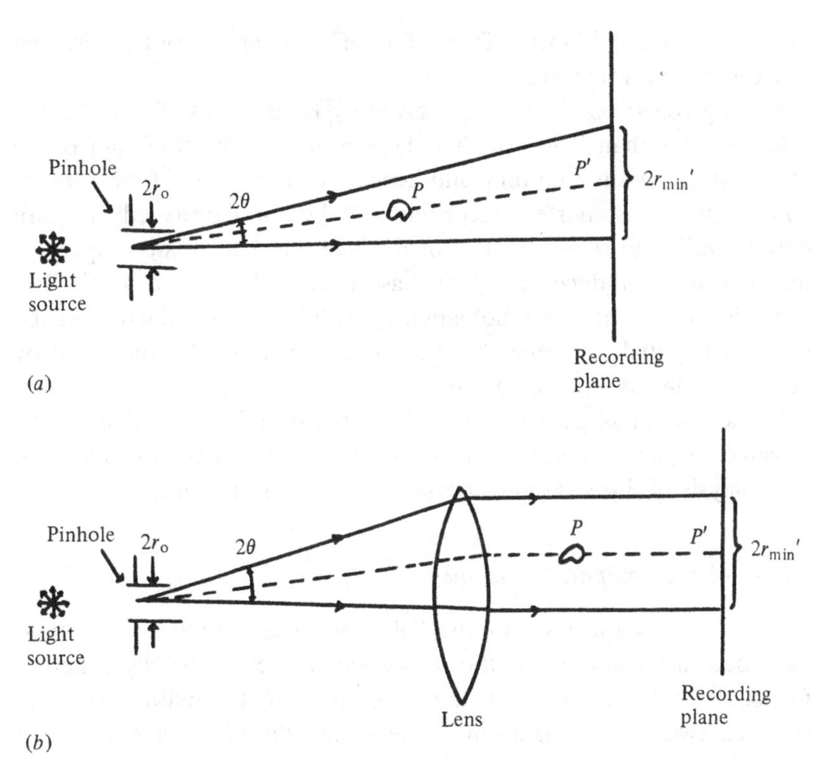

Fig. 4.7 Diagram for determining the spatial coherence requirements in in-line Fraunhofer holography (*a*) when the light from the pinhole is used as such, and (*b*) when a lens is used to collimate the beam.

particles, as in Figure 4.7(*a*), or the lens, as in Figure 4.7(*b*), then θ is less than 0.0025 rad. For a high pressure mercury arc lamp with $\lambda_o = 0.5461\ \mu m$, r_o as large as $50\ \mu m$, $\theta = 0.0025$ rad, $|\mu_s|$ given by Equation (4.38) is greater than 0.75. The well-accepted criterion for $|\mu_s|$ is that it should be more than $1/\sqrt{2} = 0.707$. Thus a $50\ \mu m$ radius pinhole is good enough in the above example. For a typical value of $100\ W/cm^2$ at $\lambda_o = 0.5461\ \mu m$ and $\Delta\lambda_o = 0.0050\ \mu m$ we have a useful power of 8 mW or 80×10^3 erg/s. A typical emulsion, Agfa 10E56, suitable for this wavelength region requires only about $20\ erg/cm^2$. Using a wavelength filter to reduce the above spectral width will further reduce the useful power. However, the above mentioned power may be used in many in-line holographic applications with small subject volume cross-sections if a suitable laser source is not available. It is worth mentioning here that for off-axis applications, the nominal beam spread (θ) requirements are as high as 15° and pinhole radii as small as $r_o = 0.5\ \mu m$ are required even with available power at $\Delta\lambda_o = 5$ nm.[18] Thus, for off-axis applications, thermal sources are not practical.

For a given θ, λ_o, and r_o, $|\mu_s|$ given by Equation (4.38) can have a value greater than 0.707 so that there is practically minimum or no effect on hologram recording and hence reconstruction. If the value is considerably less, perfect recording will be for a reduced aperture corresponding to reduced value of θ.[6] The effect of reduced aperture might result in reduced resolution as given by Equation (4.24). The reduction in the effective hologram aperture against reduction in the degree of spatial coherence has been experimentally demonstrated by Asakura, Matsushita and Mishina.[16]

In fact, when single transverse mode performance of a laser is not possible, it can be considered as an incoherent source. In that case, the analysis of this section is useful even for a laser source.

4.5.2 *Temporal coherence*

Even if the laser or the light source is spatially coherent as discussed in Section 4.5.1, the source should be temporally coherent for successful recording of the hologram. If the path difference between two interacting beams is less than the coherence length of the source then the interference pattern will have practically the same contrast as defined by Equation (4.32). As the path difference increases, the contrast reduces, ultimately resulting in loss of the interference pattern. For a single frequency gas laser the coherence length can be as high as 1 km.[18] For the more common pulsed ruby laser in far-field applications, it could be as low as 80 μm in multi-mode operation or as high as 1.4 m in the single frequency Q-switch operation.[19] In this section we determine the temporal coherence requirements in the in-line Fraunhofer case of interest.

Figure 4.8 describes the recording geometry. The object point P is at a distance R from the optical axis. Therefore, if the line joining the source S and P meets the recording plane at the point P', then according to Equation (3.4), the distance $O'P'$ is $m_o R$ where m_o is the magnification from the recording configuration alone. Suppose Q_1 and Q_2 define points in the recording plane such that the distance Q_1Q_2 defines the aperture of a hologram sufficient from resolution and recordability considerations. This distance can be conveniently defined as $2r'_{min}$ where r'_{min} is defined in equation (4.27) or (4.36). The maximum path difference between the object and the reference beams at the upper portion of the aperture is therefore

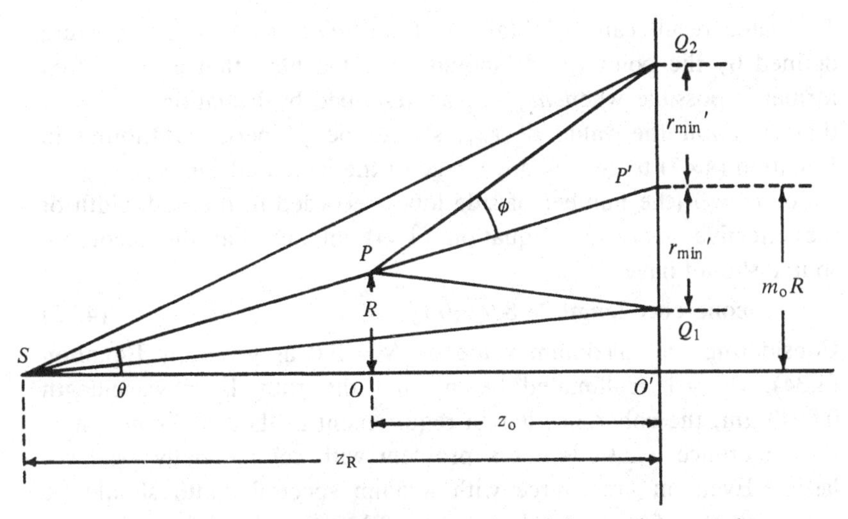

Fig. 4.8 Diagram for the determination of the maximum path difference between the object and the reference beams in in-line holography. (After Ref. 21.)

$$\delta_{oR} = \overline{SP} + \overline{PQ_2} - \overline{SQ_2} \tag{4.39}$$

Applying first order approximation

$$\begin{aligned} \overline{SP} &= [(z_R - z_o)^2 + R^2]^{1/2} \\ &\approx z_R - z_o + \frac{R^2}{2(z_R - z_o)}. \end{aligned} \tag{4.40}$$

Similarly

$$\begin{aligned} \overline{PQ_2} &= [z_o^2 + (m_oR + r'_{min} - R)^2]^{1/2} \\ &\approx z_o + \frac{(m_oR + r'_{min} - R)^2}{2z_o} \end{aligned} \tag{4.41}$$

and

$$\begin{aligned} \overline{SQ_2} &= [z_R^2 + (m_oR + r'_{min})^2]^{1/2} \\ &\approx z_R + \frac{(m_oR + r'_{min})^2}{2z_R}. \end{aligned} \tag{4.42}$$

Substituting values from Equations (4.40), (4.41), (4.42) and (3.4) in Equation (4.39) gives

$$\delta_{oR} \approx r'^2_{min}/2m_oz_o. \tag{4.43}$$

Substituting the value of r'_{min} from Equation (4.27) and using $N = \lambda_oz_o/d^2$ as the number of far-fields defined in Equation (4.31), Equation (4.43) becomes

$$\delta_{oR} \approx (1 + m)^2N\lambda_o/2m_o. \tag{4.44}$$

The same result can be obtained at the lower side of the aperture defined by the point Q_1. It should be noted here that a larger film format is possible when $m_o > 1$, as described by Equation (4.37). In that situation the value of r'_{min} should be properly substituted in Equation (4.43) to get full advantage of the increased film size.

For $m = 3$ (the number of side lobes recorded in the half width of the effective aperture), Equation (4.44) implies that the recording source should have

$$\text{coherence length} > 8N\lambda_o/m_o. \quad (4.45)$$

Considering the maximum value of $N = 100$ as given by Equation (4.34), $m_o = 1$ (collimated beam) and the ruby laser wavelength 0.6943 μm, the coherence length requirement is about 0.5 mm. Thus, the coherence length is not a problem with commercially available lasers. Even an arc source with a 5 nm spectral width should be sufficient at a few far-fields.[9] Successful application of a miniature Q-switched ruby laser operated by a NiCd battery has also been reported.[20]

4.6 Object velocity and exposure time

The main advantage of particle field holography is that it can cope with transient microobjects in a volume. Thus it is very important to determine the pulse duration requirements for a given object velocity. Consequently for a given pulse duration, a certain light energy is required for a given sensitivity of the recording medium. In this section we first consider transverse and longitudinal object displacement effects and then describe the combined or general motion. Finally, a more suitable geometry for higher allowable displacements is described for known displacement directions.

Considering the central diffraction maxima and m side lobes sufficient for resolution, Equation (4.27) can be written as

$$r'_{min} = (1 + m)\lambda_o|z_o|/d, \quad (4.46)$$

where d is the object diameter. Equations (4.28) and (4.46) yield the minimum spacing of the fine interference fringes as

$$\Delta r'_{min} = \frac{dm_o}{1 + m}. \quad (4.47)$$

Assuming that the object motion during exposure must not smear this fine interference pattern, the allowable subject movement will be discussed in this section. The main results are also discussed in Ref. 21.

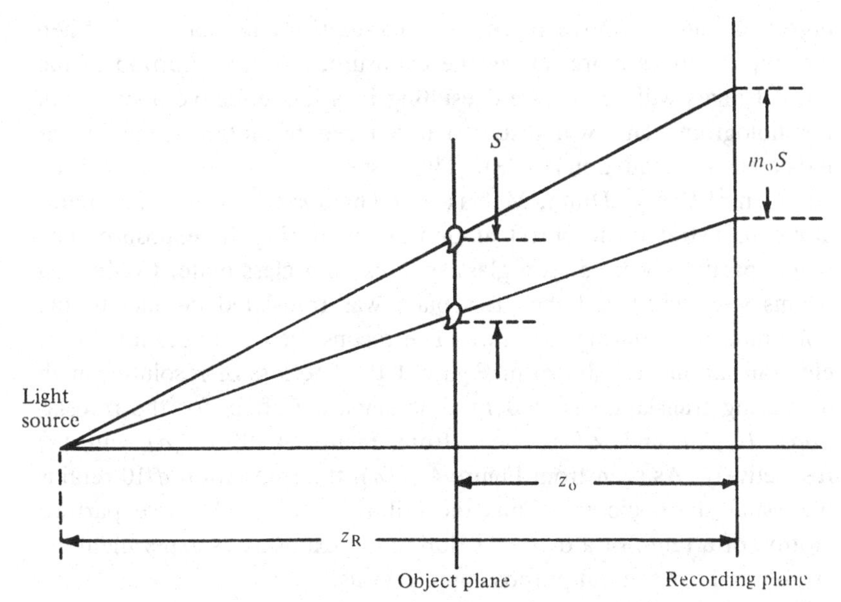

Fig. 4.9 Schematic diagram (not to scale) to determine the movement of the interference fringes at the recording plane against a transverse movement S of the object. The entire fringe system is shifted by the amount m_oS at the recording plane. (After Ref. 21.)

4.6.1 *Transverse movement*

If during the exposure, the object moves an in-plane (ξ, η) distance S then the corresponding movement of the pattern in the recording plane, according to Equation (3.5), would be m_oS. This situation is represented in Figure 4.9. This means that r' is also changed by m_oS, i.e. the entire fringe structure is moved by this distance. For a continuously moving object during the exposure time, these fringes will be continuously moving. However, if the maximum fringe movement is much less than the minimum required fringe spacing given by Equation (4.47), then the image resolution will remain good. Otherwise there will be a smearing effect.

As a criterion, we can assume that the fringe movement m_oS is not more than one-tenth of the minimum required fringe spacing for the *smallest* object. Equation (4.47) then yields:

$$S < d_{\min}/10(1 + m), \tag{4.48}$$

where $d_{\min}$ is the diameter of the smallest object to be recorded. If we consider the central diffraction $(m = 0)$ maxima sufficient for common applications,[6,7] we obtain the common conclusion[3,8] that the object should not move more than one-tenth of its diameter. When the object moves more during the exposure, a greater portion of the outer fringes will be smeared resulting in a less effective aperture of the hologram. This will result in a reduced resolution capability as indicated by Equation (4.48). This aspect has been discussed by Brenden.[20] Crane, Dunn, Malyak and Thompson[22] have experimentally demonstrated the object motion effects during the exposure. The object particles were 50 μm glass spheres on a glass plate. Collimated beams were used and the glass plate was translated parallel to the hologram plane during exposure. The reconstructed images for different translations are shown in Figure 4.10. The loss of resolution with increasing translation is evident. The smearing effect for the translations $d/2$, d and $2d$ is clear from Figure 4.10(c), (d) and (e) respectively. As seen from Figure 4.10(a), the translation $d/10$ retains the usual sharp picture. Thus the criterion of an allowable particle motion of a tenth of a diameter during the exposure is experimentally established for general purpose applications.

For a 5 μm diameter object the allowable subject movement given by Equation (4.48) is 0.5 μm. With a Q-switched ruby laser of 15 ns pulse duration, the allowable object velocity is about 33 m/s. This is more than adequate for many practical situations. Agfa 10E75 emulsion requires about 44.5 erg/cm^2 exposure (see Section 5.5.1). To expose a 15 cm × 10 cm area, one would require only about 0.7 mJ energy in a pulse. This is easily obtained in commercially available ruby oscillators.

4.6.2 *Longitudinal movement*

For allowable object movement in the longitudinal direction, Figure 4.11 represents the situation. Here, when the object moves a distance Δz_0 along the longitudinal direction, the center of the diffraction pattern changes from $m_0 R$ by $\Delta m_0 R$ distance, where Δm_0 is the change in the magnification due to the movement of the object. Also the fine fringe frequency given by Equation (4.14) depends not only on r' and m_0 but also on z_0. Thus, with respect to the optical axis, the position of an interference fringe in the recording plane is

$$r = m_0 R \pm r', \tag{4.49}$$

where the positive and negative signs are for upper and lower fringes respectively with respect to the center of the diffraction pattern.

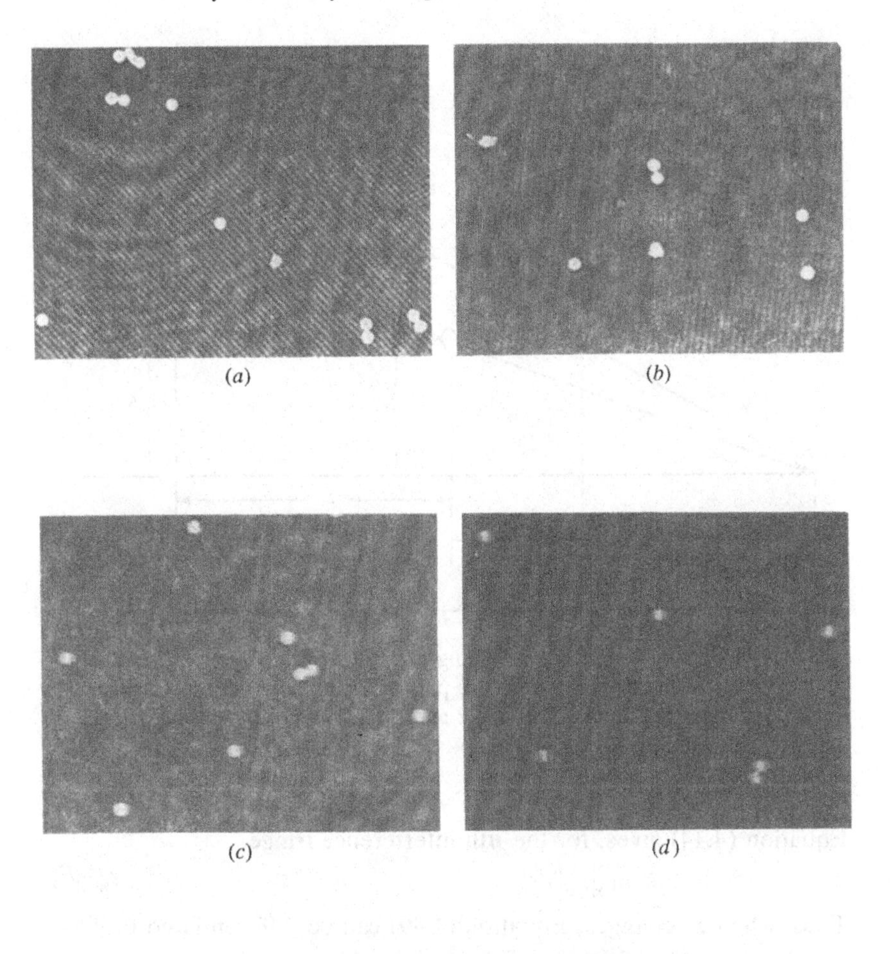

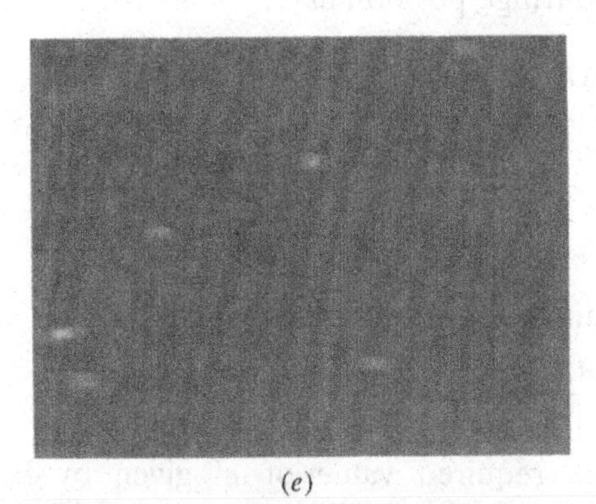

Fig. 4.10 A reconstructed image from a hologram of 50 μm glass microspheres. The particles were translated horizontally by (*a*) $d/10$ (*b*) $d/5$ (*c*) $d/2$ (*d*) d and (*e*) $2d$ during the exposure. (Crane *et al.*[22])

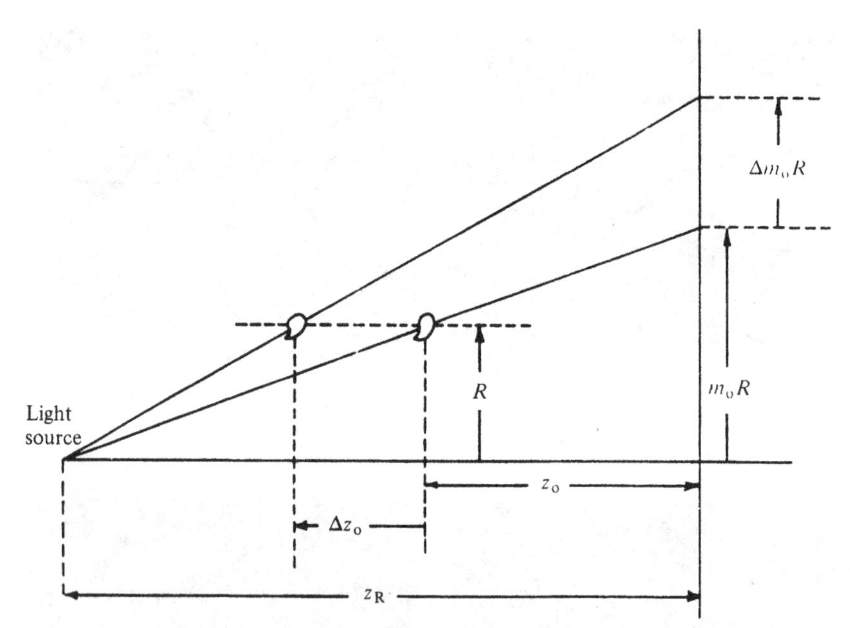

Fig. 4.11 Diagram representing the effect of longitudinal movement Δz_o of the object. The entire pattern at the recording plane is shifted by the amount $\Delta m_o R$ and the frequency of the fine fringes is also altered. (After Ref. 21.)

Equation (4.14) gives, for the nth interference fringe,

$$r' = (2n\lambda_o m_o |z_o|)^{1/2}. \tag{4.50}$$

Thus, when z_o changes, Equation (4.49) can be differentiated to yield the change in the fringe position as

$$\delta_r = R\Delta m_o \pm \frac{r'}{2}\left(\frac{\Delta|z_o|}{|z_o|} + \frac{\Delta m_o}{m_o}\right) \tag{4.51}$$

The relationship for m_o given by Equation (3.4) can be used to obtain

$$\Delta m_o = m_o^2 \Delta z_o / z_R, \tag{4.52}$$

which on substituting into Equation (4.51) gives

$$\delta_r = \frac{R m_o^2 |\Delta z_o|}{|z_R|} \pm \frac{m_o r' |\Delta z_o|}{2|z_o|} \tag{4.53}$$

At the minimum required value of r' given by Equation (4.46), Equation (4.53) becomes

$$\delta_r = \frac{R m_o^2 |\Delta z_o|}{|z_R|} \pm \frac{m_o \lambda_o (1 + m)|\Delta z_o|}{2d}. \tag{4.54}$$

Dividing by the minimum fringe spacing from the resolution requirements given by Equation (4.47), we obtain the fractional fringe change as

$$\frac{\delta_r}{\Delta r'_{\text{min}}} = \frac{Rm_o(1+m)|\Delta z_o|}{|z_R|d} \pm \frac{\lambda_o(1+m)^2|\Delta z_o|}{2d^2}. \tag{4.55}$$

Obviously, this fraction should be as small as possible, say less than 0.1, so that the fringe pattern is not smeared. Due to the positive and the negative signs in RHS of Equation (4.55), the upper and lower sides of the effective hologram will be affected differently. In general, one area in the hologram may be wiped out whereas the other may remain perfect. However, if both the terms in the RHS of Equation (4.55) are individually kept small, the hologram will be practically perfect. The maximum fractional fringe change will be

$$\left.\frac{\delta_r}{\Delta r'_{\text{min}}}\right|_{\text{max}} = \frac{Rm_o(1+m)|\Delta z_o|}{|z_R|d} + \frac{\lambda_o(1+m)^2|\Delta z_o|}{2d^2}. \tag{4.56}$$

Let us have a careful look at Equation (4.56). The quantity Rm_o/z_R, with the help of Equation (3.4) and geometry of Figure 4.8, is nothing but $\tan\theta$. Similarly, Equation (4.46) yields $(1+m)\lambda_o/d = r'_{\text{min}}/z_o$, which, under the paraxial approximation is $\tan\phi$. Angle ϕ is shown in Figure 4.8. Therefore Equation (4.56) can be written as

$$\left.\frac{\delta_r}{\Delta r'_{\text{min}}}\right|_{\text{max}} = \frac{(1+m)|\Delta z_o|\tan\theta}{d} + \frac{(1+m)|\Delta z_o|\tan\phi}{2d}. \tag{4.57}$$

Let us consider the situation when the object movement Δz_o is the allowable transverse movement S; S is given by Equation (4.47) for allowable $m_oS = \Delta r'_{\text{min}}/10$. Equation (4.57) then becomes

$$\left.\frac{\delta_r}{\Delta r'_{\text{min}}}\right|_{\text{max}} = \frac{\tan\theta}{10} + \frac{\tan\phi}{20}. \tag{4.58}$$

Even with θ and ϕ hypothetically as high as 45° each, the fractional change given by Equation (4.58) is only 0.15. Thus, the effect of longitudinal object motion, even with the divergent beam case, can be generally neglected when the criterion for the transverse motion is met.

4.6.3 *Combined effect*

The effects of transverse and longitudinal object motions have so far been described separately. Their combined effect can be simply obtained from the discussions of Sections 4.6.2 and 4.6.3. The

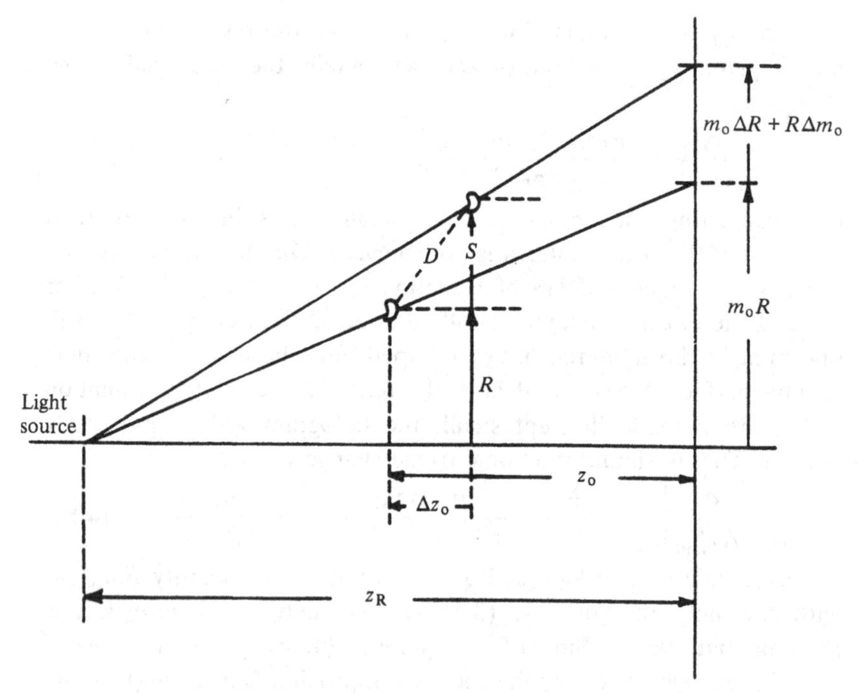

Fig. 4.12 Diagram to describe the effect of the object motion of magnitude D in a general direction. (After Ref. 21.)

situation is illustrated in Figure 4.12. Here the object distance R from the optical axis as shown in Figure 4.11 is not a constant. Thus, when RHS of Equation (4.49) is differentiated, we obtain the change in the fringe position as

$$\delta_r' = m_0 \Delta R + \delta_r, \tag{4.59}$$

where δ_r is given by Equation (4.51), ΔR is nothing but the transverse subject movement S and $m_0 \Delta R$ is therefore the corresponding fringe movement on the recording plane.

Therefore the maximum fringe movement is the addition of the maximum movements corresponding to longitudinal and transverse motion components individually. The net maximum fractional fringe change will therefore be the sum of RHS of Equation (4.56) and $m_0 S/\Delta r'_{\text{min}}$.

4.6.4 *Geometry for high velocity particles*

It is clear now that common in-line Fraunhofer holography with pulsed lasers such as Q-switched ruby is adequate in many

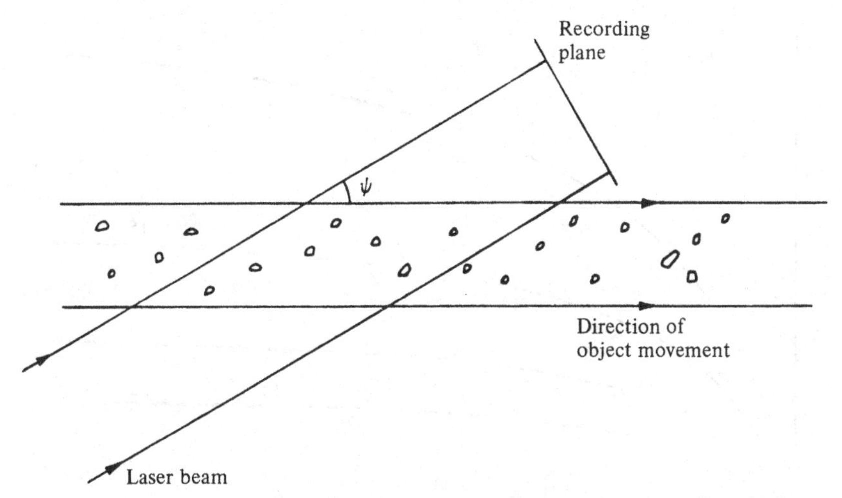

Fig. 4.13 Schematic diagram for recording geometry to allow higher object displacements. (After Ref. 21.)

transient situations. However, hypervelocity particles pose a new problem. At 20 km/s, a particle would move 300 μm during a 15 ms pulse. Obviously this motion is large for the successful recording of holograms of microobjects. Special techniques such as velocity synchronized Fourier transform holograms (see Chapter 11) can be used in these situations. These techniques are far more complex than the in-line approach. We know that the longitudinal subject movement is far less critical than the transverse movement. Thus, for known motion direction, higher object velocities can be allowed from geometrical considerations. If the object velocity is not a problem, the approach will allow a broad laser pulse. As seen in Figure 4.13, the recording geometry can be arranged so that the movement is mainly longitudinal with respect to the recording plane. The collimated beam case is considered here for simplicity. The motion will have transverse and longitudinal components and their respective effects on the hologram fringes can be determined for the optimum value of ψ. If D is the subject movement, then the longitudinal and transverse movements are $D \cos \psi$ and $D \sin \psi$ respectively. Considerations for Equation (4.48) for $m \neq 0$, Equation (4.56) and Equation (4.59) can be used to get the maximum fractional fringe change (for the collimated beam case) as

$$\delta_r'' = \frac{D(1+m)\sin\psi}{d} + \frac{D\lambda_0(1+m)^2\cos\psi}{2d^2}. \tag{4.60}$$

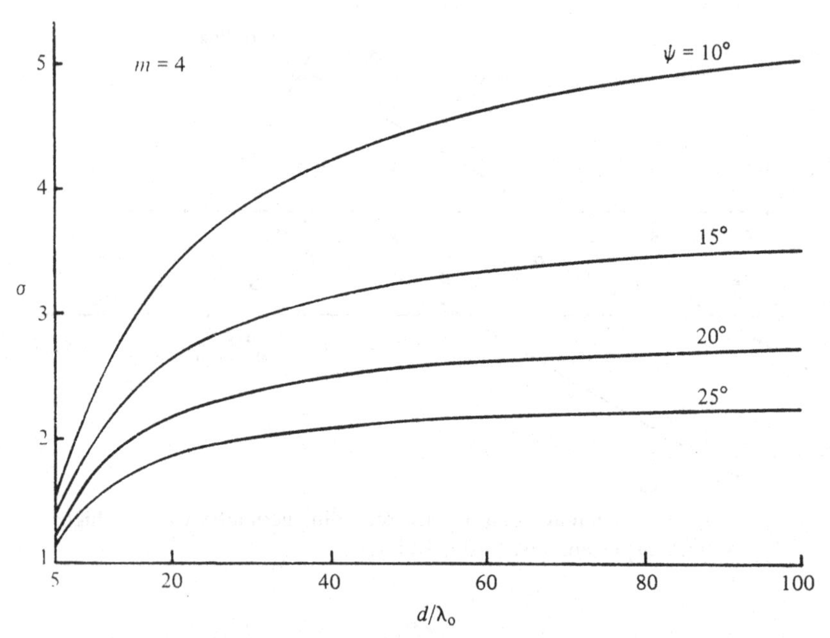

Fig. 4.14 Plot of σ against d/λ_o for some values of ψ and $m = 4$. (After Ref. 21.)

As discussed earlier in Section 4.6.2, the term $(1 + m)/d$ is generally large compared to $\lambda_o(1 + m)^2/2d^2$. Thus, if the value of ψ is kept near 0°, the value of δ_r'' can be reduced to a minimum. Due to physical obstructions, $\psi = 0$ may not be practically possible or convenient. However, by keeping ψ low, considerable gain in the allowable subject movement is possible. Let us consider the maximum allowable value of δ_r'' to be 0.1, Equation (4.60) can be used to obtain the maximum subject movement D_{max} as

$$D_{max} = 0.1d\sigma/(1 + m), \quad (4.61)$$

where

$$\sigma = [\sin\psi + \lambda_o(1 + m)\cos\psi/2d]^{-1} \quad (4.62)$$

Since $0.1d/(1 + m)$ is the previously accepted allowable subject movement, the allowance is increased if σ is greater than unity.

In Figure 4.14, σ has been plotted against d/λ_o for some values of ψ and $m = 4$. It is found that for higher d/λ_o and lower ψ, maximum values of σ are obtained. For large d/λ_o, σ approaches cosec ψ. It is interesting to note that for a given ψ, σ is larger for large particle diameters d. Also the maximum allowable object movement is defined in terms of the object diameter as given by Equation (4.16).

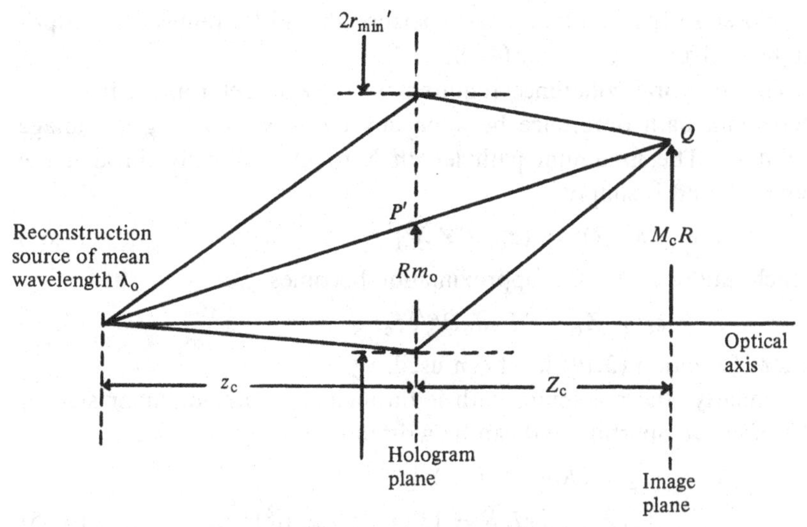

Fig. 4.15 Schematic diagram (not to scale) for determining source coherence requirements during the reconstruction.

Thus the net object displacement can be significantly increased by the geometrical approach, particularly for larger objects.

4.7 Coherence requirements during reconstruction

In Section 4.5 we have studied spatial and temporal coherence requirements in in-line far-field holography at the recording stage. Similar requirements can be determined during the reconstruction. The reconstruction arrangement is represented in Figure 4.15. The reconstruction source of mean wavelength λ_c is situated on the optical axis at the distance z_c from the hologram. The object point is at P (Figure 3.1) whose diffraction pattern is centered at the point P' in the recording plane. The distances of P and P' from the optical axis are known to be R and $m_o R$ respectively. According to Equation (3.33), the virtual image point is at a distance $M_c R$ from the optical axis where M_c is the transverse magnification of the conjugate image. Equation (3.19) therefore gives that the reconstruction source, points P' and Q are on a straight line as shown in the figure. The van Cittert–Zernike theorem[17,18] can again be used to determine the spatial coherence requirements of the source, as discussed in Section 4.5.1. The discussion of Section 4.5.1 shows that z_c should be more than 200 cm for a 100 μm diameter source aperture in the case of $\lambda_o = 0.5461$ μm. Obviously if the source is to be closer to the hologram, the aperture should be reduced appropriately as given by equation (4.38).

The temporal coherence requirements can be determined from the maximum path difference between any two rays reaching the image point Q. The minimum path length Σ to Q is directly through the point P' and is simply

$$\Sigma \approx [(M_c R)^2 + (z_c + Z_c)^2]^{1/2}, \tag{4.63}$$

which, under first-order approximation becomes

$$\Sigma \approx z_c + Z_c + M_c m_o R^2/2z_c, \tag{4.64}$$

where Equation (3.19) has been used.

Similarly, the maximum path is through the outer or inner side of the effective aperture and can be written as

$$\Sigma' = [z_c^2 + (Rm_o \pm r'_{\text{min}})^2]^{1/2} + \{Z_c^2 + [M_c R - (Rm_o \pm r'_{\text{min}})]^2\}^{1/2}, \tag{4.65}$$

where the positive and negative signs correspond to the upper and the lower paths respectively as shown in Figure 4.15. Under first-order approximation and using Equations (3.16) and (3.19), Equation (4.65) becomes

$$\Sigma' \approx z_c + Z_c + \frac{M_c m_o R^2}{2z_c} + \frac{nr'^2_{\text{min}}}{m_o z_o}. \tag{4.66}$$

Equations (4.64) and (4.66) yield the maximum path difference

$$\delta_\Sigma = \Sigma' - \Sigma \approx nr'^2_{\text{min}}/2m_o z_o. \tag{4.67}$$

Except for the effect of the wavelength change ($n \neq 1$), this is basically the same as the requirement at the recording stage described by Equation (4.43).

As discussed in the Section 4.5.2, the temporal coherence requirements are therefore not a problem with commercially available lasers and even an arc source with a filter could be used. Anyway CW lasers such as HeNe with sufficient spatial and temporal coherence are readily available. The preceding analysis is basically to learn the requirements no matter how relaxed they are found to be.

4.8 Diffraction-limited resolution and depth-of-focus

The integral limits in Equation (3.8) are considered to be infinite meaning an infinite size of the hologram. This is physically impossible. Even for a large size recording plate, the actual effective hologram aperture for a microobject is limited by the emulsion resolution requirements, coherence considerations, etc. The infinite hologram size assumption leads to the idealistic irradiance distribution in the reconstruction as given by Equation (3.33) irrespective of the object size. The physically finite hologram aperture acts as a lens and there is a certain aperture-limited resolution. Even for a point object, there will be certain minimum image size governed by the diffraction-limited irradiance distribution. Similarly, a point source will yield a focal uncertainty near the image. The aperture related resolution and depth-of-focus are discussed in this section. The effect of random background noise or speckle noise is also outlined.

4.8.1 *Lateral resolution*

Referring to Figure 4.4, the normalized diffracted irradiance due to the circular aperture of radius r' at the image plane at distance Z_c is given by the Airy distribution (see for example Ref. 23).

$$I(R') = \left[\frac{J_1(2\pi r' R'/\lambda_c Z_c)}{2\pi r' R'/\lambda_c Z_c}\right]^2, \tag{4.68}$$

where R' is the radial distance in the image plane from the center of the image point. J_1 represents the Bessel function of order 1. This distribution is basically the intensity diffraction pattern of the circular aperture. As implied from Figure 4.5 there is a bright central spot and luminous rings of rapidly decreasing intensity. The radius of the central diffraction spot corresponds to $g \approx 3.83$, which in terms of R' comes out to be

$$2R'|_{\text{diffraction spot}} \approx \frac{3.83\lambda_c Z_c}{\pi r'} \simeq \frac{1.22\lambda_c Z_c}{r'}. \tag{4.69}$$

Therefore even a point object will result in an image spot diameter as given by Equation (4.69). Obviously this spot size can be called the minimum resoluble image size. The spot radius is also the separation between two points that can be resolved according to the Rayleigh resolution criterion.[24] This criterion states that two point sources can be resolved when the maximum of the illumination from one source coincides with the first minimum of the illumination produced by the other. The criterion is changed with coherence considerations and also with the geometry of the imaging system.[25] However, the changes in the resolution limits are only marginal[25] for our purpose here in establishing the order of the resolution capability. As seen earlier in the chapter, the image quality is affected by many practical considerations. Therefore, detailed discussions leading to marginal changes in the criterion are beyond the scope here.

4.8.2 *Depth-of-focus*

Equation (4.68) describes the intensity variation in the image plane. Similarly, the intensity variation along the axis near the image is described by the normalized variation[23]

$$I(\Delta Z_c) = \left[\frac{\sin(\pi r'^2 \Delta Z_c / 2\lambda_c Z_c^2)}{\pi r'^2 \Delta Z_c / 2\lambda_c Z_c^2}\right]^2. \tag{4.70}$$

The variation $(\sin g)^2/g^2$ has been plotted in Figure 4.16. Assuming about 20% loss in the intensity with respect to that at the center is permissible, the allowable argument of the sine function is about 0.8. This consideration yields the focal tolerance as:

$$\begin{aligned} \Delta Z_c|_{\text{allowable}} &\sim \pm \frac{0.8 \times 2}{\pi}\left(\frac{Z_c}{r'}\right)\lambda_c \\ &\sim \pm \frac{1}{2}\left(\frac{Z_c}{r'}\right)^2 \lambda_c. \end{aligned} \tag{4.71}$$

The total depth-of-focus is thus $(Z_c/r')^2\lambda_c$. Equations (4.69) and (4.71) can now be combined crudely, to get the well-known result that for a system that can resolve the diameter d, the depth-of-focus is d^2/λ, where λ is the wavelength of the light used.

4.8.3 *Speckle noise*

Different fine unwanted surface reliefs, refractive index variations, grains in the emulsions, etc. in the processed hologram act as diffusers. We will outline here the combined effect in a general sense as a coherent noise. At any plane in the space, these random path variations in the emulsion combine constructively and destructively to form a random granular pattern called laser speckles. The contrast of these patterns will depend on the scattered noise intensity. The size of these speckles at the reconstructed image plane is of interest here. Recalling Figure 4.4, the typical speckle diameter σ_s in the image plane is[26]

$$\sigma_s \approx \frac{\lambda_c Z_c}{r'}, \tag{4.72}$$

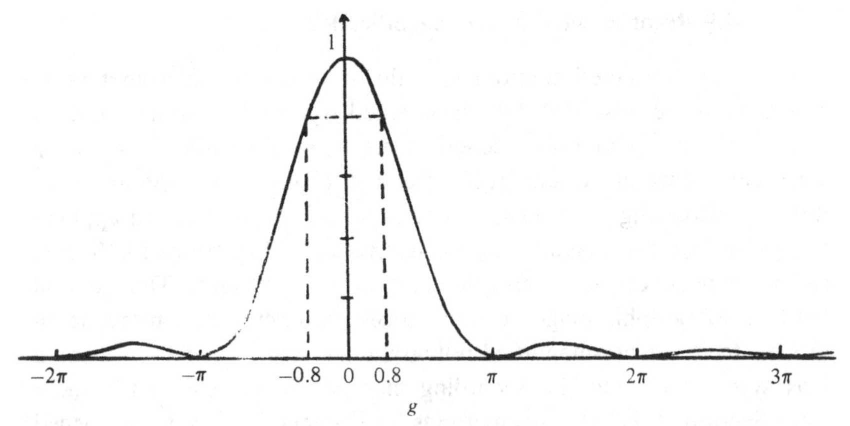

Fig. 4.16 Plot of the function $(\sin g)^2/g^2$ against g. The value of the function is 0.8 at $g \sim 0.8$.

where r' here is the radius of the illuminated section of the hologram during reconstruction. Without going into speckle theory in too much detail, it can be stated that this diameter corresponds to the diffraction-limited spot diameter given by Equation (4.69). In physical terms, no structure finer than the aperture-limited resolution capability can be clearly seen. This leads to the similarity of the aperture-limited diffraction spot diameter and the typical speckle diameter.

In practical in-line Fraunhofer holography, the actual effective film radius needed for an individual microobject is a fraction of a centimeter (see Section 4.4). On the other hand, a much larger film cross-section is generally recorded to cover a test-section volume. The entire recorded cross-section is again generally illuminated by the reconstruction beam during the analysis. Therefore the speckle size is generally very small as compared to size of the microobjects under study. For this reason as well as low scattering noise in modern holographic emulsions, speckle noise is generally not a problem.[27]

Suppose a weak diffuser is introduced within the reconstruction path[28] near the final image plane (i.e. near the magnified image plane produced by a lens). Fine speckles will then be produced at the image plane. These speckles are created by scattering of the light which reaches the diffuser. If the diffuser is rotated, these speckles will move and/or change in a complex manner.[29] The average effect will then yield a smoother background and easier detection of the image therein.[28] Thus, the random background noise effect can be reduced by a rotating diffuser.

4.9 Reconstructed image magnification

Reconstructed microimages down to a few micrometers in size cannot be resolved by an unaided eye. Thus some kind of magnification is generally needed. Lensless magnification, i.e. using spherical recording and/or reconstruction beams, with the same or different wavelengths is possible and is utilized in specialized applications. This kind of magnification is described by Equations (3.18) and (3.24) respectively for conjugate and primary images. The general lensless holographic magnification process is described in more detail in Chapter 6. A pre-magnified hologram can also be recorded using a lens system between the recording medium and the object volume (see Section 5.2). These methods are generally used in special circumstances and even so have practical limitations regarding the magnification. Consequently, they can be used as such (i.e. with no further magnification during the reconstruction process) for larger objects only. Magnification at the reconstruction stage is thus often required. These methods are discussed in this section.

4.9.1 *Microscopic observation*

One of the straightforward solutions is to examine the real images using a microscope. There are actual applications of this, for example by Pavitt, Jackson, Adams and Bartlett[27] and Bartlett and Adams[28] in the in-line method and by Briones, Heflinger and Wuerker[30] and Heflinger, Stewart and Booth[31] in the off-axis approach. However, microscopic viewing is not very convenient for the eye and so this approach is not very common in particle field holography.

4.9.2 *Screen projection*

The images can be projected onto a screen.[28] As seen in Figure 4.17, the lens at a distance u from the reconstructed scene plane forms an image at the screen at a distance v. The magnification m_l due to the lens is thus

$$m_l = v/u. \tag{4.73}$$

The lens aperture must be carefully selected for proper resolution capability. Equation (4.69) can be used to obtain the minimum

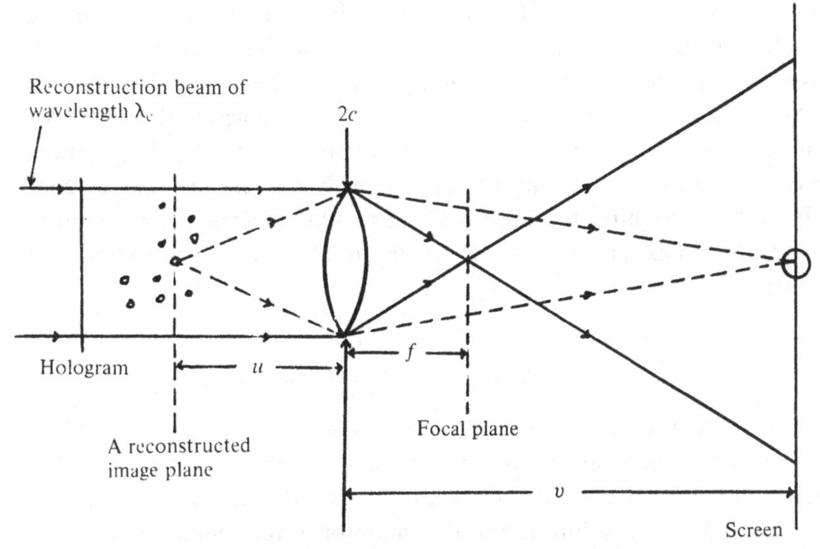

Fig. 4.17 Diagram illustrating screen projection of a reconstructed image plane. A collimated reconstruction beam is shown but a spherical beam can also be used. The screen is replaced by the image tube for closed circuit television observation method.

diffraction spot diameter $2R^{\min}_{\text{screen}}$ on the screen as

$$2R^{\min}_{\text{screen}} \approx 1.22\lambda_c v/c \tag{4.74}$$

The corresponding resolution capability $2R^{\min}_{\text{rec. image}}$ is therefore

$$\begin{aligned} 2R^{\min}_{\text{rec. image}} &= 2R^{\min}_{\text{screen}}/m_1 \\ &\approx \frac{1.22\lambda_c u}{c} = 2.44\text{F}\lambda_c\left(\frac{1+m_1}{m_1}\right), \end{aligned} \tag{4.75}$$

where the lens formula

$$\frac{1}{u} + \frac{1}{v} = \frac{1}{f} \tag{4.76}$$

has been used and $F = f/2c$ is the lens $f\#$ number ($f\#$). For large $m_1 \gg 1$ needed in this operation Equation (4.75) becomes

$$2R^{\min}_{\text{rec. image}} \approx 2.44F\lambda_c. \tag{4.77}$$

Section 4.8 deals with hologram aperture-limited resolution of the reconstructed images. The lens resolution given by Equation (4.75) should match with that described in Section 4.8 to retain the capability at the recording stage.

For a HeNe laser at $\lambda_c = 0.6328\ \mu\text{m}$ and $F = 2.8$, $2R^{\min}_{\text{Rec. image}}$ given by Equation (4.77) is about 4 μm. This should be more than sufficient for common particle field holography.[7] Now, to visualize clearly a small object, the magnification should be sufficiently high. Let us assume that the minimum measurable size on the screen is 1 mm. A 10 μm size reconstructed image then has to be magnified by at least 100 times! Such large magnifications will require a very high power reconstruction laser beam for sufficient brightness on the screen. Hologram bleaching for improved diffraction efficiency (see Section 5.5.3) becomes particularly desirable in the screen projection approach.[28]

4.9.3 *Closed circuit television system*

The problem due to the large magnification requirement in the screen projection method can be simply solved by replacing the screen by the image tube of a closed circuit television system. A typical 0.5 mm resolution on the monitor screen and an internal electronic magnification of 30 greatly relieve the lens magnification (m_l) requirement. With the lens magnification as 10 and the electronic magnification as 30, a 10 μm diameter reconstruction will appear as 3 mm diameter image on the monitor screen. That is why the closed circuit television method was adopted during very early stages of particle field holography[11] and its use remains very common. In fact, for larger objects the electronic magnification can be sufficient and the real image can be observed in a lensless manner. Laser beam intensity requirements and related controls during reconstruction using a closed circuit television system are discussed in Section 5.8.

5

Practical considerations

In Chapter 4, we discussed the optics necessary for system designing to record and reconstruct an in-line Fraunhofer hologram. A single microobject model and the simplest mode of recording and reconstruction were considered. The technique often finds limitations in practical situations. For example, if the density of the microobjects in the test-section is high, no reference beam might be present in the transmitted light. Thus, there is a practical limit on the density for the successful recording and hence reconstruction. In some other situations, the ideal position of the recording medium may not be practical. This, for example, could be inside a combustion chamber. The scene in that situation can be relayed to a convenient location using lenses. A number of these practical limitations and possible ways to circumvent them are considered in this chapter.

Techniques to enhance the image quality such as special recording and/or processing of the hologram, spatial frequency filtering methods, etc. are also described.

5.1 Density of microobjects

In Chapter 3 we saw that an in-line hologram is formed by the interference between the directly transmitted light and the diffraction pattern produced by the microobject. For a single microobject, there is no problem because the reference beam is present almost everywhere. When the number of microobjects increases in a given cross-sectional area, the effective presence of the reference beam will reduce and hence the hologram recording process will be affected. In fact, this is the simple way to describe the situation. For a high density of microobjects the simple representation of the source light given by Equation (3.1) is not valid. Initially along the propagation

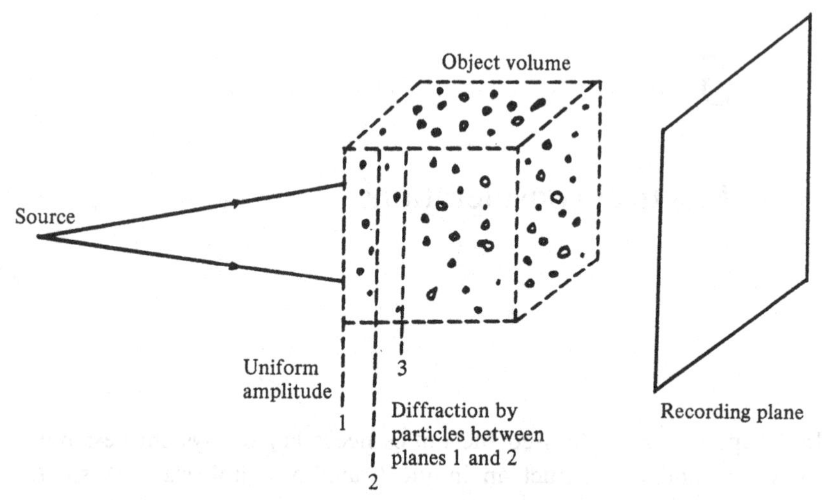

Fig. 5.1 Schematic diagram for representing the in-line Fraunhofer hologram recording process when the density of micro-objects is high.

direction, say up in plane 2 in Figure 5.1, Equation (3.1) may be valid. Afterwards the incident light can first be characterized by Equation (3.2) and then by an unknown complex function. More complexity is expected if the Fraunhofer condition is not satisfied. Thus, the objects after plane 2 are illuminated by a complex unknown wavefront unlike the simple one described by Equation (3.1). Similarly at the other planes along the way, the light amplitude becomes more and more complex. In a limiting situation, this kind of coupling from one plane to another in the object volume becomes so strong that the individual object model of the irradiance distribution at the recording plane becomes invalid. In an extreme situation of very high density microobject, only a speckle pattern is formed on the recording plane without any individual pattern due to a single microobject.

Thus, a sufficient amount of light must pass without modulation. A commonly accepted practical limit for in-line holography is that at least 80% of the light through the cross-sectional area must be undiffracted.[1–3] In terms of practical density, Figure 5.2 represents density vs transmission of various monodisperse opaque particles. In practice, however, there is a size range present and the quality of the chamber wall surfaces gives a base undiffracted transmission. Thus, the best limit must be obtained experimentally in a particular situation.

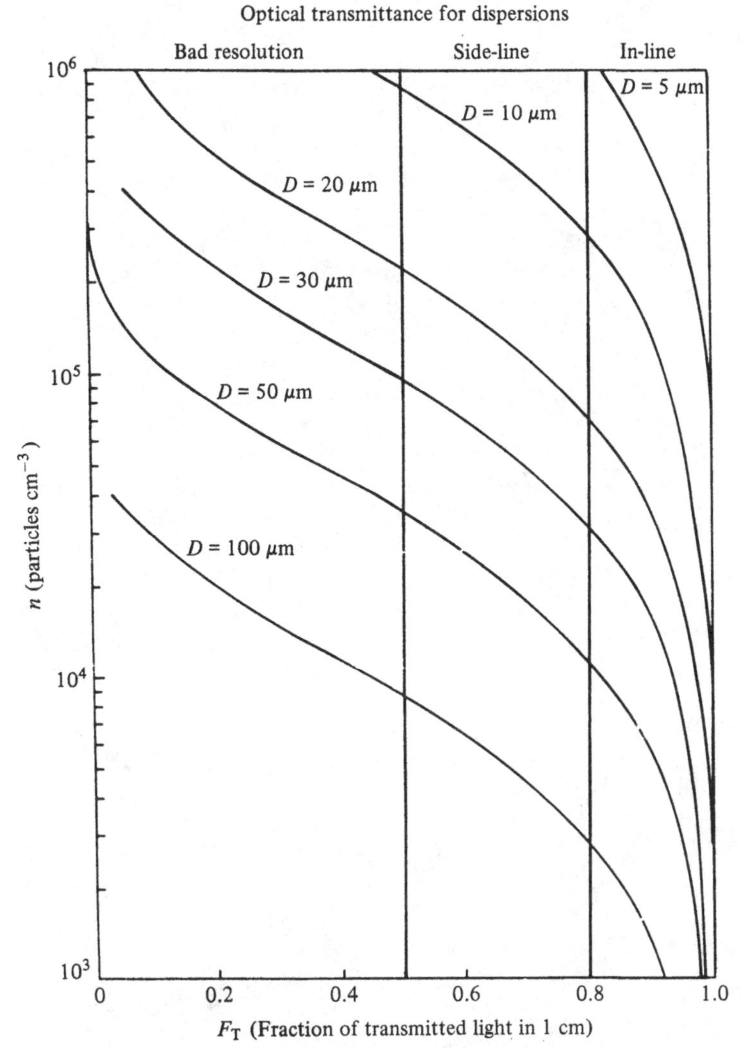

Fig. 5.2 Variation of density of monodisperse opaque particles against optical transmittance. (Witherow[2].)

The effect of increasing density on the appearance of the hologram is shown in Figure 5.3. Figure 5.4 shows typical reconstructed images from these holograms. The formation of good diffraction patterns, as seen in Figure 5.3(*a*) indicates that the density of microobjects is not too high. This can be seen using a microscope objective right after the fixing of the emulsion.

Off-axis holography generally allows more density, as indicated in Figure 5.2. The off-axis approach is discussed in detail in Chapter 11.

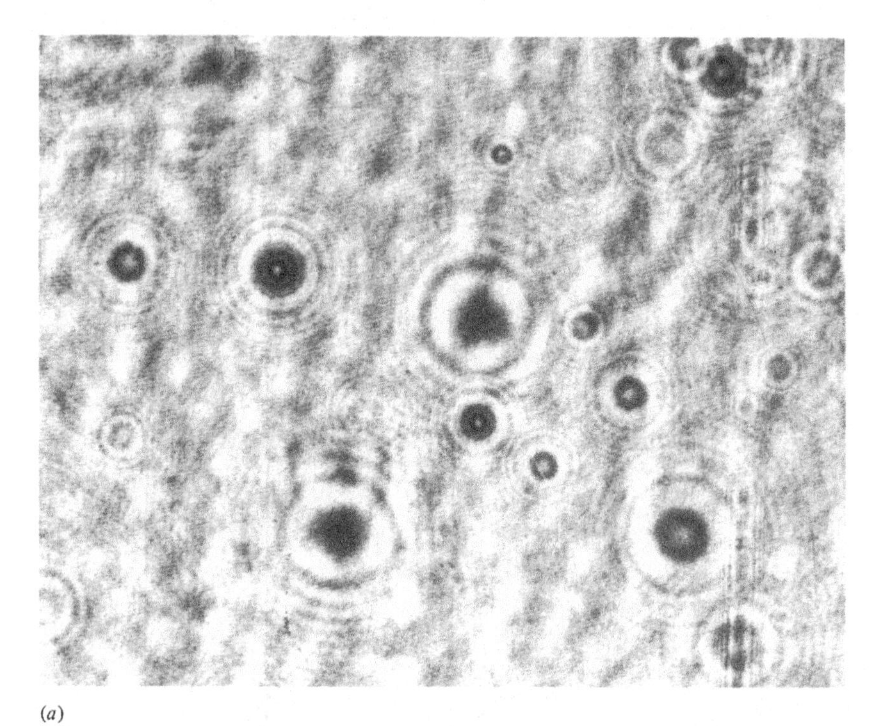

(a)

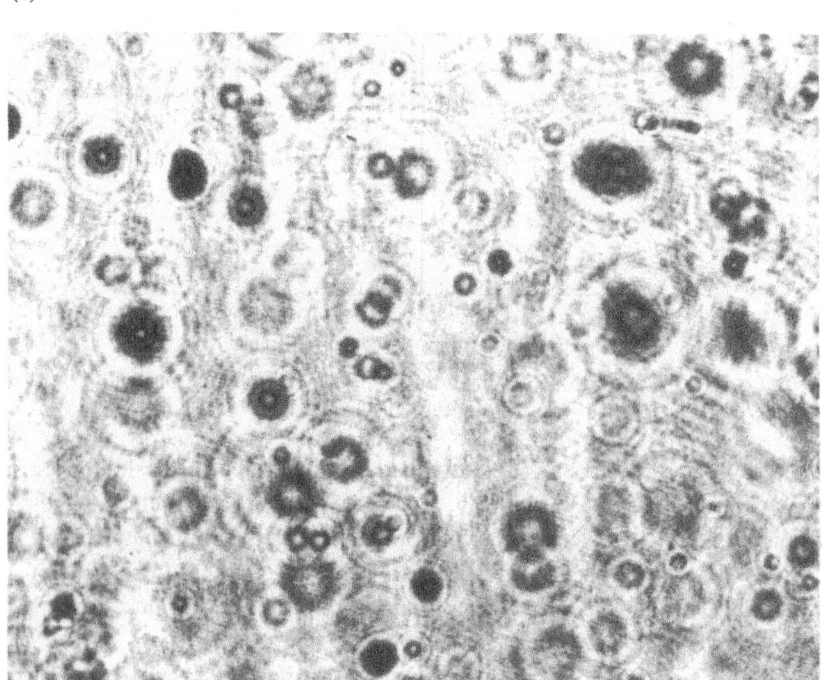

(b)

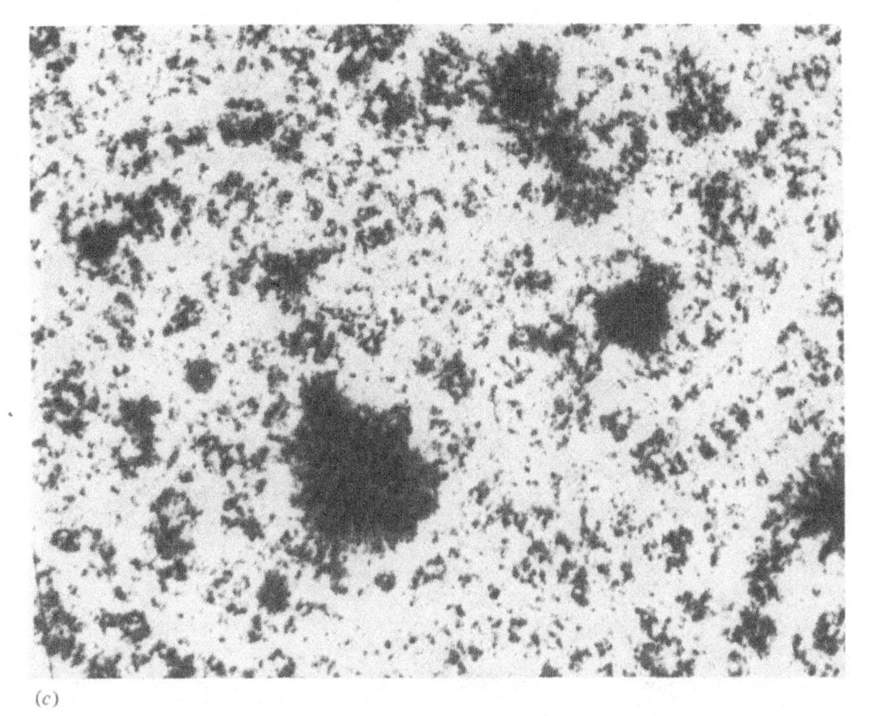

(*c*)

Fig. 5.3 Holograms of air bubbles in water: (*a*), (*b*) and (*c*) respectively correspond to progressively increasing bubble density.

5.2 Application of relay lenses

So far in our discussion about in-line Fraunhofer holography, we have considered a lensless recording. However, lenses have been used between the object and the recording planes for a variety of reasons. The main advantages of the lenses are object pre-magnification, reduced film resolution requirements, relaying the scene volume at a suitable location, and even de-magnifying the scene for large microobjects. In this section the role of these lens systems is discussed.

5.2.1 *Single lens*

These lenses are generally used to pre-magnify the sample volume so that a lower resolution film can be used. As seen in Section 4.2, successful recording of the in-line Fraunhofer hologram with a collimated beam requires film resolution of the order 1000 lines per millimeter for a 5 μm diameter object. This requirement can

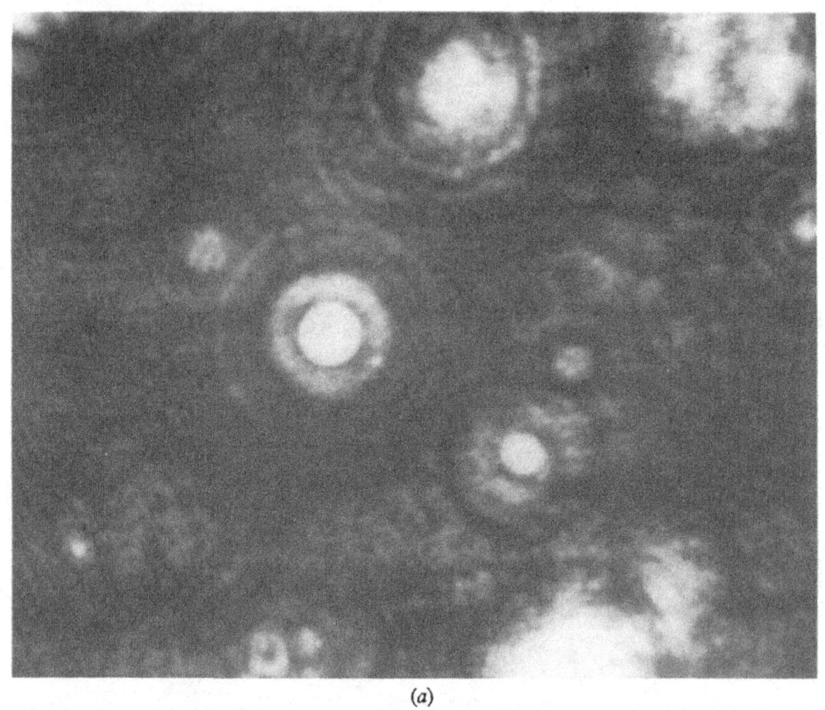

(a)

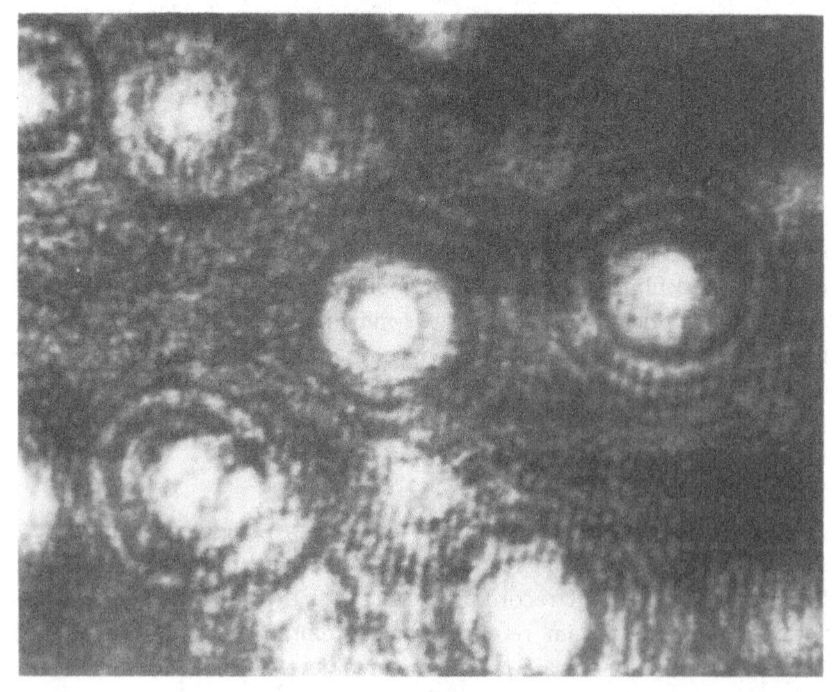

(b)

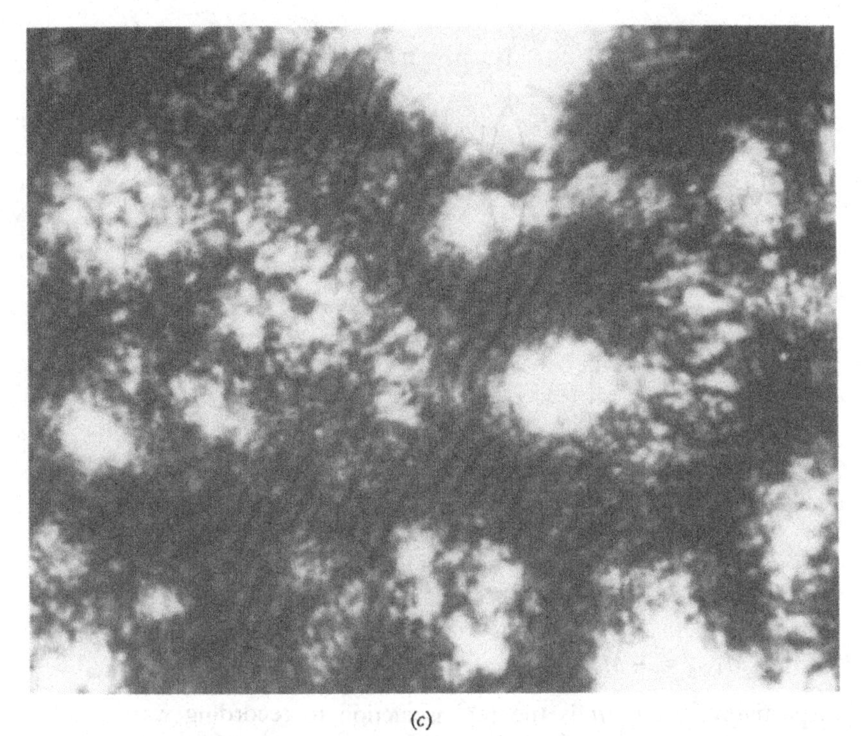

(c)

Fig. 5.4 Reconstructed images: (*a*), (*b*) and (*c*) correspond to the holograms of Figures 5.3(*a*), (*b*) and (*c*) respectively.

be relaxed by pre-magnification.[4–8] A typical system utilizing a lens is represented in Figure 5.5. The generally divergent beam (or the collimated beam in a particular case) illuminates the scene. A lens images this scene and a hologram is recorded in a plane as seen in the figure. The simplest way to describe the magnification process is that for the hologram recording plane (at distance v from the lens), there exists an imaginary object plane (at distance v/m_{H} backwards from the lens). Here m_{H} is the transverse magnification due to the lens only at the hologram recording plane. Thus the situation is like recording the hologram at the object (of the hologram) plane and magnifying it m_{H} times. Obviously, the image is inverted as shown in the figure, but that hardly matters for the purpose.

If, as shown in Figure 5.5, z_{o} and z_{R} are the object and recording source distances respectively from the imaginary hologram plane, then the transverse and longitudinal magnifications of the conjugate image are known (Section 2.3.1) to be

$$M_{\mathrm{c}} = \frac{m_{\mathrm{H}}}{1 - m_{\mathrm{H}}^2 z_{\mathrm{o}}/n z_{\mathrm{c}} - z_{\mathrm{o}}/z_{\mathrm{R}}} \tag{5.1}$$

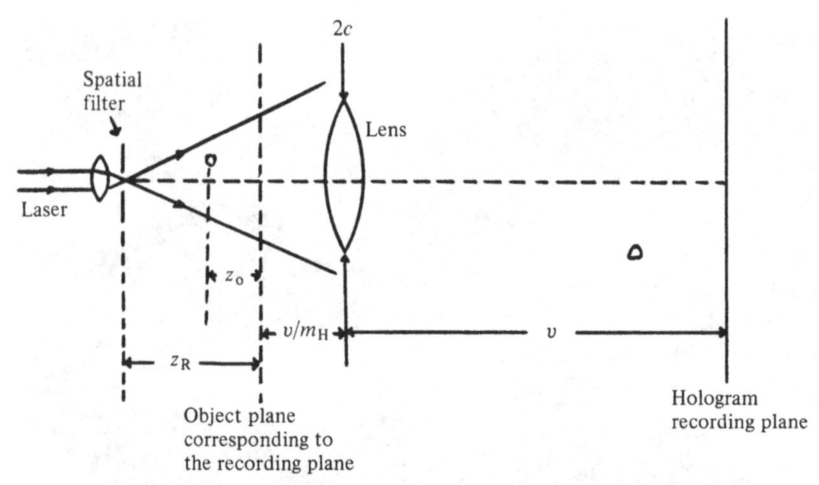

Fig. 5.5 Diagram representing the arrangement of recording an in-line Fraunhofer hologram using a lens.

and

$$M_{\text{long}} = -M_c^2/n \tag{5.2}$$

respectively, where n is the reconstruction to recording wavelength ratio and z_c is the reconstruction source distance from the hologram (Figure 3.2). For collimated recording and reconstruction beams ($z_R = z_c = -\infty$), Equations (5.1) and (5.2) become

$$M_c = m_H \tag{5.3}$$

and

$$M_{\text{long}} = -m_H^2/n \tag{5.4}$$

respectively. It is interesting to note that the magnifications are independent of the particle position.

Care should be taken in the proper positioning of the lens. If the microobjects are both sides of the input diffraction plane, then the images will be both sides of the recording plane. Upon reconstruction, the conjugate images from one side of the objects from the recording plane will be mixed with the primary images of the objects from the other side. The extra images will make concentration studies difficult.[7] Thus, the input diffraction plane should be beyond the point where the microobjects are present. As a result the image space will also contain the microobjects from only one side of the recording plane.[7,9]

5.2.2 *Single lens resolution requirement*

As seen in Section 5.2.1, the lens basically magnifies the plane in the object space to be recorded as a hologram in the image space. The lens must be capable of coping with the fine details in this plane. As discussed earlier in Section 5.2.1, a typical 1000 lines per millimeter resolution is required in the object space. The lens should meet or exceed this requirement. Although this requirement is for 5 μm diameter objects, this situation is common in far-field holography. Obviously for the case of *only large* objects, the resolution requirement is relaxed.

To relate the resolution requirements with the *f#* of the lens, the resolution in the object space is given by the well-known Rayleigh limit as[10] (see also Section 4.8)

$$R_o \approx \frac{1.22(v/m_H)\lambda_o}{2c} \tag{5.5}$$

where $2c$ is the lens diameter. Using the general lens formula

$$\frac{1}{u} + \frac{1}{v} = \frac{1}{f}, \tag{5.6}$$

Equation (5.5) yields

$$R_o \approx 1.22(1 + 1/m_H)\lambda_o F \tag{5.7}$$

where F is the *f#* of the lens given by

$$F = f/2c. \tag{5.8}$$

For the ruby laser wavelength $\lambda_o = 0.6943\ \mu m$, $m_H = 5$ and the 1000 lines per millimeter resolution requirement demands F given by Equation (5.7) to be about 1. Thus for smaller objects, a careful lens selection is needed.

5.2.3 *Double lens (telescopic) system*

As seen in Section 5.2.1, a lens can be used to magnify the pattern to be stored as a hologram. One or more lenses can be added in a similar fashion, although the resolution, depth of field, etc. are determined by the first lens. However, there is a special case of two imaging lenses separated by the sum of their first lengths. This type of telescopic arrangement gives constant magnification over the range of object positions when a collimated beam is used. As noted in Section 5.2.1, even the single lens yields constant magnification in the in-line case. Thus, the telescopic arrangement is of more significance

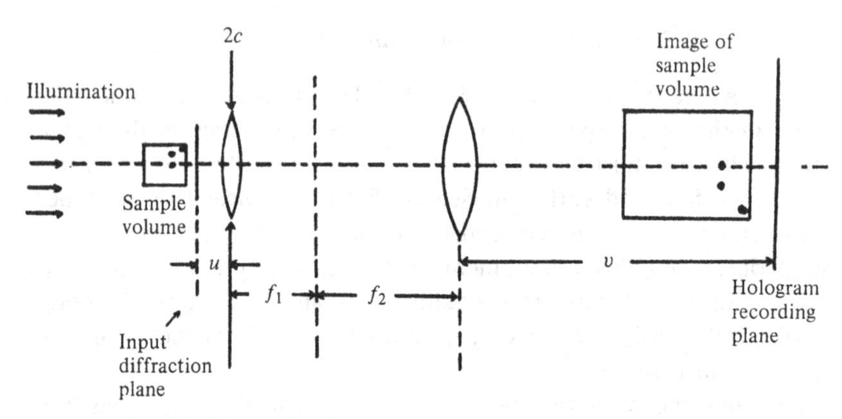

Fig. 5.6 Schematic diagram showing an in-line Fraunhofer holographic system using collimated light and two lenses separated by the sum of their focal lengths.

in the off-axis method to be discussed in Chapter 11. Nevertheless, the telescopic arrangement can be used in the in-line method.[1,11–13] The system is represented in Figure 5.6. To understand the imaging process, let us consider an object at back distance u from the first lens of focal length f_1. Using Equation (5.6), the image distance for the first lens comes out to be $f_1u/(u - f_1)$. Thus, the object distance for the second lens of focal length f_2 (separated from the first lens by the distance $f_1 + f_2$) is $f_1 + f_2 - f_1u/(u - f_1)$. Using Equation (5.6) again we obtain the final image distance v from the second lens as

$$v = -(f_2/f_2)^2u + f_2 + f_2^2/f_1 \tag{5.9}$$

Similarly, the net transverse magnification can be determined in two steps, i.e. calculating v/u twice and multiplying with each other to get

$$m_H = -f_2/f_1 \tag{5.10}$$

where the negative sign means image inversion. Thus the image is magnified by f_2/f_1 times which is a constant and independent of the object position. If the image and object planes correspond to the hologram recording and the input diffracted plane (i.e. the object plane for the recording plane) respectively, then Equations (5.3) and (5.4) can again be used for final holographic magnifications.

According to Equation (5.9), if u is large then v tends to become negative, i.e. no special advantage of the lens system of the image relay is obtained. Also, large u means large lens diameter ($2c$) for resolution requirements. For $u < f_1$, v given by Equation (5.9) is greater than f_2. This is the practical working range of this kind of two lens relay system.[10]

The resolution in the object space is again[10]

$$R_o \approx 1.22u\lambda_o/2c \tag{5.11}$$

which for $u < f_1$ is at least $1.22\lambda_o\ f\#$. Thus a proper $f\#$ lens should be used so that the fine interference fringes at the input diffraction plane are meaningfully recorded in the hologram. As discussed at the end of the Section 5.2.2, a lower $f\#$ is required for smaller microobjects.

5.2.4 *Demagnification of large particles*

As seen in Section 4.3, the far-field distance is equal to square of the object diameter divided by the wavelength of the light used. For large particles, say a 400 μm diameter object and ruby laser wavelength ($\lambda_o = 0.6943\ \mu$m), the far-field distance is as high as 23 cm. Since the object should be a few far-fields away at the time of recording, similar distances have to be coped with during the reconstruction. These distances are prohibitively large for larger particles. Since film resolution requirements are very relaxed for such large particles (Section 4.2), a demagnification of the pattern will not pose any recording problems in general. The demagnification can be carried out by single or double lens methods. In the single lens method (Section 5.2.1) as well as in the double lens method (Section 5.2.3), the total longitudinal image space is shrunk ($m_H < 1$) to m_H^2 times the distance otherwise ($m_H = 1$). Here m_H is the transverse demagnification of the hologram. As stated above, a demagnification of 5 will not pose other problems for large particles but it will reduce the image–hologram distance by 25 times! Thus, image–hologram distance can be kept at conveniently small values.[14,15]

5.3 Test-section with single window opening

The common in-line holography utilizes passing the laser beam through the particle volume and recording the hologram after such transmission. However, there might be situations when only one window is available uninterrupted. An example is bubble chamber holography where the laser beam cannot be passed through due to the yoke of the magnet.[12] In this situation the laser beam can be reflected back through the test-section as shown in Figure 5.7 and a

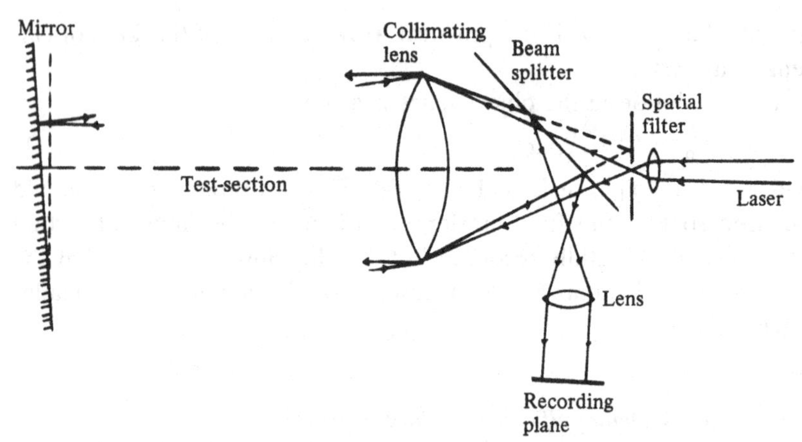

Fig. 5.7 Schematic diagram of in-line holography when the laser beam enters and exits through the same window. This is possible using a mirror.

hologram is conveniently recorded. The telescopic system using two lenses (Section 5.2.3) is shown in the figure but the single lens (Section 5.2.1) or the lensless method can also be used. In the single lens method, for example, the small collimating lens just before the recording plane can be removed. In that situation the beam collimating lens acts as the single relay lens as well. In the lensless method, the beam splitter can be placed between the mirror and the main collimating lens as shown in Figure 5.8.

The very slight tilt of the mirror in each case is to avoid part of the beam going back to the laser cavity and accidently damaging it. The beam splitter can be replaced by a mirror near the spatial filter plane.[12] However, the mirror must be placed in a non-overlapping region of the light from the spatial filter and the light reflected back through the test-section. Since this non-overlapping region is very small for small mirror tilts, the beam splitter arrangement is simpler. On the other hand, only a fraction of the available laser power is finally used to record the hologram in the beam splitter arrangement. This aspect should be considered in designing the system when the available laser power is very limited in a particular case.

The light passes twice through the same sample volume and therefore virtually produces two holograms of the same object at different distances. Consequently, two images of each microobject are reconstructed. These images are separated along the optical axis and so do not overlap. In fact this situation can be used to advantage. As seen in Figure 5.9, the equivalent effect of the mirror is like recording

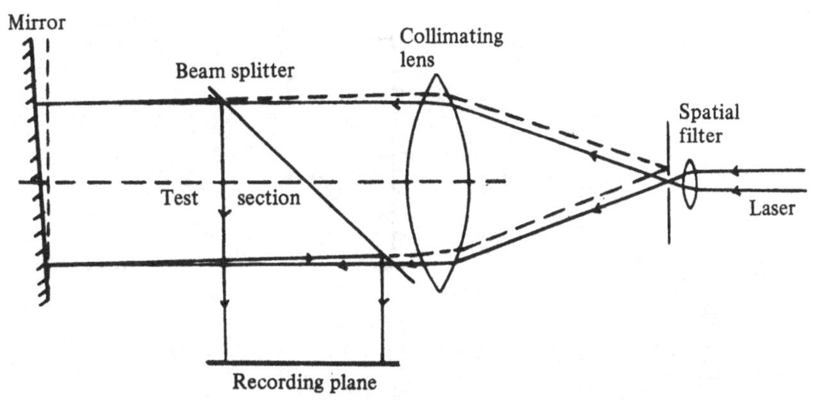

Fig. 5.8 Diagram representing the one window in-line recording method without relay lens. The collimating lens can be removed if the divergent beam is to be used for the recording.

two holograms of the same object from two object–recording plane distances. The difference is significant for objects not very close to the mirror. Thus, the closer (to the hologram) image can be used for smaller objects for good resolution and the farther one for the far-field condition. In this way the allowable particle size range can be increased. As discussed in Section 4.3, this range is limited.

5.4 Very small objects

Small particles demand special attention in particle field holography. Firstly, when the particle is small as compared to the wavelength of the light being used, the object acts as a point source. Secondly, even when the object is a few times larger than the wavelength so that it can no longer be considered as a point source, there are certain errors and limitations. These special features are addressed in this section.

5.4.1 *Submicrometer particles*

The general formulation in Chapters 3 and 4 assumes that the object size is large compared to the wavelength of the light used. For visible light such as that from a *Q*-switched ruby laser, the object size should be at least 1 μm. Researchers have put this limit in the region of 1–5 μm.[2,5–7,10,16–23] In that situation, the Fraunhofer diffraction pattern contains information about the object cross-section. On the other hand, submicrometer objects will act as point scatterers.[6,24] For

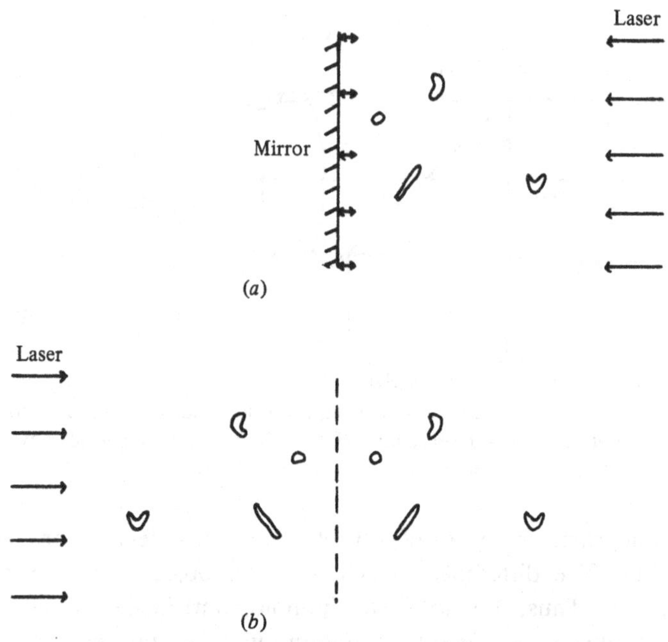

Fig. 5.9 Diagram illustrating the effect of the light passing twice through the same object volume: (*a*) the actual physical situation and (*b*) the equivalent effect for the conventional reconstructed image analysis.

simplicity let us consider a plane wave incident on such particles as shown in Figure 5.10. An object point at a distance z_o from the recording (x, y) plane will cause scattering. The light intensity distribution at the recording medium can be determined from the limiting case of Equation (4.14). Considering opaque objects, the imaging part in RHS of Equation (4.14) becomes zero. Thus the irradiance distribution is

$$I(r) \cong B[1 - \alpha \sin(k_o r^2/2|z_o|)], \tag{5.12}$$

where

$$r^2 = x^2 + y^2 \tag{5.13}$$

and α relates the amplitude of the spherical wave as compared to the background, presently a plane wave. In fact $\alpha = 1$ corresponds to the ideal sine-wave zone plate. In the present case of submicrometer objects, α is generally much less than unity.

Comparing Equations (4.14) and (5.12), the side lobe structure is absent here. This structure gives information about the object shape and size for larger objects. In the present situation, only the position (z_o) is recorded in the hologram described by Equation (5.12).

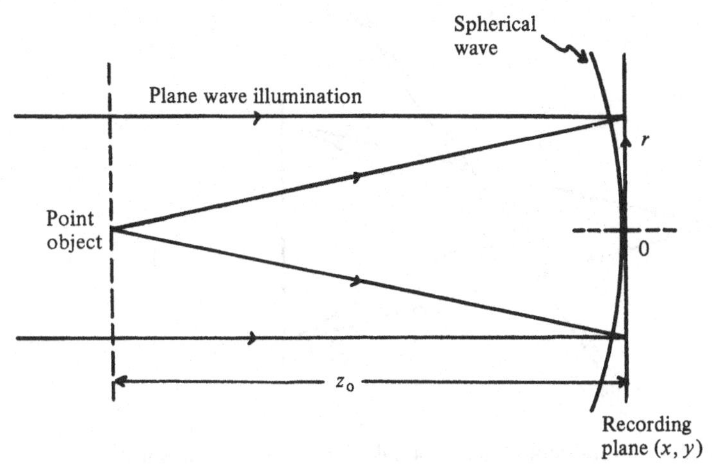

Fig. 5.10 Diagram showing the recording of an in-line hologram of a point object with plane wave illumination.

The fringe spacing can be determined from the argument of the sine term in RHS of Equation (5.12). The phase difference between the nth and the $(n + 1)$th fringe is 2π. Also r is generally large as compared to the fringe spacing. So we obtain the fringe spacing Δr as

$$\Delta r \cong \lambda_o |z_o| / r \tag{5.14}$$

The system resolution depends on the hologram aperture size as discussed in detail in Section 4.2. However, the film resolution requirement increases for increasing r. Thus there is an effective aperture if the minimum Δr at the maximum r on the film cannot be resolved. The reconstruction image size will depend on the system resolution. The image intensity is thus the impulse response of the system at the image point.[6,24] For submicrometer particles with visible light holography, the image will generally be larger than the object size.[6,24] The image therefore is detected in position without accurate size information.

Thompson and Zinky[24] have determined the positional accuracy of such images. The depth of field resolution is the impulse response diameter squared divided by $2\lambda_c$. An impulse response diameter (or the system resolution image diameter) of a few micrometers can easily be obtained with common recording emulsions in modern holography. The depth of field resolution is therefore a few micrometers again.

Thus the particle can be detected without accurate analysis of its size. However the detection capability can still be used in velocimetry. Two or more light pulses can be given to record the same

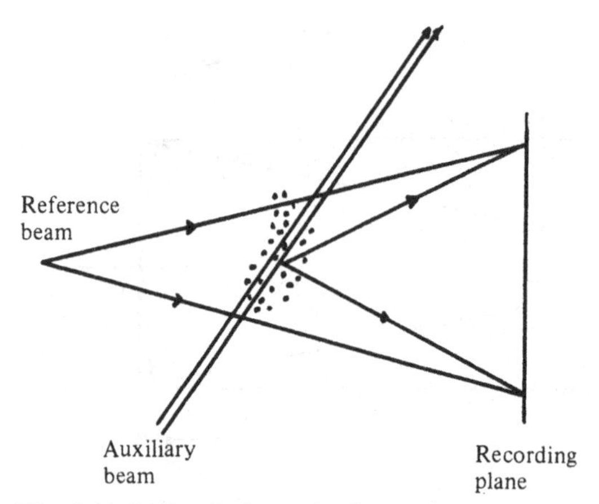

Fig. 5.11 In-line holography for submicrometer particles using an intense auxillary beam for enhancing the object beam intensity. The already very poor object to reference beam intensity ratio can thus be increased. Reference and auxiliary beams are derived from the same laser source.

hologram to reconstruct the time history of the object position. For example, in gas dynamics studies, submicrometer particles can be used as tracers due to their small inertia. Particle velocity measurements will be discussed in more detail in Chapter 10 but an experimental arrangement by Royer[25] is of particular interest here. The amount of light diffracted by these particles at the hologram plane is very small resulting in small values of α in Equation (5.12). This results in poor contrast of the hologram fringes and their recordability. As shown in Figure 5.11, an intense auxiliary beam derived from the same laser can be used to illuminate the particles. Effectively this means increasing the object beam intensity. The basic process is still in-line but with adjustable reference to object beam intensity ratio.

The use of radiation from a *Q*-switched ruby laser at $\lambda_o = 0.6943\,\mu m$ is very common in particle field holography. It generally provides more than sufficient power to expose common recording materials for holography in a time of about 15 ns and still retain excellent spatial and temporal coherence requirements. However, in the context of small particles, a smaller wavelength can be used to avoid the object behaving like a point source. For example, an argon laser at $0.5145\,\mu m$[26,27] can be used. An ultraviolet ruby laser at $0.35\,\mu m$[28] and a dye laser at $0.590\,\mu m$[29] have also been used in the in-line method. Since the wavelength ratio of these lasers as compared to 0.6943 μm is not very high, only a moderate advantage is expected. Fortunately, a variety of suitable ultraviolet pulsed lasers such as nitrogen, neodymium:YAG, neodymium:glass, dye, are commercially available with wavelengths as low as 0.19 μm. With these lasers, the resolution can be increased. For a given object diameter d, the far-field distance d^2/λ_o and the allowable sample volume depth are also increased (Section 4.3). With smaller diameter objects, the film resolution requirements are also changed (Section 4.2) and must be taken into account.

Hickling[30,31] has generated artificial Fraunhofer holograms of submicrometer particles and successfully reconstructed images. Using classical Mie solution, holograms of spherical liquid droplets can be computed and produced using a cathode ray tube camera linked to the computer. The method can be used for spherical liquid droplets in the size range of 0.5–20 μm.

5.4.2 *Electron-beam and x-ray holography*

The discussion of Section 5.4.1 concludes that the ultimate lower limit for the accurate size analysis of particles is basically linked to the wavelength of the light being used. Thus, for very small particles, a suitably small wavelength is needed. Electron beams are historically linked to holography for application to electron microscopy. The original interest in nanometer domain resolution has not been achieved. However, electron beams can be used for the size analysis well below the micrometer region. This is evident from the work of Tonomura, Fukuhara, Watanabe and Komoda[32] where holograms of opaque gold particles of 10 nm diameter on a carbon film are recorded using $\lambda_o = \lambda_{el} = 0.0037$ nm. The object distance z_0 was kept at 80 far-fields. A relay lens was used for pre-magnification of the hologram. The reconstruction was performed using a HeNe laser at $\lambda_c = 632.8$ nm. The electron micrograph, image hologram and the reconstructed image are shown in Figure 5.12. Although the reconstructed image is slightly inferior to the electron micrograph image, it is an excellent example of the far-field holography of very small objects. Developments in electron holography have been reviewed by Tonomura.[33]

It is difficult to image a living cell with an electron microscope due to sputtering on the specimen, dehydration, fixing, staining, etc. These processes can also alter the specimen's structure. X-ray holography allows live specimens and also avoids the problem of structure

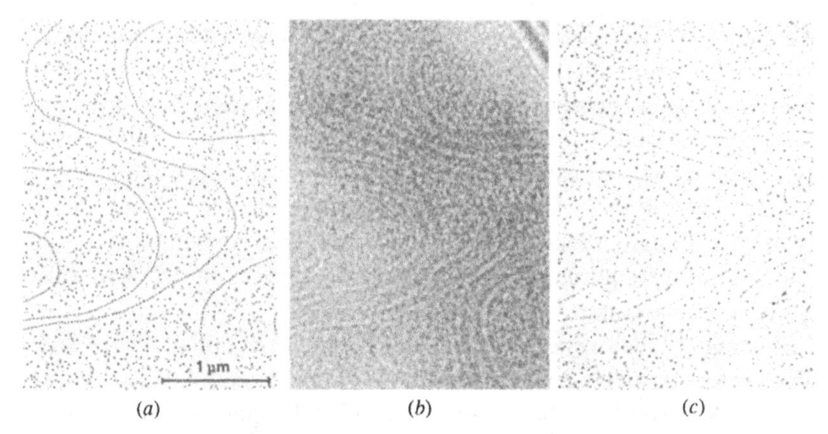

Fig. 5.12 (*a*) Electron micrograph image, (*b*) hologram, and (*c*) reconstructed image of 10 nm diameter gold particles on a carbon film. Recording and reconstruction were performed by electron beam and HeNe laser respectively. (Tonomura *et al.*[32])

alteration. El-Sum and Kirpatrick[34] and El-Sum[35] were able to generate visible reconstruction of a thin wire from an x-ray diffraction pattern recorded 20 years earlier. Baez[36] provided a theoretical feasibility study and concluded that x-ray holography is possible. A history of x-ray holography and its status has been reviewed more recently.[37] Associated problems and design considerations have also been discussed.[37,38] Transverse resolution of 1–2 μm is possible with exposure times in the range of 3–100 min.[37] The limitation is mainly due to the lack of a strong x-ray source with sufficient spatial and temporal coherence and also due to limited film resolution. The diffraction limit inherent with visible light during the reconstruction is also a consideration. With new generation of soft x-ray sources, using hologram pre-magnification, Fourier transforms geometry, high resolution resist detectors, etc. the situation is improving.[39–41]

5.4.3 *Hologram aperture related effects*

In Chapter 3 the discussion regarding the general theory of in-line Fraunhofer holography assumes an infinite hologram (x, y) aperture. This is not physically possible. There is an effective aperture due to the resolution limit of the emulsion, as discussed in Chapter 4. The homogeneity of the physical recording medium should also be considered. The homogeneous diameter for Agfa 8E75 and Afga 10E75 plates is as small as between 10 and 25 mm.[20] The infinite integration limits of Equation (3.8) or x, y limits of the integral of Equation (3.13) should therefore be replaced by appropriate finite values. These kinds of finite aperture considerations result in an image intensity distribution and not a sharp edge like that described by Equation (3.33). These effects are discusssed in detail in Section 5.10. With the conventional method of measuring the image diameter from inflexion points, there is significant error when just a few side lobes of the diffraction pattern are recorded in the hologram. Thus, for smaller objects, the finite aperture effects are important because for a given aperture size, the number of side lobes to be recorded is small. Wavefront aberrations also become important for smaller objects because the effective hologram aperture needed is larger for smaller objects. Chapter 7 is devoted to third-order wavefront aberrations with special reference to in-line Fraunhofer holography.

The hologram recordability also depends on the contrast of the high frequency interference fringes. For small objects, the contrast unfavorably becomes smaller. The contrast also depends on the object shape. These kinds of interference fringe related contrast effects are described in Chapter 8.

All these hologram aperture related effects have special reference to smaller size objects. These must be considered carefully when accurate size analysis rather than just detection is required.

5.5 Factors associated with recording materials and their processing

There are several particular aspects of the recording conditions and materials dealing with in-line holography of microobjects which need to be considered. Special exposure and processing conditions are required for optimum image to background intensity ratio, signal to noise ratio, etc.

5.5.1 *Exposure requirements with thin absorption emulsions*

In common off-axis holography with amplitude holograms, the average exposure should generally be to yield an amplitude transmittance of the order of 0.5 for optimum image quality.[42] The exact value varies with the particular recording medium and processing conditions. For example, Agfa 10E75 plates require an average amplitude transmittance of 0.45 corresponding to the optical density

0.7.[43,44] By common photographic standards, the holograms therefore look underexposed, i.e. they appear lighter than a typical photographic negative.

In the case of in-line Fraunhofer holograms, the effect of the background term must be taken into account[45] for a high image to background intensity ratio. The situation can be explained by the linear model.[46] Let us rewrite Equation (3.33) for the irradiance distribution in the conjugate image plane:

$$I(\mu, \nu) = I_o|1 + \Gamma A^*(\mu/M_c, \nu/M_c)|^2. \tag{5.15}$$

For an opaque object, A is unity inside the object cross-section and zero otherwise. The image to background intensity ratio σ is therefore

$$\sigma = (1 + \Gamma)^2 \tag{5.16}$$

From Equations (3.3) and (3.6), it can be determined that the background amplitude transmittance (i.e. the transmittance without the object in the cross-section) τ_o is

$$\tau_o = \tau_b - K(B/m_o)^2. \tag{5.17}$$

Equations (3.10), (3.11), (5.16), and (5.17) can be combined to obtain

$$\sigma = (\tau_b/\tau_o)^2. \tag{5.18}$$

A typical exposure vs amplitude transmittance curve for an Agfa 10E75 plate is plotted in Figure 5.13. Slopes at different points were plotted to determine the incident τ_b and ultimately σ. The image to background intensity ratio σ is also plotted in Figure 5.13. The maximum value of σ is approximately 14.5 at the average exposure of about 4.45 $\mu J/cm^2$. This corresponds to an average amplitude transmittance τ_o of about 0.09 or a photographic density of about 2.1. This means less than 1% intensity transmittance. From photographic or off-axis holographic standards, the situation reflects an overexposure.

The idealistic curves of Figure 5.13 are for example only. The exposure vs amplitude transmittance curve changes significantly with processing conditions, base density, etc.[43] Our model also assumes an opaque object whereas in practice the object may have some transmittance. The influence of diffraction patterns of other objects, minute incoherent background exposures, etc. is also not included. That is why experimental optimum densities with Agfa 10E75 plates are found to be as diverse as 1.2[47,48], 1.6–1.7[49], 1.7[50] and 2[46,51]. Even higher densities (2–2.5) are reported for other recording materials such as Mikrat VR emulsion.[20]

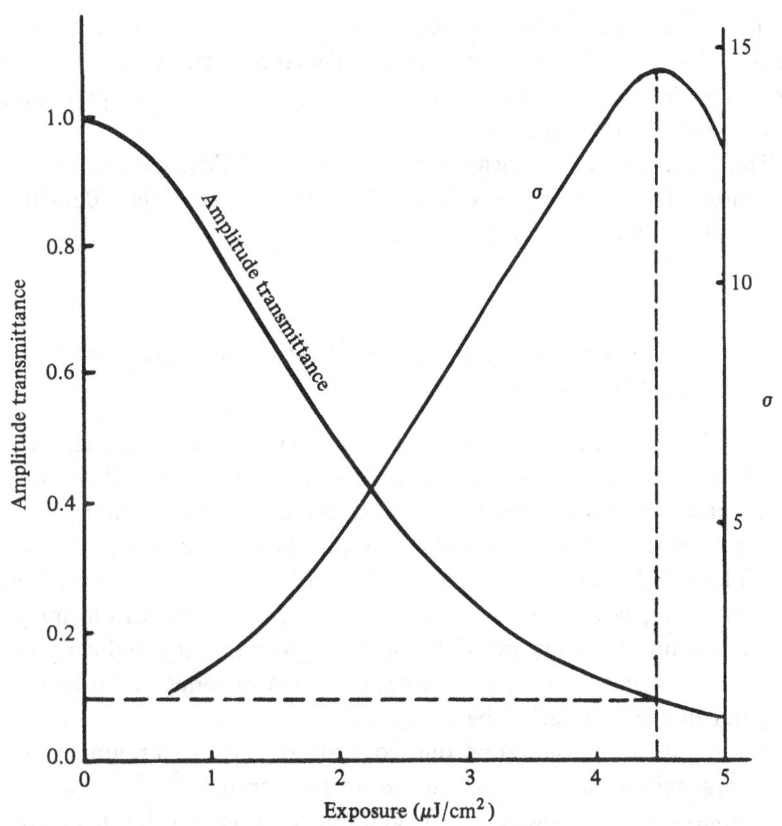

Fig. 5.13 Exposure vs amplitude transmittance curve of an Agfa 10E75 plate and a calculated plot for σ using Equation (5.18). The maximum image to background intensity ratio σ is at the amplitude transmittance $\tau_o = 0.09$.

For high $(\tau_b/\tau_o)^2$, the value of τ_b should be as large as possible at a given value of τ_o. This means a high γ or the slope of the D–$\log E$ curve. The γ of the emulsion can be controlled by the development procedure.[43] For a given optical density or the amplitude transmittance, one way to increase γ is to reduce the amount of exposure and increase the development time.[43] Thus, to produce images of high visibility in the dark background, high γ of the order of 6–9 has been recommended for Agfa 10E75 plates.[49,50]

Even if the exact exposure vs amplitude transmittance curve is available for a particular recording and processing condition, the linear model leading to Equation (5.18) or Figure 5.13 is valid for small exposure variations. In practice a significant portion of the curve may be used. The optimum exposure can then be obtained by numerical analysis with variable τ_b and K in Equation (3.6) and hence Equation (3.8). Theoretical calculations by Özkul[52] yield optimum image contrast as $D = 1.2$ ($\tau_o = 0.25$) for a particular object–hologram range.

Thus, the optimum density depends on the particular experimental situation. The only possible generalization is that higher densities than in the off-axis case are needed.

5.5.2 *Special processing of thin absorption holograms and signal to noise ratio*

Usually the processing of the hologram recording suggested by the film or plate manufacturer is sufficient. These as well as other recommended procedures are well documented in common holography literature. However, variations can improve the image quality in in-line holography by better edge definition. As discussed in Section 5.5.1, developing the emulsion at high γ improves the image to background intensity ratio. Thus, for a given density, reducing the exposure and increasing the development time to some extent gives a brighter image in a darker background.

The signal to noise ratios due to scattering by silver grains and inhomogeneities have been discussed in the literature.[5,20,53] However, in common practice, the loss of resolution is not caused by this factor using modern holographic materials such as Agfa 10E75 plates.[53] The common high contrast developers produce adequate image quality, but they also produce a coarse silver grain. These grains increase scattering and contribute to image noise. In special applications such as automated quantitative analysis of holographic reconstructions, there is always the need to improve the signal to noise ratio leading to better edge definition. In this connection, Dunn and Walls[47,54] have found an improved processing condition for absorption holograms. The development process at 20 °C is as follows

1. (1) Pre-development in 200 mg/l benzotriozole/distilled water solution (as an antifogging agent) for 5–10 s.
2. (2) Development with concentrated Neofin Blue for approximately 1.5 min to achieve an optical density of about 1.2.
3. (3) Washing in running water for 2 min.
4. (4) Fixing in 20% solution of G334 for 2 min.
5. (5) Washing in running water for 3 min.
6. (6) Drying in 90% solution of methanol for 1 min.

This process produces clear well-defined images with low exposures whereas, the conventional process yields poor and low contrast images.[54] Grabowski[23] has also observed the superiority of Neofin Blue developer over standard Agfa-Gevaert developer G3p from the signal to noise point of view.

5.5.3 *Thin phase (bleached) holograms*

The use of thin phase holograms particularly by bleaching is well known for improved diffraction efficiency. The maximum diffraction efficiency of thin phase holograms is about 34% as against only about 6% in the case of thin amplitude or absorption holograms.[55] However, the use of several baths changes the gelatin structure[23] resulting in a poor signal to noise ratio in in-line holography. For this reason, the use of bleached holograms is not common in the in-line method. Since the bleaching process produces noise there are suggestions that these holograms should be avoided.[49,50] Thus, to take advantage of improved diffraction efficiency such as when the images are to be projected onto a screen,[56] special processing methods to reduce the noise are required. Dunn and Walls[47] developed an optimum bleach procedure in this connection. In this, pre-development, washing after the development, and the final drying procedures are the same as those for the absorption process described in Section 5.5.2. The development duration of Section 5.5.2 is increased to obtain a density greater than 2. The fixing and washing afterwards are replaced by bleaching in a 2 l solution of deionized water, 300 mg phenosafranine, and 150 gm ferric nitrate. These processes resulted in well-defined reconstruction images from both phase and absorption in-line holograms. Figure 5.14 shows microdensitometer traces across phase and absorption in-line hologram reconstructions of a 100 μm diameter wire at different hologram–image distances. In each case, the central column represents the position of the image while the side peaks correspond to out-of-focus virtual images and background noise.

There is a low signal to noise ratio in the case of phase holograms, due to large side peaks. The sharp edge definition is also apparent, defined by large negative peaks. On the other hand, the edge definition in the absorption hologram is poorer but the overall signal to noise ratio is better. Therefore, it is concluded[47] that absorption holograms are more appropriate for overall detecting and sizing of images particularly at high particle concentrations and with automatic

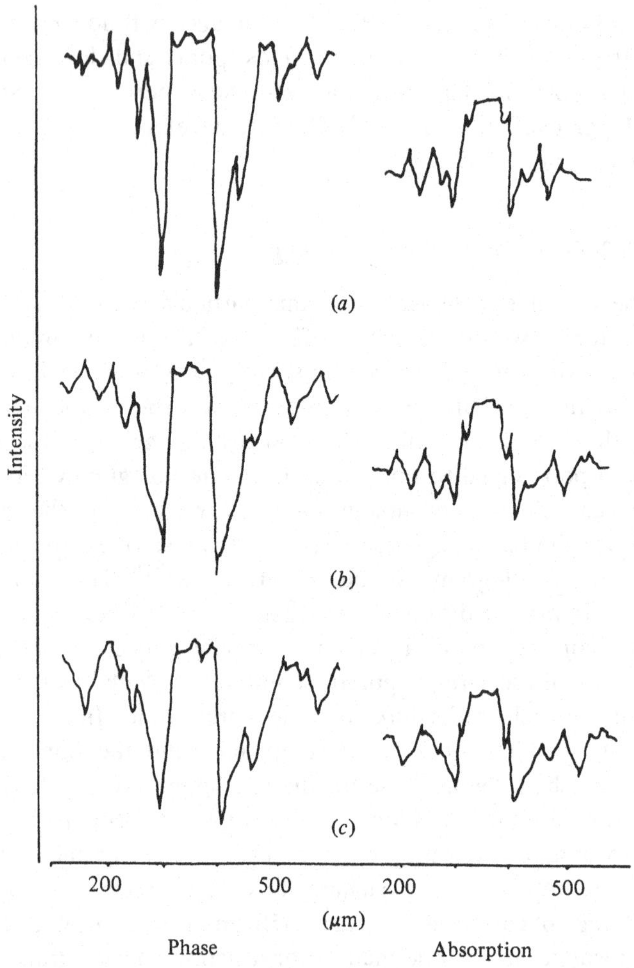

Fig. 5.14 Microdensitometer traces across reconstructed images of a 100 μm diameter wire from phase and absorption in-line holograms. The distance Z_c between the hologram plate and the image corresponds to (*a*) 21 cm, (*b*) 43 cm and (*c*) 59 cm (Dunn and Walls[47].)

image analysis systems. Phase holograms, due to their better edge definition, can be used in specialized applications, where particle shape determination is very important.

For very small objects, such as 4 μm spherical spores,[23] the low signal to noise ratio of phase holograms results in difficult image detection. Zero-order beam filtering (Section 5.6.2) in the back focal plane of the image magnifying lens gives a sufficient signal to noise ratio and a very homogeneous background.[23]

The general conclusion about the phase holograms is that they are useful for specialized cases only as mentioned in this section.

5.5.4 *Real time recording and measurement*

The commonly used photographic emulsion takes time to process before it can be used for the image reconstruction. The recording of an event with simultaneous image reconstruction is desired for time saving and *real time* data processing. Cartwright[57,58] demonstrated this possibility using a liquid crystal light valve (LCLV). The valve contains several thin layers of electrodes, liquid crystals, a photoconductor, a dielectric mirror and face plates. The device changes the state of polarization of a light beam passing through it. The polarization changing property can be controlled by the intensity of another beam. As shown in Figure 5.15, the magnified (due to the limited resolution capability of the device) in-line Fraunhofer hologram is formed on the *write side* of the device. The intensity variation then controls the local polarization rotation capability of the other beam (reconstruction).

The reconstruction beam, if polarized perpendicular to the plane of the paper, is entirely reflected by the polarizing beam splitter to the *read side* of the LCLV. After passing through the liquid crystal, the reconstruction beam is reflected back by a dielectric mirror in the LCLV. As stated above, the polarization rotation is related to the local intensity of the write beam.

The light with an unchanged polarization component is ultimately reflected back in the original reconstruction beam path. The orthogonal component is directly transmitted by the polarizing beam splitter forming the reconstructed image. An Ar^+ laser at 0.5145 μm for writing and reconstruction was used.[58]

Pre-magnification of the hologram becomes very important because of the poor resolution (~10 lines/per millimeter) capability of the LCLV device. Another drawback is that LCLV cannot store the hologram.

5.6 Spatial frequency filtering

The role of the selective filtering in the frequency plane is well known in modern optics.[59–64] These methods have been used or proposed for applications in particle field holography. One application deals with selective filtering out of parallel line objects (particularly beam tracks in bubble chamber experiments). This kind of filtering is possible during recording as well as at the reconstruction stage. Another application is high pass filtering at the reconstruction stage to reduce the effect of the directly transmitted beam for better image detectability. This process basically involves a tiny opaque mask at the center of the focal plane of the image magnifying lens. Low spatial frequencies are cut off and hence slow irradiance variations in the image are filtered out highlighting the edges.

Special spatial filtering approaches at the recording stage are also used to enhance the contrast of the diffraction pattern and also of the fine interference fringes. These will be considered in Chapters 8 and 9.

5.6.1 *Filtering of parallel line objects*

The method has been proposed particularly in connection with in-line Fraunhofer holography of bubble chambers. Beam tracks, usually appearing as a set of parallel lines, generally form an unwanted background. The useful information is in the event tracks. The removal of the beam tracks from the image will improve the quality of the event tracks. More importantly, this will reduce the data handling and processing capability requirements of the image analysis systems.[65,66]

A typical spatial filtering technique at the recording stage is illustrated in Figure 5.16. With a plane collimated incident beam, the form of the amplitude transform in the frequency plane becomes independent of the object distance from the lens.[60] This is very useful in holography where the objects are commonly distributed in space. There is another interesting aspect that the Fourier transform of all these parallel lines will pass through the origin in the frequency plane. Their amplitudes will overlap each other and the envelope will form a line perpendicular to the direction of the original lines. The Fourier transform of these lines can be filtered out using an opaque strip with a hole in the center (to pass the reference beam). The line objects are thus removed in the holographic recording. Other than this selective filtering, the combination is a two lens relay system as described in Section 5.2.3.

Suppose the parallel lines in the object planes are defined by the equation

$$\eta = m\xi + b_i, \tag{5.19}$$

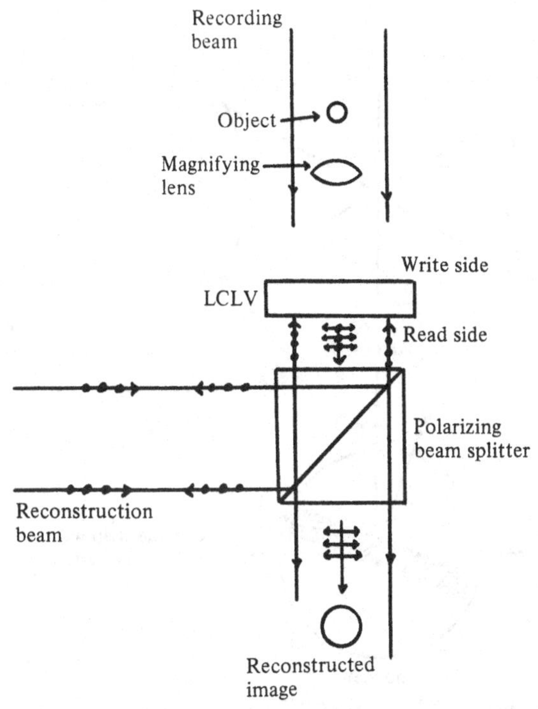

Fig. 5.15 Real time recording and reconstruction of in-line Fraunhofer holograms using a LCLV device. The actual device is made of several thin layers of different materials along with the liquid crystal.

where m is the common slope and b_i is the intercept at the η-axis of the ith line. In the filtering plane (α, β), the Fourier transform of these lines is along the perpendicular direction defined as[65]

$$\beta = -\alpha/m. \tag{5.20}$$

Notice that this transform always passes through the origin. In Figure 5.16, the line defined by Equation (5.20) is represented by the u-axis. The line defined by the v-axis is parallel to the original line objects represented by Equation (5.19). Let us consider a single line object of diameter d and length l. The normalized amplitude of the Fourier transform or the distribution in the filtering plane (α, β) is the well-known Fourier transform of the rectangular aperture:[59–64]

$$\psi(u, v) = \operatorname{sinc}\left(\frac{\pi du}{\lambda_o f}\right) \operatorname{sinc}\left(\frac{\pi lv}{\lambda_o f}\right), \tag{5.21}$$

where $\operatorname{sinc}(\chi) = \sin\chi/\chi$. For more than one line object parallel to each other, the patterns in the filtering plane will overlap each other.

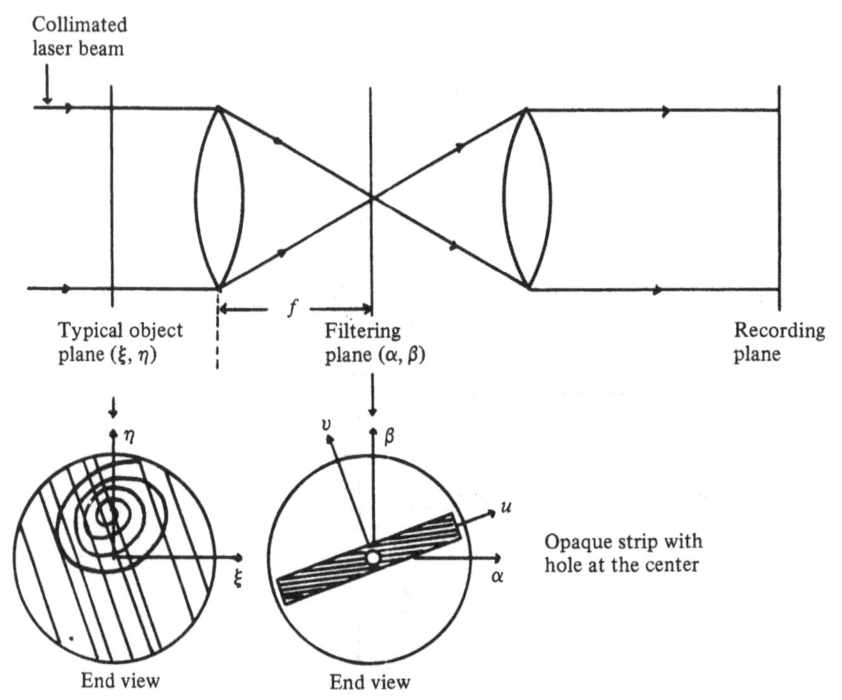

Fig. 5.16 Schematic diagram of the optical system for recording an in-line Fraunhofer hologram while filtering out the effect of the parallel line objects. The collimated incident beam is better because then the absolute value of the amplitude distribution at the filtering plane becomes independent of the object–lens distance.

Since the length l is generally very large compared to the track thickness d, the distribution is mainly spread along the u-direction. Along the v-direction, the distribution will diminish very rapidly. The value l for that purpose is the actual track length or the diameter of the collimated beam cross-section – whichever is less. The width of the opaque strip should be at least $2\lambda_0 f/l$ to filter out the central maximum line of the intensity. Murata, Fujiwara and Asakura[66] suggest observing the Fourier spectra with a microscope for suitable filter placement.

The filtering operation can similarly be performed at the reconstruction stage. The arrangement is basically similar to the one represented by Figure 5.16 except that the object space is replaced by the recorded hologram and the reconstructed images. The arrangement is shown in Figure 5.17.

At this stage, it is important to notice that Murata *et al.*[66] have considered a double lens telescopic system as described in Section

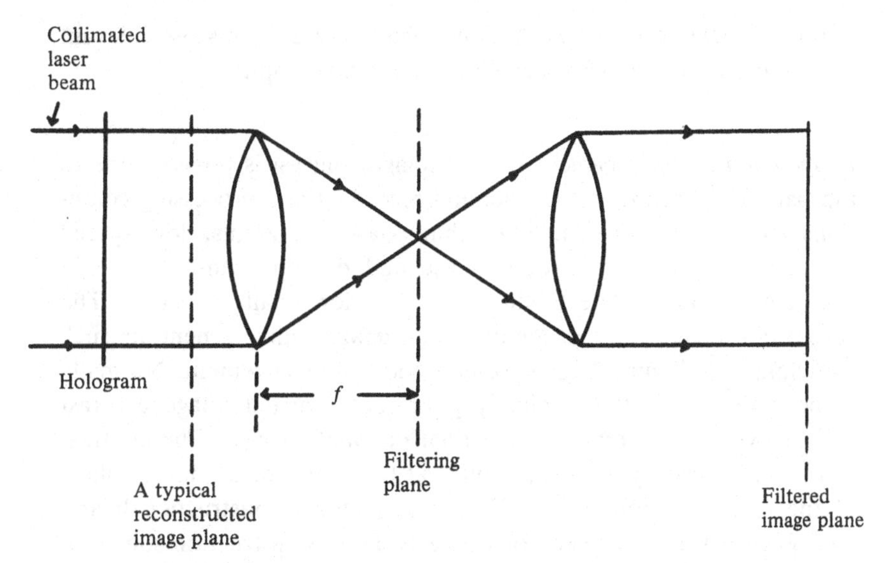

Fig. 5.17 Schematic diagram for in-line holography filtered at the reconstruction stage.

5.2.3. The same type of system is represented in Figures 5.16 and 5.17. The filtering operation is performed at the focus of the first lens where the wanted and unwanted spectra are formed. The role of the second lens is generally to perform relay to obtain constant magnification. With collimated recording and reconstruction beams, constant magnification is achieved even with the single lens system. This fact is described in the Section 5.2.1. Thus, the filtering operation can be performed even with a single lens. The filtering plane should be the focal plane of the relay lens during the recording stage or the magnifying lens during the reconstruction.

5.6.2 *High pass filtering during reconstruction*

As discussed in Chapter 3, the directly transmitted light and the virtual image do not seriously affect the real image in practical far-field holography. Nevertheless, an optical dc filter can be used at the reconstruction stage to reduce the effects of the background, the out-of-focus image, dust and scratches, etc. in the reconstruction. The low spatial frequencies at the filter plane are caused by nearly paraxial radiation. Their effects can be removed by an opaque filter at the frequency plane to cut off the zero frequency. The role of high pass filtering in coherent imaging is well discussed in optics literature

– for example Refs. 67–71. In this section we shall discuss relevant results and conclusions applicable to in-line holography.

Image contrast enhancement A poor image contrast is possible due to the particular recording material and recording and processing conditions (see Section 5.5.3). With thin phase holograms, low spatial frequencies of the background noise and directly transmitted light also create image detectability problems for small objects.[23] The zero-order beam can be removed for image enhancement in such situations.[23,72] Figure 5.18 represents such an arrangement. Normally without the filter, the originally poor reconstructed image contrast will likewise be observed in the final distribution also. The contrast can be enhanced by an opaque filter at the center of the image plane of the reconstruction source. For a collimated reconstruction beam, this plane is the focal plane of the lens. The reconstruction source at S on the optical axis is imaged at S' on the optical axis at a distance Z from the magnifying lens. For a collimated reconstruction source, the distance Z is the focal length f of the lens. The distribution in the filtering plane (α, β) is the well-known Airy pattern (see Section 4.8) whose normalized irradiance is:

$$I(\gamma) = \left[\frac{J_1(2\pi c\gamma/\lambda_c Z)}{\pi c\gamma/\lambda_c Z}\right]^2 \tag{5.22}$$

where

$$\gamma = (\alpha^2 + \beta^2)^{1/2} \tag{5.23}$$

and c is the lens radius. J_1 represents the Bessel function of order one. This kind of distribution is plotted in Figure 4.5. The major portion of the energy is contained inside the first dark ring described by:

$$2\pi c\gamma/\lambda_c Z = \pm 3.83, \tag{5.24}$$

corresponding to the spot diameter

$$2\gamma|_{\text{diffraction spot}} \cong 1.22\lambda_c Z/c. \tag{5.25}$$

For the collimated reconstruction beam, the diameter becomes

$$2\gamma|_{\text{diffraction spot}} \cong 2.44\lambda_c F, \tag{5.26}$$

where $F = f/2c$ is the $f\#$ of the lens. The opaque filter diameter should be at least the diffraction spot diameter given by Equation (5.25) or Equation (5.26) to remove the background intensity effectively. Complete elimination of the zero-order beam is possible[23] by using a filter diameter about four times larger than the diffraction

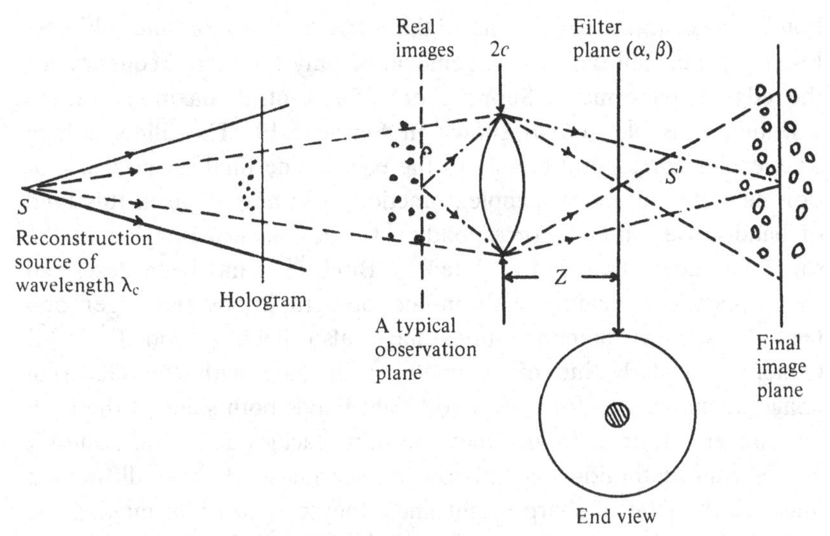

Fig. 5.18 Diagram showing contrast enhancement of the reconstructed images by dc filtering. The distance Z becomes the focal length f for the collimated reconstruction beam case.

spot diameter. The required filter size for the purpose is generally very small. With $\lambda_c = 0.6328$ μm and $F = 2.8$, the diffraction spot diameter is only about 4.3 μm.

Edge line holography In the image contrast enhancement process discussed earlier in this section, the main emphasis is to remove the directly transmitted light. This results in a very thin opaque filter at the plane where the image of the reconstruction source is formed. The filter size corresponds to the diffraction spot diameter of the reconstructed source at the plane due to the lens aperture. The image of the microobject (such as a bubble) also forms a Fourier transform at the frequency plane. Suppose the zero frequency of this transform is blocked out by an opaque filter. The higher frequencies which are passed correspond mainly to the edges. The resultant images will contain a dark background with a visible distribution near the edges only.

The collimated reconstruction beam has the unique advantage that the Fourier transform in the frequency (focus) plane is independent of the lens–object (presently reconstructed image) distance.[60] A reconstructed image of a line object of diameter d and length l will have its Fourier transform in the frequency plane like one given by Equation (5.21). In Section 5.6.1, the aim was to remove the entire

Fourier transform to eliminate the effects of these line objects. Presently we shall discuss the removal of only the zero frequency for the edge enhancement. Suppose only the central maximum of the distribution is blocked as shown in Figure 5.19. The allowed high frequencies correspond mainly to the edges. The final image distribution near the edge is a complex function. The amplitude in this kind of band-passed image, corresponding to the classical imaging of the slit,[69] has been discussed in detail by Birch.[70] It has been described for applications dealing with in-line holography by Trolinger and Gee.[73] Practical demonstrations have also been provided.[1,56,73,74] Generally a dark line of symmetry coinciding with the Gaussian image of the edge is formed. Two bright bands both sides of the dark line are also formed in an otherwise dark background. The accuracy of the edge definition depends on the sharpness of these diffraction lines. In the case of sharp bright lines, they can also join making the central dark edge disappear. The center line of the band will then define the edge.

For a sharp diffraction edge line, the zero frequency in the filter plane should be blocked (as shown in Figure 5.19) and as large a number of side bands (30 or more) as possible should pass through.[69,70,73] As indicated by $\mathrm{sinc}\,(\pi du/\lambda_o f)$ in Equation (5.21), large dimensions in the filtering plane are needed for very small object diameters (d). The width of the blocking filter is

$$D = 2\lambda_c f/d \tag{5.27}$$

For circular objects of diameter d, the two-dimensional Fourier transform in the frequency plane is

$$I(\gamma) = \frac{J_1(\pi d\gamma/\lambda_c f)}{\pi d\gamma/2\lambda_c f} \tag{5.28}$$

The diameter of the central blocking circular filter corresponds to the first zero of the Bessel function. The argument of the Bessel function is about ± 3.83 at the first zero. Thus, the filter diameter D, like Equation (5.25) is obtained as

$$D \cong 2.44\lambda_c f/d \tag{5.29}$$

Again the filtering operation is physically not possible for very small objects where the blocking filter diameter (and hence the total diameter of the filtering plane) may become too large. A way to explain this aspect is as follows. For $\lambda_c = 0.6328\ \mu\mathrm{m}$, $d = 10\ \mu\mathrm{m}$, D given by Equation (5.29) is about $0.154f$. About 30 times this

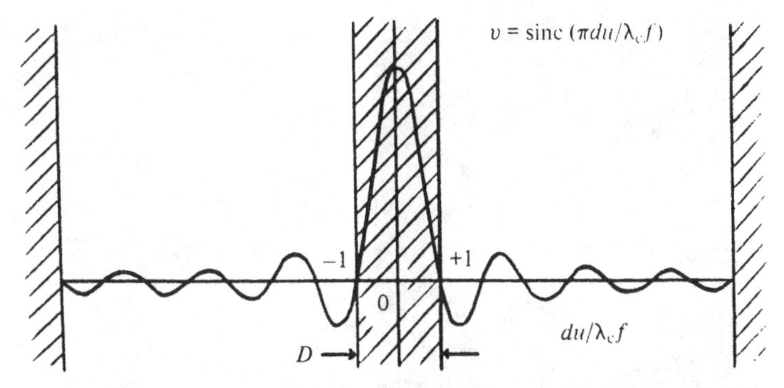

Fig. 5.19 Schematic diagram for the filter parameters in edge line holography of line objects during reconstruction. The coordinate system (u, v) is in the focal plane (α, β) of the magnifying lens – as shown in Figure 5.16.

diameter needed in the filtering plane is about $4.6f$. For a lens $f\#$ as small as 1, the filtering plane diameter is thus 4.6 times the lens diameter! Also the available theoretical discussion[69,70,73] about the edge differentiation is for sharp edged objects (the reconstructed image in the present context). Due to finite hologram aperture effects, sharp edged reconstructions are anyway not present in the case of small objects (see Section 5.10). The spatial filtering will consequently not produce the sharp edged line. Therefore, the differentiation is said to be impossible for small objects – those with a diameter of the order of 10 μm.[23]

Since the filter removes the paraxial radiation, z-imaging sensitivity is improved.[74] Again, with a magnifying lens and a closed circuit television system, this is not a practical problem.

Thus, filtering for the enhanced edge line is practical only for larger objects. Even then, the optimum filter dimensions are different for the different sizes and shapes of microobjects commonly present in an experimental situation. One possible application area is in in-line holography of large objects where the far-field condition cannot be satisfied. However, the filtering action of the television system clips the optical energy below a certain level. Consequently, the edge location capability in the unfiltered case can be as good as that in the filtered case.[73] This situation is seen in Figure 5.20 where the reconstructions from an in-line hologram of a .22 caliber bullet in flight are shown. Figures 5.20(a) and (b) show unfiltered and filtered images respectively from a lensless closed circuit television monitor. Figures 5.20(c) and (d) show the corresponding enlarged images.

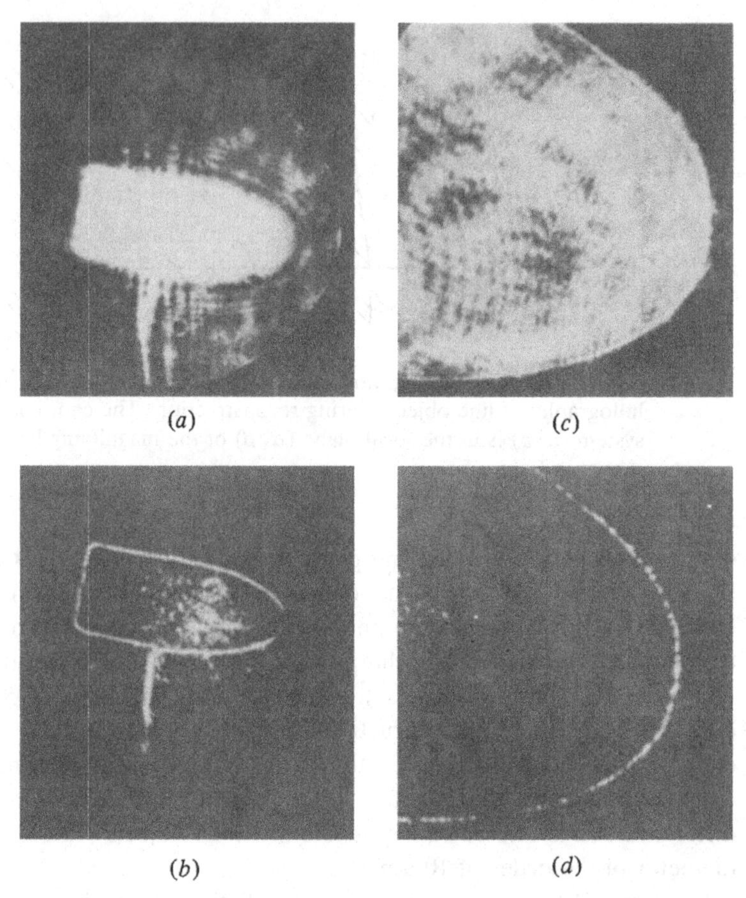

Fig. 5.20 Reconstructed image of a .22 caliber bullet recorded in flight by the in-line method using a Q-switched ruby laser: (*a*) unfiltered; (*b*) bandpassed filtered; (*c*) enlargement of the nose unfiltered; (*d*) enlargement of the nose filtered. (Trolinger and Gee.[73])

Combined effect for a distribution As seen earlier in this section, generally smaller filter sizes are needed for image contrast enhancement. For edge line holography, the optimum filter size depends on the object size as given by Equations (5.27) and (5.29). A limited cross-section in the filtering plane is physically available. Thus edge line holography is not practical for very small objects. In a practical situation generally a size and shape (i.e. one- and two-dimensional objects) range is present. For a fixed filter size, the situation might mean image contrast enhancement for smaller objects and edge line holography for large objects. This kind of situation is shown in Figure 5.21. Notice that the larger objects show edge lines whereas many small objects have a significantly increased signal to noise ratio after the high pass filtering.

5.7 Exposure controls during recording

Due to the limited dynamic range of emulsions used in holography, accurate control of exposure becomes very important. For off-axis holography using Agfa10E75 NAH plates, Phipps, Robertson and Tamashiro[44] determined the required exposure accuracy within ±3.5%. With many laser systems, this precise control of exposure is not easy.

For in-line holography such a precise control is not necessary. As indicated in Figure 5.13, the optimum exposure of 4.45 $\mu J/cm^2$ yields an image to background intensity ratio (σ) about 14.5. A ± 10% variation in the exposure does not alarmingly reduce the ratio σ.

On the other hand, pulsed lasers such as ruby often used in particle field holography have significant pulse-to-pulse energy variation. The effect is very pronounced when the laser is operated near the threshold and the pulse repetition rate is on the high side. The output energy variation in such a situation becomes very frustrating leading to underexposure in one case and overexposure in the other. TEM_{00} output from commercially available ruby lasers is generally more than 0.02 J. According to Figure 5.13, an Agfa10E75 plate, which requires 4.45 $\mu J/cm^3$, will need only about 445 μJ to expose a 100 cm^2 area. This need is only about 1/45 of the available output of the 0.02 J. Thus, as such, the laser has to be operated near the threshold for proper exposure resulting in the exposure uncertainty stated above.

The problem can be partly or entirely solved by beam expansion, decreasing the size of the TEM_{00} mode selector and the spatial filter aperture, using lower speed emulsions, or any combination of above techniques. Hologram pre-magnification (Section 5.2) also reduces laser intensity per unit cross-sectional area. However, the use of neutral density filters at the output of the laser beam to cut off the major portion of the available laser power is very common. Applications of these filters are found in the literature from the very beginning of the particle field holography.[6,16]

With the above mentioned considerations in mind, exposure in the proper range is generally obtained. As seen in Figure 5.13, significant exposure variations are allowed for a high image to background intensity ratio. Some exposure variation can be handled during or even after the development. In the safe darkroom light, one can

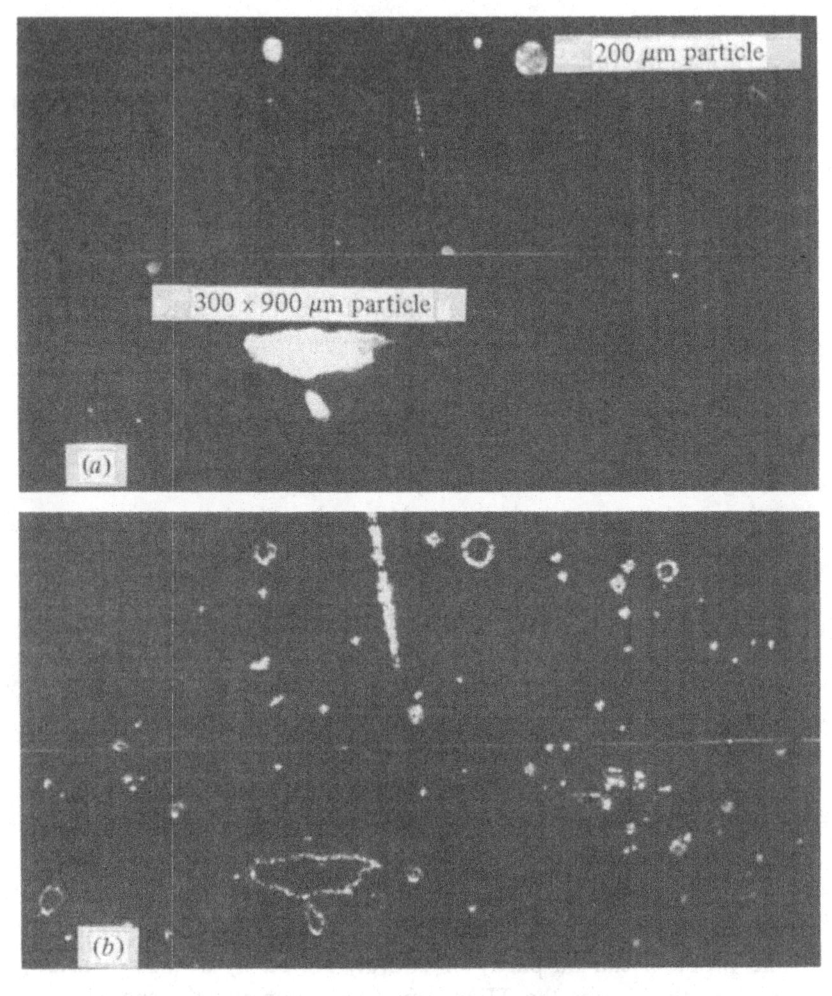

Fig. 5.21 High pass filtering in holography: (*a*) a reconstructed image plane; (*b*) filtered image of (*a*). (Trolinger.[1])

detect underexposure or overexposure at the early stages. The development time can then be reduced or increased from the normal duration by some extent. Although not very commonly used, reprocessing of non-optimally exposed but properly developed holograms is also possible.[44]

5.8 Controls during reconstruction

After the successful hologram recording, the magnified images are generally viewed on the monitor of a closed circuit television system. The camera–monitor system has a certain dynamic range. There is a certain minimum (threshold) detectable intensity seen in the form of the monitor brightness. There is also a saturation level of incident intensity beyond which the monitor brightness remains constant. The saturation to threshold intensity ratio of the picture tube alone in generally of the order of 100–1000.[75] As seen in Figure 5.13, the optimum image to background intensity ratio is much below the typical dynamic range. Since the whole aim is to visualize the image clearly, a television system with brightness and contrast controls is very helpful. The best way is to keep the background intensity on the picture tube slightly above the threshold situation. The image, whose intensity is several times more than that in the background (see Section 5.5.1), will then be clearly seen.

The background intensity reaching the picture tube is a function of the reconstructed laser intensity reaching the hologram, the background amplitude transmittance of the hologram and the lens magnification of the image. Depending on the sensitivity of the picture tube the requirement of the intensity reaching it will vary. The saturation light intensities vary from 0.005 $\mu w/cm^2$ for an Isocom to about 0.59 $\mu W/cm^2$ for a standard Vidicon at the HeNe laser wavelength of 0.6328 μm.[75] As discussed in Section 5.5.1, the optimum background intensity transmittance is about 0.01 for Agfa10E75 plates. Suppose a 5 mW laser beam is used to illuminate a 100 cm^2 area of the hologram. The transmitted background intensity is therefore 0.5 $\mu W/cm^2$ as such or 0.005 $\mu W/cm^2$ if the reconstruction is magnified to 10 times using a lens. For the Isocon tube, this is already the saturation intensity! For a Newvicon tube, the saturation intensity requirement is 0.18 $\mu W/cm^2$.[75] Considering the typical saturation to threshold intensity ratios, even the Newvicon tube is too sensitive particularly when the laser beam is expanded to less than 100 cm^2 area and the lens magnification is less than 10 in the above example.

In these situations of excess picture tube sensitivity, the available laser power can be reduced by a neutral density filter, a beam splitter, using a smaller pinhole of the spatial filter, a slight misalignment of the spatial filter, excess beam expansion, etc. With the commonly used polarized laser output, a Polaroid sheet can be used at the laser output. Rotation of the sheet will transmit a variable laser power. In fact, the variable laser power gives a quick adjustment in case of slight to moderate hologram exposure variations. The laser intensity adjustment is also needed when the lens magnification is changed.

On the other hand, there are laser intensity losses at the spatial filtering stage, reflection at different lens and hologram surfaces, etc. In these situations, low sensitivity tubes such as the standard Vidicon may not do so well with a low power reconstruction laser. Then, besides using higher power reconstruction lasers, the lowest needed lens magnification (see considerations in Section 4.9.3), minimum required (at a time, from resolution considerations as discussed in Section 4.4) beam expansion, and larger size aperture in the spatial filter are helpful.

Liburdy[76] has proposed the use of two closed circuit television systems simultaneously. The image on the monitor of the first system is reimaged with the help of the second system. The final image on the monitor of the second system is a function of controls on both the systems. Overall, it becomes easier to obtain an acceptable image quality.

5.9 Transparent objects

In common holography, phase imaging via holography was introduced by Gabor *et al.*[77] and Tanner[78–80]. Generally, the off-axis method is used to store the phase information ultimately yielding fringes like those obtained by holographic interferometry. These fringes can be related to refractive index variations and then to physical properties such as mass density, temperature, strain, etc.. Such methods and applications are well discussed by Vest.[81]

In Fraunhofer holography, phase objects such as droplets and bubbles are routinely encountered. Thompson[15] reconstructed a shock wave showing the wavefront. Boundary layers in water tunnels have also been observed by several workers.[82–4]

Critical studies of the phase structure from the hologram and the reconstruction have been performed by Lomas[85] for tapered glass fibers. Far-field holography of phase objects has been discussed in some detail by Cartwright, Dunn and Thompson,[86] Cartwright,[58] Prikryl,[87] Prikryl and Vest,[88] and Lu and Meng.[89] The analogy between the coherent images and those holographically obtained has also been described.[58,86,88] Paterson[90] used the role of the energy transmitted through the bubble to distinguish bubbles from solid particles.

The transparent object case can be described using the available background from Chapter 3. In the far-field approximation, the normalized irradiance distribution during the reconstruction is

$$I(\mu, \nu) = |1 + \Gamma A^*(\mu/M_c, \nu/M_c)|^2$$
$$= 1 + 2\Gamma \operatorname{Re}[A(\mu/M_c, \nu/M_c)]$$
$$+ \Gamma^2|A(\mu/M_c, \nu/M_c)|^2. \quad (5.30)$$

We have omitted the field from the virtual image. However, the virtual image term can also be included.[58,86] Generally, the irradiance is discussed assuming opaque objects implying 'A' is unity over the object cross-section and zero outside. Semi-transparent or phase objects can simply be described by proper definition of A.

High values of Γ are generally desired (see Section 5.5.1) and in that case, Equation (5.30) can be written as

$$I(\mu, \nu) \cong \Gamma^2|A(\mu/M_c, \nu/M_c)|^2. \quad (5.31)$$

For collimated recording and reconstruction beams, M_c becomes unity resulting in:

$$I(\mu, \nu) = |A(\mu, \nu)|^2 \quad (5.32)$$

where Γ is omitted being a constant. Equation (5.32) is sometimes[89] used to describe the image of a phase object.

The recording irradiance distribution for the semi-transparent objects is similarly known (see Section 4.1) with proper definition of A. Specific discussion about the image is possible for a known A distribution. Two particular cases are considered here. One is one- or two-dimensional objects with uniform amplitude transmittance and phase. The other is spherical microobjects.

5.9.1 *One- or two-dimensional uniform objects*

The case of a simple uniform bar has been considered by Cartwright, Dunn and Thompson.[86] The function 'A' for one- or two-dimensional objects in this case is

$$A(\xi, \eta) = [1 - \alpha \exp(i\delta)]A_G(\xi, \eta)$$
$$= (1 - 2\alpha \cos\delta + \alpha^2)^{1/2} \exp(-i\theta)A_G(\xi, \eta), \quad (5.33)$$

where

$$\theta = \tan^{-1}\left(\frac{\alpha \sin\delta}{1 - \alpha\cos\delta}\right), \quad (5.34)$$

α is the uniform amplitude transmittance of the bar or the lamina and δ is the constant phase. As shown in Figure 5.22, $A_G(\xi, \eta)$ defines the geometrical boundary of the object, i.e.

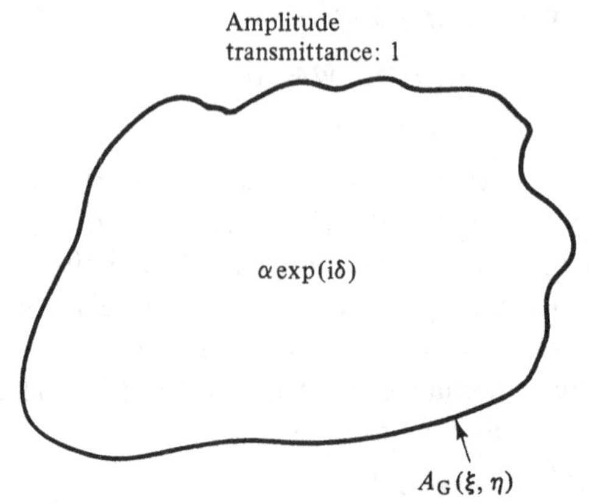

Fig. 5.22 Diagram representing a laminar phase object. The object boundary is described by the geometrical function $A_G(\xi, \eta)$ whose value is unity inside the object cross-section and zero outside. The complex amplitude transmittance of the object is $\alpha \exp(i\delta)$ and that of the surrounding area is 1.

$$A_G(\xi, \eta) = \begin{cases} 1 & \text{object cross-section} \\ 0 & \text{otherwise} \end{cases} \tag{5.35}$$

In the conventional model of opaque objects, $\alpha = 0$ and hence $A(\xi, \eta)$ becomes $A_G(\xi, \eta)$.

As an example, let us consider the collimated beam recording. The normalized irradiance distribution in the hologram for a two-dimensional object cross-section (see Section 4.1) becomes

$$I = \left| 1 + \frac{i(1 - 2\alpha \cos \delta + \alpha^2)^{1/2}}{\lambda_o |z_o|} \times \exp\left[\frac{ik_o(x^2 + y^2)}{2|z_o|} - \theta\right] \tilde{A}_G\left(\frac{x}{\lambda_o |z_o|}, \frac{y}{\lambda_o |z_o|}\right) \right|^2 \tag{5.36}$$

Similarly, for a one-dimensional object (such as a long thin bar)

$$I = \left| 1 - \frac{(1 - 2\alpha \cos \delta + \alpha^2)^{1/2}}{\lambda_o |z_o|} \times \exp\left(\frac{ik_o x^2}{2|z_o|} - \frac{\pi i}{4} - \theta\right) \tilde{A}_G\left(\frac{x}{\lambda_o z_o}\right) \right|^2. \tag{5.37}$$

Besides the fringe shifts due to change in the phase terms, the amplitude of the modulation is multiplied by $(1 - 2\alpha \cos \delta + \alpha^2)^{1/2}$ in either case. As we know, the limited modulation limits the recording range (see Section 4.3). The additional modulation here can be of advantage. For example, for a pure phase object of $\alpha = 1$ and $\delta = \pi$, the additional modulation is 2.

The form of the reconstruction described by Equation (5.30) can be specifically discussed here. With the help of Equation (5.33), it becomes

$$I(\mu, \nu) = 1 + [2\Gamma(1 - 2\alpha\cos\delta + \alpha^2)^{1/2} + \Gamma^2(1 - 2\alpha\cos\delta + \alpha^2)]A_G(\mu, \nu) \quad (5.38)$$

Substituting the value of $A_G(\mu, \nu)$ from Equation (5.35) provides the irradiance distribution. The simple solution of $I(\mu, \nu) = [1 + \Gamma(1 - 2\alpha\cos\delta + \alpha^2)^{1/2}]^2$ inside the image and 1 outside is not always true. In practice, spikes at the edges are formed. These spikes (edge ringing) are due to the phase step and finite resolution capability, and can be explained by the impulse response of the system.[86,89]

5.9.2 *Spherical objects*

This case is very often encountered in practical far-field holography in connection with liquid droplets and gas bubbles in liquids. Besides the diffraction of the recording beam by the object cross-section, refraction through the sphere and reflection from its surface must be considered. Based on the scattering properties of air bubbles in water,[91] it is known that an interference pattern is formed within the image.[90] The pattern is due to the interference between the diffraction image and image of the light passing through the bubble during the recording. In that respect the pattern can be used to distinguish bubbles from opaque particles.[90,92,93]

Holography of semi-transparent spherical microobjects has been discussed in detail by Prikryl and Vest.[88] Lu and Meng[89] provided explicit solutions for spherical microobjects with a higher refractive index than that of the surrounding medium. The microobject then acts as a highly aberrated positive lens. Neglecting multiple reflections, the situation is represented in Figure 5.23. $n_o < n_s$ can also be considered but total internal reflection then has to be accounted for.

In opaque objects, only diffraction by the center plane (ξ, η) of the sphere is considered. In the non-opaque case, reflected and transmitted fields are also to be added. In the Fraunhofer diffraction calculation, the equivalent plane object in the center plane (ξ, η) should be obtained. This can be done by projecting back the transmitted and reflected fields in the (ξ, η) plane. The Huygens–Fresnel Principle (see Chapter 3) can then be used to determine the reconstructed field. If the amplitude transmittance of the object plane

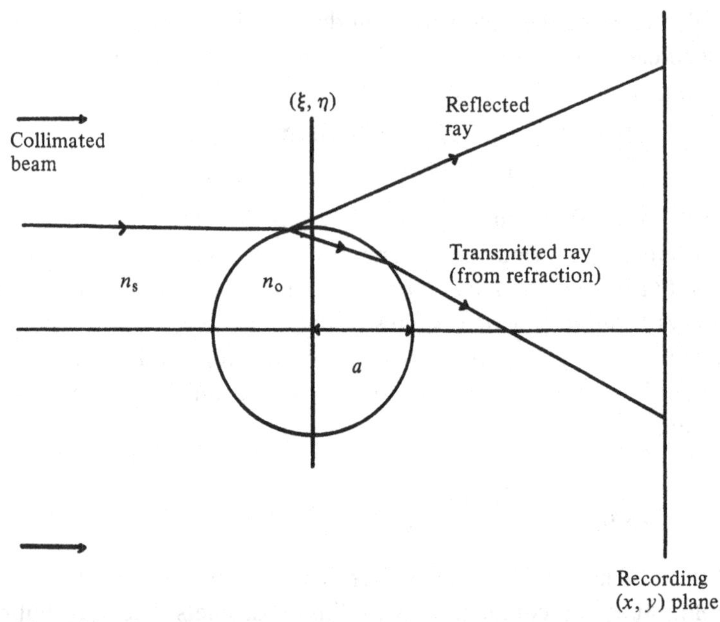

Fig. 5.23 Diagram showing reflected and transmitted rays when collimated light is incident on a semi-transparent sphere. The refractive index n_o of the sphere material is greater than that n_s of the surrounding medium showing the positive lens effect.

is $1 - A(\xi, \eta)$, then we know the image irradiance can be determined from $A(\mu, \nu)$, using say Equation (5.32). In the present case

$$A(\xi, \eta) = 1 - \operatorname{circ}(r/a) - f(r), \tag{5.39}$$

where the circ function is defined in Section 3.4. The function $f(r)$ can be written as[89]

$$f(r) = t(r)\exp[\mathrm{i}\phi_1(r)] + w(r)\exp[\mathrm{i}\phi_2(r)], \tag{5.40}$$

where $t(r)$ and $\phi_1(r)$ are the amplitude transmittance and the phase respectively of the refracted field projected back at the (ξ, η) plane; $w(r)$ and $\phi_2(r)$ are similar quantities corresponding to the reflection. Obviously, only rays with incident angles equal to or larger than 45° will reach the plane. The intensity of the real image is then

$$\begin{aligned} I(R) = |A(\mu, \nu)|^2 &= \operatorname{circ}(R/a) - 2\operatorname{circ}(R/a)t(R) \\ &\quad \times \cos[\phi_1(R)] + 2t(R)w(R)\cos[\phi_1(R) - \phi_2(R)] \\ &\quad + t^2(R) + w^2(R), \end{aligned} \tag{5.41}$$

where the term containing $w(r)\operatorname{circ}(R/a)$ has been omitted because the reflected field is zero when $R < a$. The contour $\operatorname{circ}(R/a)$ defines the usual image of opaque objects. The circular interference fringes

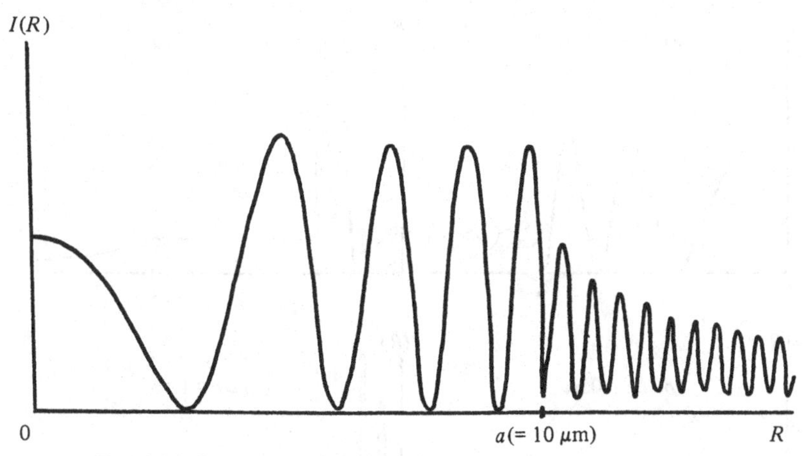

Fig. 5.24 Computer plot of the intensity distribution at the real image plane of a spherical water droplet of 10 μm radius in air. $\lambda_o = \lambda_c = 0.6328$ μm is used. (Lu and Meng.[89])

indicated by the cosine terms in RHS of Equation (5.41) are due to the object transparency and reflectivity. There are two groups of cosine fringes. The interference between the contour image and the one corresponding to the refracted field forms concentric fringes in the region $R < a$. This corresponds to the $\cos[\phi_1(R)]$ term. For $R > a$, another group of concentric fringes is formed, governed by the $\cos[\phi_1(R) - \phi_2(R)]$ term. The fringes are due to interference between reflection and refraction images. Figure 5.24 is a computer plot of Equation (5.41) for a 10 μm water droplet in air when a HeNe laser ($\lambda_o = \lambda_c = 0.6328$ μm) is used.

The 'planar object' can also be considered to be slightly different but parallel to the central plane of the sphere. The image at out-of-focus planes can thus be studied.[89] Computer plots of such out-of-focus image intensities are shown in Figure 5.25.

At the focus, the envelope of the intensity maxima is almost constant within the diffraction image. Outside the diffraction image, the fringes are finer and the envelope decreases as R increases.

In out-of-focus images, the region of fringes becomes restricted near the center only (Figure 5.25(c), (d)). This corresponds to the region of focus of the 'spherical lens' and the refracted beam is focused or centered in a small region.

Care must be taken to distinguish these fringes from those resulting from the finite aperture effects (see Section 5.10). The analysis in this section assumes that the aperture-limited resolution of the system is

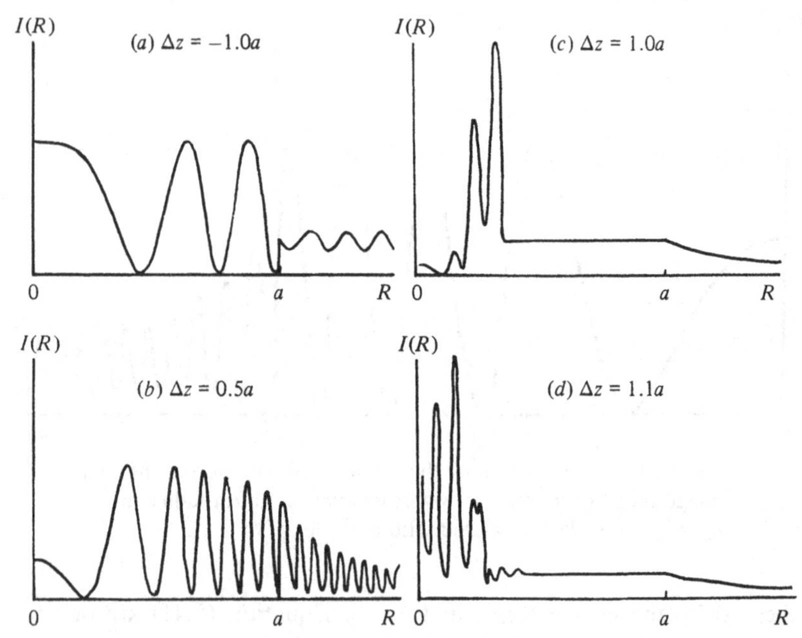

Fig. 5.25 Computer plots of intensities at some out-of-focus images of the same droplet as in Figure 5.24. The amount of misfocusing Δz is represented in terms of the droplet radius a. (Lu and Meng.[89])

capable of distinguishing the fine fringes within and outside the image.

A discussion of the role of the recording arrangement (in-line with negative or positive emulsion; off-axis with plane wave or diffuse illumination) is available.[88] The in-line method with a positive emulsion and the off-axis method require careful selection of numerical apertures. The most useful way is to use the in-line holography with negative emulsions. Fortunately, this approach is common in practice.

5.10 Finite aperture effects

The image in Chapter 3 is evaluated assuming infinite hologram aperture. The actual hologram aperture is, however, limited by its physical size, effective aperture due to limited film resolution capability, actual reconstruction beam cross-section, etc. The integral limits over the recording plane should therefore be appropriately finite. The image form with infinite integral limits has been considered by Robinson,[94] Belz[95,96] and others[23,52,58,97–101]. The study by

Özkul[52] includes the effect of film non-linearities on the aperture-limited in-line Fraunhofer holography.

The image form in the finite aperture case can be evaluated using appropriately finite limits of x and y in the integrals in Section 3.2. In this section, only the conjugate image term will be considered. In practical in-line Fraunhofer holography, this term describes the image intensity distribution (see Section 5.9). However, other terms can likewise be considered. For simplicity, let us consider collimated beam recording and reconstruction ($z_c = z_R = -\infty$) with the same wavelength ($\lambda = \lambda_o = \lambda_c$). The integral I_2 corresponding to the conjugate image, under the far-field approximation for two-dimensional object cross-sections, becomes:

$$I_2(\mu, v) \propto \int_{y=y_{\min}}^{y=y_{\max}} \int_{x=x_{\min}}^{x=x_{\max}} \int_{\eta=-\infty}^{\eta=+\infty} \int_{\xi=-\infty}^{\xi=+\infty} A^*(\xi, \eta) \left\{ -\frac{ik}{|z_o|} [(\mu - \xi)x + (v - \eta)y] \right\} d\xi \, d\eta \, dx \, dy \tag{5.42}$$

where $x_{\min}$, $y_{\min}$ are the minimum values of the x, y coordinates respectively. Similarly $x_{\max}$ and $y_{\max}$ define their maximum values. For one-dimensional objects, a similar expression can be written as

$$I_2(\mu) \propto \int_{\xi=-\infty}^{\xi=+\infty} \int_{x=x_{\min}}^{x=x_{\max}} A^*(\xi) \exp\left[-\frac{ik}{|z_o|} (\mu - \xi)x \right] d\xi \, dx. \tag{5.43}$$

The distributions can also be written as convolutions of A^* with a suitably defined impulse response of the system.[99]

5.10.1 *Opaque circular cross-section object*

For a circular cross-section opaque object of diameter d and effective hologram aperture ρ we obtain:[23,95–7]

$$I_2(\alpha) \propto \int_0^{\alpha_k} J_1(R') J_0(\alpha R') \, dR' \tag{5.44}$$

where $\alpha = 2R/d$ is a normalized coordinate in the image plane. J_0 and J_1 are the zero- and first-order Bessel functions of the first kind. The integral limit α_k is $\pi\rho d/\lambda z_o$. As discussed in Section 4.2, α_k can also be written as $(1 + m)\pi$ where m is the number of side lobes recorded and utilized for the reconstruction. The behavior of $I_2(\alpha)$ has been extensively discussed.[23,95–7]

In Figure 5.26, the variations of $|I_2(\alpha)|^2$ has been plotted for some numerical situations. With an infinite hologram aperture ($\alpha_k = \infty$),

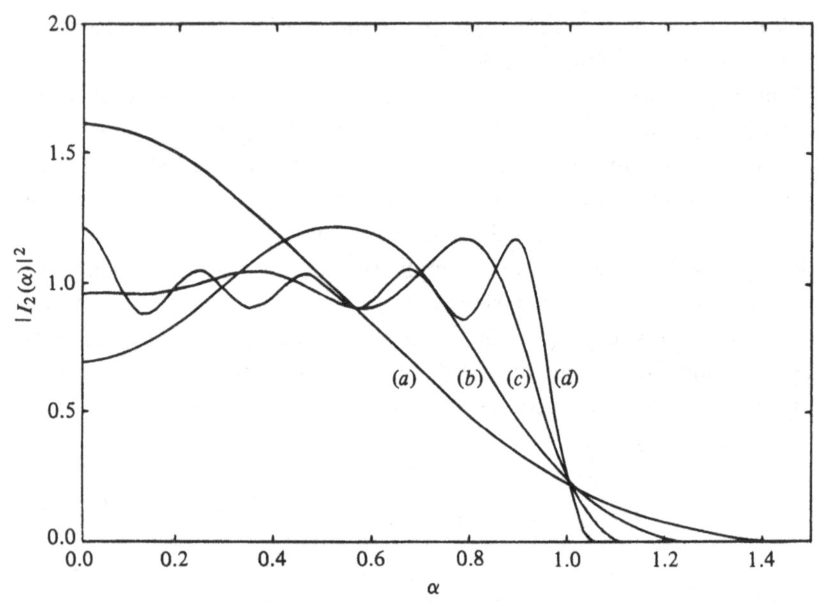

Fig. 5.26 Image intensity distribution given by square of RHS of Equation (5.44): (*a*) $\alpha_k = 3$; (*b*) $\alpha_k = 6$; (*c*) $\alpha_k = 15$; (*d*) $\alpha_k = 30$.

$I_2(\alpha)$ is a step function. For the finite case, the variation, as seen from Figure 5.26, yields a fringe-like pattern. The ideal edge is at $\alpha = 1$ but that is not reflected from the finite aperture curves. Thus, error in the size analysis is possible with the conventional image-edge or inflexion point detection.

From the behavior of $|I_2(\alpha)|^2$, it has been found[95-7] that the error is zero when the image width is measured where the edge intensity is 25% of the center intensity. However, exact intensity measurement is not common in practical particle field holography. Also, the above conclusion is for circular cross-section opaque objects. Consequently, the approach of considering the image inflexion point of apparent edge remains common. Generally the actual size is larger than predicted this way. The error is reduced for increased α_k or m. Practically, $m > 3$ for the smallest microobject yields acceptable errors.[23]

5.10.2 *Opaque line objects*

For a one-dimensional cross-section object (such as a long thin wire) of diameter $2a$ and effective hologram width 2ρ, Equation

(5.43) can be solved. First, the integration over x (within limits $\pm \rho$) can be performed to obtain a sinc function ultimately yielding:

$$I_2(\mu) \propto \left[\frac{S_i(A) - S_i(B)}{\pi}\right], \quad (5.45)$$

where the sine integral is defined as

$$S_i(g) = \int_0^g \frac{\sin t \, dt}{t}, \quad (5.46)$$

and

$$A = \frac{k(\mu + a)\rho}{|z_0|} = \left(\frac{\mu}{a} + 1\right)(1 + m)\pi, \quad (5.47)$$

$$B = \frac{k(\mu - a)\rho}{|z_0|} = \left(\frac{\mu}{a} - 1\right)(1 + m)\pi, \quad (5.48)$$

m is the number of side lobes recorded (see Section 4.2). From Equation (5.45), the function

$$I\left(\frac{\mu}{a}, m\right) = \left[\frac{S_i(A) - S_i(B)}{\pi}\right]^2 \quad (5.49)$$

describes the image intensity distribution. For an infinite hologram aperture ($m = \infty$), it is a step function describing the ideal image. In Figure 5.27, $I(\mu/a, m)$ has been plotted for some values of m.

The common practice is to measure the object from the sharp edge defined by the last intensity maximum (just before $\mu/a = 1$). Obviously, the measured size is smaller than the actual one defined by $\mu/a = 1$. The error can be determined from the position of the maximum. For $m > 1$ and near $\mu/a = 1$, $S_i(A)$ can be approximated as $\pi/2$. Also B and $S_i(B)$ are negative quantities in the region ($\mu/a < 1$) of interest. Thus, the maxima just before $\mu/a = 1$ corresponds to $B = -\pi$, i.e.

$$\frac{\Delta\mu}{a} \cong \frac{1}{1 + m}, \quad (5.50)$$

where $\Delta\mu$ is the difference between the apparent edge position and the ideal position ($\mu = a$). The validity of Equation (5.50) can be easily verified from Figure 5.27 for $m = 2$ and 4.

Thus, for $m = 4$ the error in the size measurement is 20%. However, the criterion for side lobes to be recorded is established for the smallest size object to be encountered. For larger objects, a larger number of side lobes is automatically recorded (see Section 4.2 and Chapter 8). Consequently, an error of 20% in the size of the smallest object is generally not bothersome.

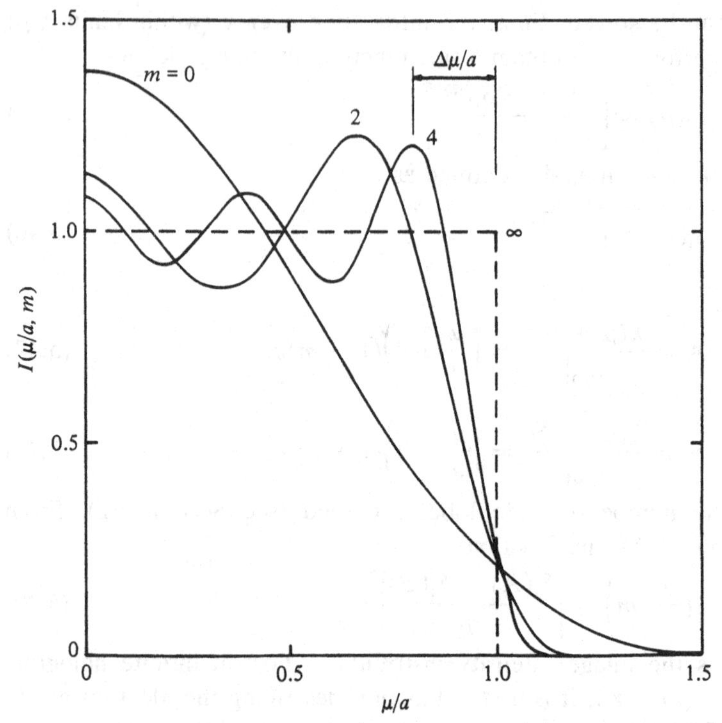

Fig. 5.27 Variation of $I(\mu/a, m)$ against μ/a for some values of m; $\Delta\mu$ is the error in the size measurement from the inflexion point. Such an error is shown for $m = 4$ in the figure. (Ref. 101.)

Nevertheless when accurate size analysis of very small objects is required, numerical processing of the image intensity distribution,[102] calibration as reported,[23] and measuring the image width at certain normalized intensity[23,95–7,101] are useful. Conventionally measured linewidths can also be corrected.[101]

5.11 Other developments

There are several less common approaches to particle field holography which are useful in particular situations. Katz, O'Hern and Acosta[103] and O'Hern, Katz and Acosta[104] designed and used a holocamera submersible in water to reduce 'scaling' errors of laboratory results applied to field cavitation phenomena. Katz[83] used a knife-edge at the focal plane of the magnifying lens during the reconstruction to reduce noise and laser speckle. Knox[105] used a holographic plate in water to demonstrate the capability of recording a larger volume at any depth in the ocean. Murata, Fujiwara and Asakura[106] proposed and demonstrated the use of a diffused recording beam in the in-line method. Bexon, Dalzell and Stainer[50] used a high power laser to produce an 'in-place' pinhole of the spatial filter of the recording beam. This helps to produce the pinhole quickly without the alignment problem. Roberts and Black[107] demonstrated infrared recording of in-line holograms using a CO_2 laser and visible reconstruction using a HeNe laser. Tyler[108] described the image coding capability of far-field holography.

When a large amount of data is to be reduced, the usual manual analysis of the reconstruction is time consuming. Automatic analysis of the image field is then desired. This process generally requires image digitization for further data processing. The early semi-automatic approach of Bexon *et al.*[49,50] requires manual focusing. Similar systems are described by Heidt and Furchert,[109] Lidl,[110] Witherow,[2] and Payne, Carder and Steward.[111] More automated systems with image depth (focusing) capability are also reported by Haussmann and Lauterborn,[102] Feinstein and Girard,[112] and Weinstein, Beeler and Lindermann.[113] A very large amount of data digitization and processing is required for such an analysis. Stanton, Caulfield and Stewart[114] and Caulfield[115] proposed a non-image plane analysis technique at only two misfocused image planes that can yield the image plane location. The technique can reduce the number of digitizing and processing operations by orders of magnitude. The optical principal of the non-image plane analysis technique is described in Section 9.2. Reflections and scatterings by sidewalls of narrow test-sections are also possible. Proper aperture stops can eliminate the unwanted light.[116] Optimizing image quality using variable intensity reconstruction beams is also possible.[117] Trolinger and coworkers[118–19] describe the role of particle field holography in crystal growth experiments in space.

6

Analysis of reconstruction

Once a hologram of an event is successfully recorded, the reconstruction at proper net magnification will yield images of the microobjects distinct in the background. As seen in Chapter 3, the image represents the object cross-section encountered at the time of recording. Generally knowledge of net transverse magnification is required for size analysis. Information about the volume magnification is also often needed for determining the density of microobjects.

The net transverse magnification is generally several hundred (see Section 4.9). This results in a limited field of view on a fixed cross-section observation screen. To cover a considerable subject or image volume, either a detector such as the television camera or the hologram on a translation stage is moved for frame by frame analysis. For the general case of a spherical recording and/or reconstruction beam, the magnifications are not straightforward.

The transverse magnification due to the camera-lens system is fixed for a given lens–screen (or vidicon) separation. This magnification can be easily determined by calibration using a resolution chart or a transparent ruler scale. However, the holographic magnification due to the spherical beams is space variant. These magnifications are described in detail in this chapter.

The analysis is generally valid for in-line as well as off-axis holography. We start with the reconstructed image position known in terms of the object, reference and reconstruction source points, and wavelengths of light sources. The needed magnifications are then discussed for modes suited to users of the technique. The conjugate image is considered as it is used almost invariably in particle field holography. The conjugate image is generally real in practical situations in the present context.

The recording and reconstruction are both initially assumed to

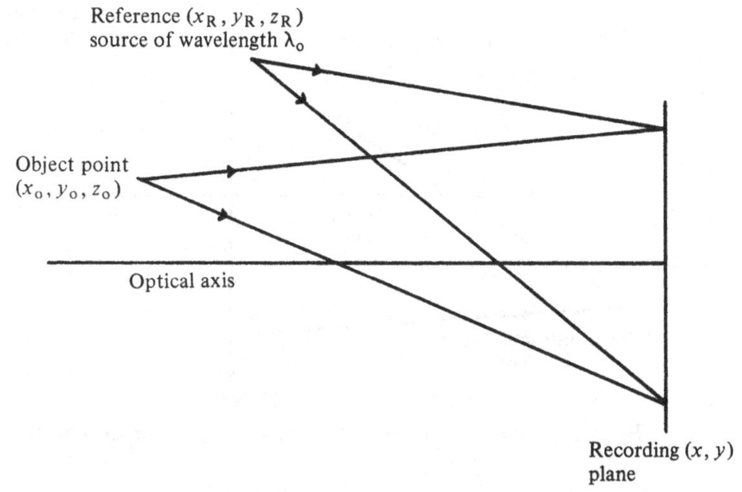

Fig. 6.1 Diagram describing the general recording arrangement.

be in air. The role of a non-unity refractive index of the medium is discussed later. Similarly, the case when the hologram is pre-magnified before the recording is described as a coordinate transformation later.

Lensless magnification in holography has been discussed by several authors.[1-8] However, the analysis of Meier[6] is very well accepted for coordinates of image points, magnifications, and third-order aberrations.

6.1 Relationships between object and image points

Figure 6.1 represents a typical recording arrangement with a spherical reference wave. With respect to the origin of the Cartesian (x, y, z) system at the center of the recording plane (x, y), the object and reference point coordinates are (x_o, y_o, z_o) and (x_R, y_R, z_R) respectively. The recording source wavelength is λ_o. The reconstruction arrangement is described by Figure 6.2. The reconstruction source of wavelength λ_c is situated at (x_c, y_c, z_c) and the conjugate image is formed at (X_c, Y_c, Z_c). The coordinates are related by[6] (see also Chapter 2).

$$X_c = -\frac{(x_c/n)(z_o/z_c) - x_o + x_R(z_o/z_R)}{1 - z_o/nz_c - z_o/z_R}, \tag{6.1}$$

$$Y_c = -\frac{(y_c/n)(z_o/z_c) - y_o + y_R(z_o/z_R)}{1 - z_o/nz_c - z_o/z_R}, \tag{6.2}$$

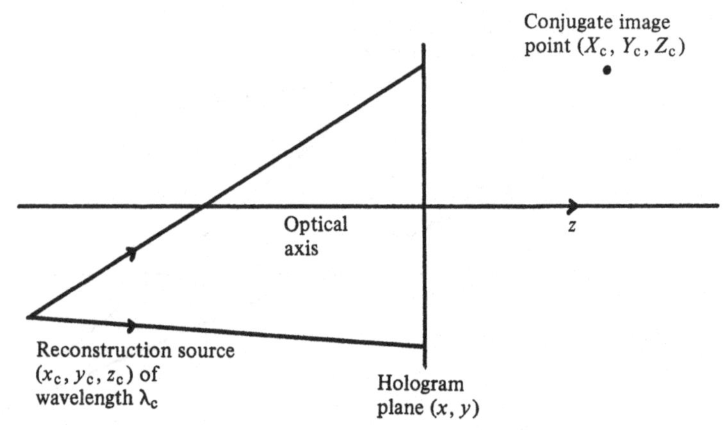

Fig. 6.2 Diagram illustrating the reconstruction process with laser light of wavelength λ_c.

and

$$Z_c = -\frac{z_o/n}{1 - z_o/nz_c - z_o/z_R}, \tag{6.3}$$

where n is the reconstruction (λ_c) to the recording (λ_o) wavelength ratio given by

$$n = \lambda_c/\lambda_o. \tag{6.4}$$

For the in-line case, Equations (6.1) and (6.2) become simpler since then $x_c = x_R = y_c = y_R = 0$. Reference, reconstruction or both can be plane waves. For a plane reference z_R, x_R/z_R and y_R/z_R become constants given by

$$z_R = -\infty, \tag{6.5}$$

$$x_R/z_R = \sin\phi_{x,R}, \tag{6.6}$$

and

$$y_R/z_R = \sin\phi_{y,R}, \tag{6.7}$$

where $\phi_{x,R}$ and $\phi_{y,R}$ are the angles that the reference beam makes with the positive direction of z-axis in (x, z) and (y, z) planes respectively. Similarly in the case of a plane reconstruction beam

$$z_c = -\infty, \tag{6.8}$$

$$x_c/z_c = \sin\phi_{x,c}, \tag{6.9}$$

and

$$y_c/z_c = \sin\phi_{y,c}, \tag{6.10}$$

where $\phi_{x,c}$ and $\phi_{y,c}$ are the angles that the reconstruction beam makes with the positive direction of the z-axis in (x, z) and (y, z) planes respectively. Therefore values of Equations (6.5)–(6.10)

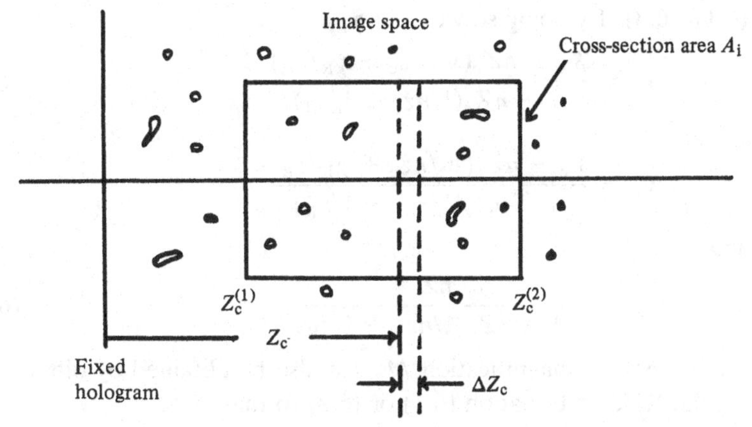

Fig. 6.3 Diagram representing image analysis with a fixed hologram and a movable camera or detector. An image volume of cross-sectional area A_i and depth $Z_c^{(2)} - Z_c^{(1)}$ is considered.

should be substituted in Equations (6.1)–(6.3) in the case of a collimated reference or reconstruction beam. We have discussed the conjugate image point. The real image is commonly analyzed. To make sure that the conjugate image is also real, a careful look into Equation (6.3) is necessary. With divergent beams, z_o, z_c and z_R are all negative. Z_c is positive for the real image. Thus, $1 - z_o/nz_c - z_o/z_R$ must be positive. This can be ensured by the proper selection of z_c and z_R for a given region of the object distance z_o.

6.2 Image analysis with fixed hologram

In this mode the hologram is fixed with respect to the reconstruction source. To cover an image volume, the detector or the television camera is moved on an x–y–z stage (see for example Section 4.9.1[9]). As seen in Figure 6.3, a fixed cross-sectional area A_i in the image space between longitudinal distances $Z_c^{(1)}$ and $Z_c^{(2)}$ from the hologram is covered. For size analysis, the transverse magnification and, for density calculations, the volume magnification will be sufficient in a practical situation. The location of the object corresponding to an image region is also useful. This mode of analysis is discussed by Vikram and Billet.[10]

6.2.1 *Object location and transverse magnification*

Corresponding to a given region in the image space, the location in the object space can be obtained by solving Equations

(6.1)–(6.3). By doing so we obtain:

$$x_o = \frac{X_c - nZ_c(x_c/nz_c + x_R/z_R)}{1 - nZ_c(1/nz_c + 1/z_R)}, \tag{6.11}$$

$$y_o = \frac{Y_c - nZ_c(y_c/nz_c + y_R/z_R)}{1 - nZ_c(1/nz_c + 1/z_R)} \tag{6.12}$$

and

$$z_o = -\frac{nZ_c}{1 - nZ_c(1/nz_c + 1/z_R)}. \tag{6.13}$$

The transverse magnification M_c can also be obtained by differentiating the RHS of Equation (6.1) or (6.2) so that

$$M_c = \mathrm{d}X_c/\mathrm{d}x_o = \mathrm{d}Y_c/\mathrm{d}y_o$$
$$= \frac{1}{1 - z_o/nz_c - z_o/z_R}. \tag{6.14}$$

Obviously for a spherical reconstruction and/or reference beam, the transverse magnification M_c depends on the object (and hence image) distance z_o. For collimated beams ($z_c = z_R = \pm\infty$), M_c becomes constant and unity.

For spherical waves, the information about z_o given by Equation (6.13) can be substituted in Equation (6.14). Thus, M_c can be obtained in terms of n, z_c, z_R and Z_c as

$$M_c = 1 - nZ_c(1/nz_c + 1/z_R). \tag{6.15}$$

The need for information about n, z_c and z_R can be avoided by knowing or calibrating the magnification for one value of Z_c. Using Equation (6.15), the magnification M_c' for another image position Z_c' is

$$M_c' = (Z_c - Z_c' + M_c Z_c')/Z_c. \tag{6.16}$$

To use Equation (6.16), at least one image position with respect to the hologram must be measured. If the magnifications at two image planes are known or calibrated then the mutual distances between the planes can be used to find the magnification at other planes. If M_c, M_c' and M_c'' correspond to the image distances Z_c, Z_c' and Z_c'' respectively, then Equation (6.15) can be used to yield:

$$M_c'' = M_c + \frac{(Z_c'' - Z_c)(M_c' - M_c)}{Z_c' - Z_c}. \tag{6.17}$$

In Equation (6.17), only the relative distances $Z_c'' - Z_c$ and $Z_c' - Z_c$ are involved.

6.2.2 *Volume magnification*

In Section 6.2.1 we have seen how transverse magnification can be determined at different longitudinal distances for proper size analysis. With the present mode of analysis, i.e. fixed hologram and the reconstruction source, suppose the depth between image distances $Z_c^{(1)}$ and $Z_c^{(2)}$ within the fixed cross-sectional area A_i is covered. The volume in the object space corresponding to this image volume should be determined for density calculations.

Let us consider an infinitesimal rectangular area $\Delta X_c \Delta Y_c$ in the image plane at the longitudinal distance Z_c. Equations (6.11) and (6.12) give that this area will correspond to the object space area given by

$$\Delta x_o \Delta y_o = \frac{\Delta X_c \Delta Y_c}{[1 - nZ_c(1/nz_c + 1/z_R)]^2}. \tag{6.18}$$

We can integrate the RHS of Equation (6.18) to relate the object-space cross-section area A_o and the image-space cross-section area A_i:

$$A_o = \iint dx_o dy_o = \frac{\iint dX_c dY_c}{[1 - nZ_c(1/nz_c + 1/z_R)]^2}$$

$$= \frac{A_i}{[1 - nZ_c(1/nz_c + 1/z_R)]^2}. \tag{6.19}$$

Now, keeping the cross-section A_i fixed, suppose an infinitesimal longitudinal depth ΔZ_c is considered. According to Equation (6.13), this depth corresponds to the object space depth Δz_o given by

$$\Delta z_o = -\frac{n\Delta Z_c}{[1 - nZ_c(1/nz_c + 1/z_R)]^2}. \tag{6.20}$$

Equations (6.19) and (6.20) yield the object volume of the cross-section with the image volume depth ΔZ_c at Z_c as

$$\Delta V_o = A_o \Delta z_o = \frac{nA_c \Delta Z_c}{[1 - nZ_c(1/nz_c + 1/z_R)]^4}, \tag{6.21}$$

where the negative sign is ignored because it does not matter for volume. However, it is assumed that $1 - nZ_c(1/nz_c + 1/z_R)$ does not change sign in the region of interest. The total object-space volume can be obtained by integrating Equation (6.21) to yield:

$$V_o = \int \Delta V_o$$

$$= \left| \frac{A_c}{3(1/nz_c + 1/z_R)[1 - nZ_c(1/nz_c + 1/z_R)]^3} \right|_{Z_c^{(1)}}^{Z_c^{(2)}}, \tag{6.22}$$

which is a simple relationship. Since the transverse magnification is to be calculated or determined anyway for the size analysis, Equations (6.1)–(6.3) can be used to write Equation (6.22) in a convenient form:

$$V_o = \left| \frac{nA_c Z_c}{3M_c^3(1 - M_c)} \right|_{Z_c^{(1)}}^{Z_c^{(2)}}. \tag{6.23}$$

Thus, the volume magnification can be simply obtained from the transverse magnification data at two longitudinal image planes defining the image volume.

Equations (6.22) and (6.23) are generally valid for spherical waves where $M_c \neq 1$. Suppose one of the recording or the reconstruction beams is collimated as described by Equations (6.5)–(6.7) or Equations (6.8)–(6.10). The analysis for volume magnification again yields Equation (6.22) with z_c or z_R as infinity depending upon the case. As a result, Equation (6.23) is still valid.

When both the beams are collimated, the integration leading to Equation (6.22) loses meaning. In that particular case ($z_R = z_c = \pm\infty$), Equation (6.21) can be written as

$$\Delta V_o = nA_c \Delta Z_c \tag{6.24}$$

and hence

$$V_o = nA_c \int \Delta Z_c = nA_c(Z_c^{(2)} - Z_c^{(1)}) \tag{6.25}$$

or

$$\text{Object-space volume} = n \times \text{image-space volume} \tag{6.26}$$

6.3 Image analysis with hologram on translation stage

In this commonly used approach, the reconstruction source and the observation plane in space are fixed. The hologram on x–y–z translation stage is moved to bring different regions of the image onto the observation plane. The general reconstruction process is described in Figure 6.4. With the general case of a spherical reconstruction beam, the longitudinal coordinate z_c is no longer fixed against longitudinal hologram movement. Instead, the distance between the reconstruction source plane and the observation plane is a fixed quantity d as shown in Figure 6.4. The transverse magnification for this type of reconstruction arrangement has been described by different workers[11–16] and the volume magnification by Vikram and Billet.[15] In this section we describe the image analysis, i.e. relating image space with the object space for movable hologram.

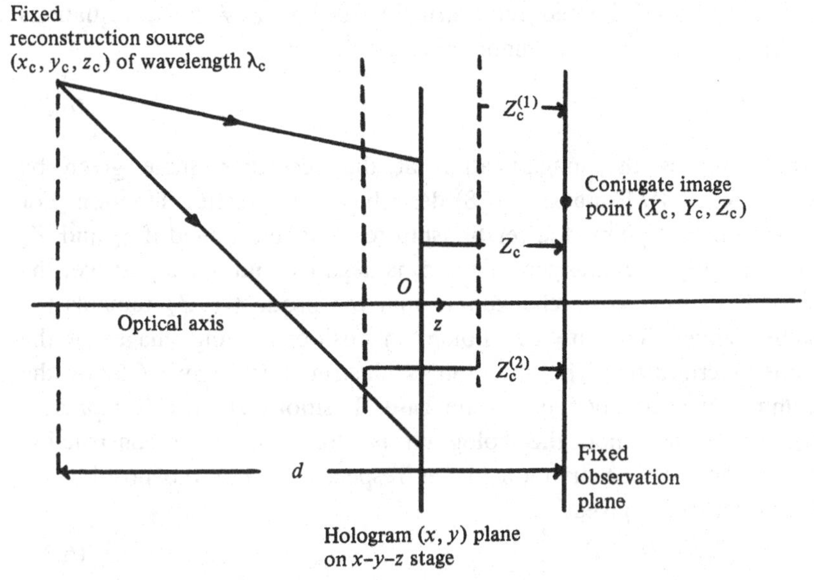

Fig. 6.4 Schematic diagram for the analysis with fixed reconstruction source and observation plane but the hologram on the moveable translation stage. (Ref. 15.)

6.3.1 *Transverse magnification and two image positions*

As seen in Figure 6.4, the reconstruction source and observation planes are fixed in space. The hologram is moved to focus microobjects on the observation screen. The observation plane can further be magnified by an auxiliary magnification (say the electronic magnification of the television system). The distance d between the observation and the source planes is fixed. The distance between the observation and hologram planes is Z_c. Thus, the variable longitudinal coordinate z_c of the reconstruction source is

$$z_c = Z_c + d, \tag{6.27}$$

where d is positive or negative if the reconstruction source is at RHS or LHS respectively of the observation plane. In the common case of the divergent beams, d is therefore negative. Equations (6.1)–(6.3) and (6.11)–(6.13) are still valid against the longitudinal hologram movement except that z_c is now governed by Equation (6.27).

For the collimated reconstruction beam case, the longitudinal hologram movement means a similar image movement in the longitudinal direction. The situation to one described in Section 6.2 even if the image plane rather than the observation plane is moving. z_c is still constant ($\pm\infty$).

For a non-collimated reconstruction beam ($z_c \neq \pm\infty$), Equations (6.3) and (6.27) can be combined to yield:

$$z_o = -\frac{nz_c Z_c}{m_o d}, \tag{6.28}$$

where m_o is the magnification at the recording stage given by Equation (3.4). Equation (6.28) describes an interesting situation. For a given object point, the relationship remains unchanged if z_c and Z_c are mutually interchanged. Since z_c is negative and Z_c is positive, the signs have also to be changed so that the product $z_c Z_c$ remains the same. Thus, there are two hologram positions giving images of the same microobject. The situation is illustrated in Figure 6.5 for the commonly used divergent beam case. Positions A and B represent the situations when the hologram is closer to the reconstruction source and the observation planes respectively. For the position A, Equation (6.27) gives

$$z_c = D + d \tag{6.29}$$

and

$$Z_c = D \tag{6.30}$$

where the hologram and the observation plane separation is a positive quantity D. According to our earlier definition for the divergent beam case, d is negative. Similarly for the position B:

$$z_c = -D \tag{6.31}$$

and

$$Z_c = -(D + d). \tag{6.32}$$

The product of $z_c Z_c$ in each case (positions A and B) is common, i.e. $D(D + d)$ meaning the same object as indicated by Equation (6.28). An interesting aspect of the two images is their magnifications. Equations (6.3) and (6.14) can be combined to obtain:

$$M_c = -nZ_c/z_o. \tag{6.33}$$

Again, Equations (6.28) and (6.33) yield

$$M_c = m_o d/z_c. \tag{6.34}$$

Position A will therefore yield higher magnification because z_c is smaller there. This kind of lensless magnification is described in detail by several workers.[7,8,11–14] for applications in particle field holography. Position A can yield high magnifications without lenses in the recording and/or reconstruction system.[8,11–14] A suitable combination of the magnification m_o and the beam divergence during the reconstruction can give as high as 1200 × magnification without lenses.[14]

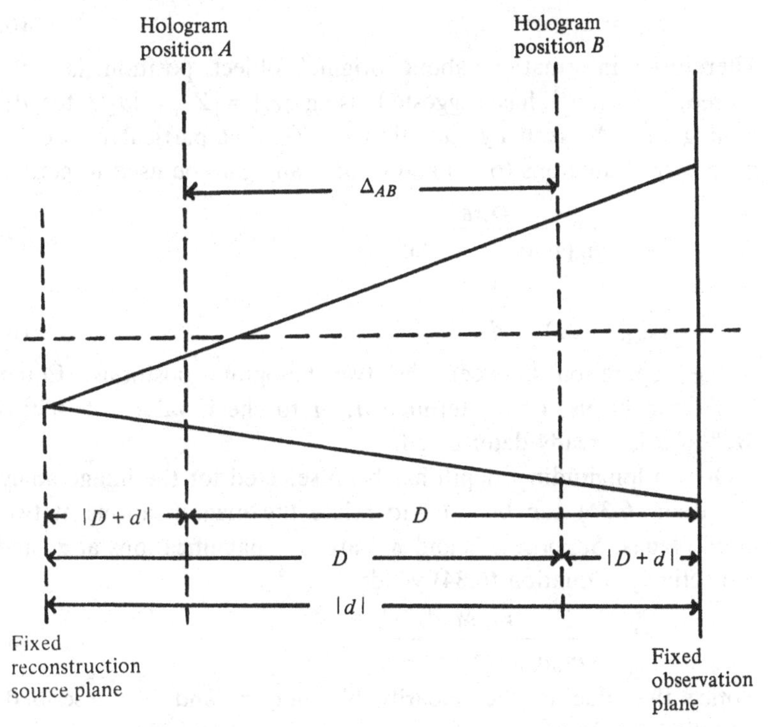

Fig. 6.5 Diagram representing two hologram positions along the optical axis yielding images of the same object. A divergent beam reconstruction is shown meaning that d is a negative quantity.

The selection of position A or B depends on the particular magnification needs. One special application of both the positions simultaneously is to determine the transverse magnification without knowledge of the auxiliary magnification.[11,12,14] Equations (6.29) and (6.34) give the magnification $M_c^{(A)}$ at the position A as

$$M_c^{(A)} = m_o d/(D + d). \tag{6.35}$$

Similarly, Equations (6.33) and (6.34) give the magnification at position B as

$$M_c^{(B)} = -m_o d/D. \tag{6.36}$$

Equations (6.35) and (6.36) can be combined to yield

$$M_c^{(A)} = m_o(1 + M_c^{(A)}/M_c^{(B)}). \tag{6.37}$$

Thus, by knowing the magnification ratio corresponding to the two image positions and m_o, the magnification $M_c^{(A)}$ can be obtained. m_o can be obtained simply by using Equation (6.28) and the definition of m_o in Equation (3.4):

$$m_o = 1 - nz_c Z_c / z_R d. \tag{6.38}$$

Therefore, information about original object position is not required.[16] Bexon[11] has suggested using $|z_c| = |Z_c| = |d|/2$ for determining m_o. As seen by Equation (6.38), that particular case is not necessary. Equations (6.35) and (6.36) can again be used to get:

$$d = \frac{\Delta_{AB}}{m_o(1/M_c^{(A)} - 1/M_c^{(B)})}, \tag{6.39}$$

where

$$\Delta_{AB} = 2D + d \tag{6.40}$$

is the separation between the two hologram positions. Equation (6.39) can be used to determine d, or to check values of $M_c^{(A)}$ and $M_c^{(B)}$ if d is already determined.

Once a longitudinal depth has been selected for the image analysis, Equation (6.34) can be used to relate the magnifications at two or more planes. Suppose M_c and M_c' are the magnifications at z_c and z_c' respectively, Equation (6.34) yields

$$M_c' = \frac{m_o' M_c d}{m_o d + M_c(z_c' - z_c)}. \tag{6.41}$$

Notice that due to the linearity between z_c and Z_c described by Equation (6.27),

$$z_c' - z_c = Z_c' - Z_c. \tag{6.42}$$

Suppose that distances d, z_c and Z_c (corresponding to various image or hologram positions) are not known or measured. Then knowledge or calibration of the magnification at two image planes with known mutual distances can be used to obtain the magnification at any third plane for the particular case of a collimated reference beam ($m_o = 1$). Equations (6.34) and (6.42) can then be used to obtain:

$$M_c'' = \left[\frac{1}{M_c} - \left(\frac{Z_c - Z_c''}{Z_c - Z_c'}\right)\left(\frac{1}{M_c} - \frac{1}{M_c'}\right)\right]^{-1}, \tag{6.43}$$

where M_c, M_c' and M_c'' correspond to the image distances Z_c, Z_c' and Z_c'' respectively.

In summary, when the hologram is moved longitudinally between the reconstruction source and the observation plane, two image positions with generally distinct magnifications are obtained. Hologram position A which is closer to the reconstruction source, yields higher magnification and may be used for lensless applications. Either the left or right half of the depth between the reconstruction and observation planes should be used for density calculations to avoid

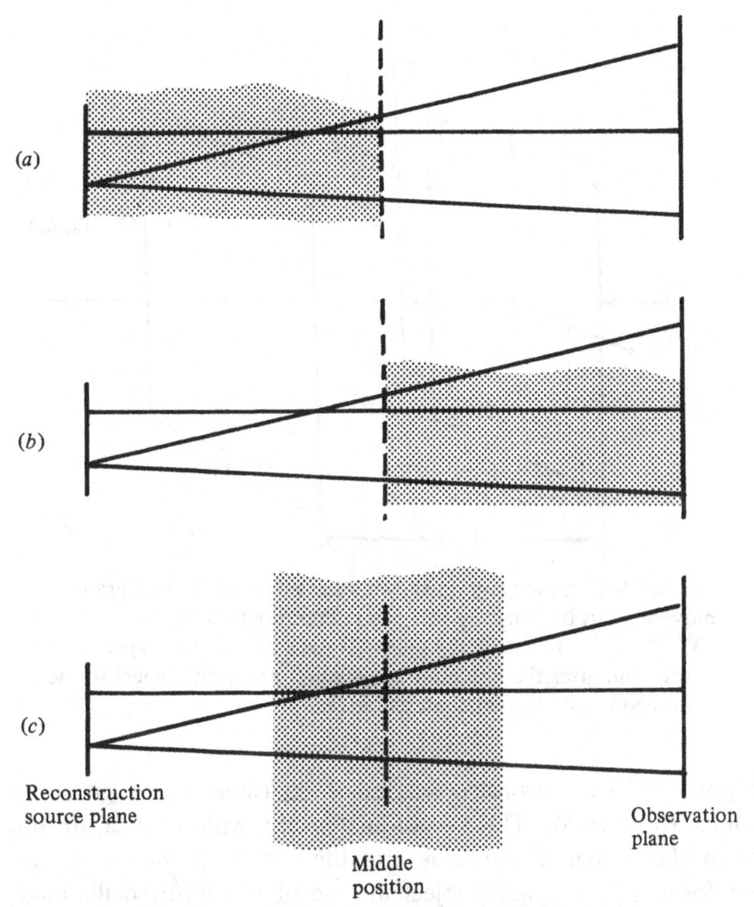

Fig. 6.6 Diagram illustrating regions for hologram movement. The shaded area corresponds to the region of hologram movement. Hologram moving (*a*) in the left half, (*b*) in the right half, and (*c*) in a region symmetrical about the center for counting the double density.

effect of two mixed images. On the other hand, a longitudinal depth symmetrical about the center can be used to count the double density of microobjects. These situations are shown in Figure 6.6.

6.3.2 *Effect of transverse hologram movement*

The hologram is moved in its plane to cover a cross-section of the space. It is important to relate the image movement to the corresponding distance in the object space. Typical object, reference, and reconstruction source coordinates are shown in Figure 6.7 in the

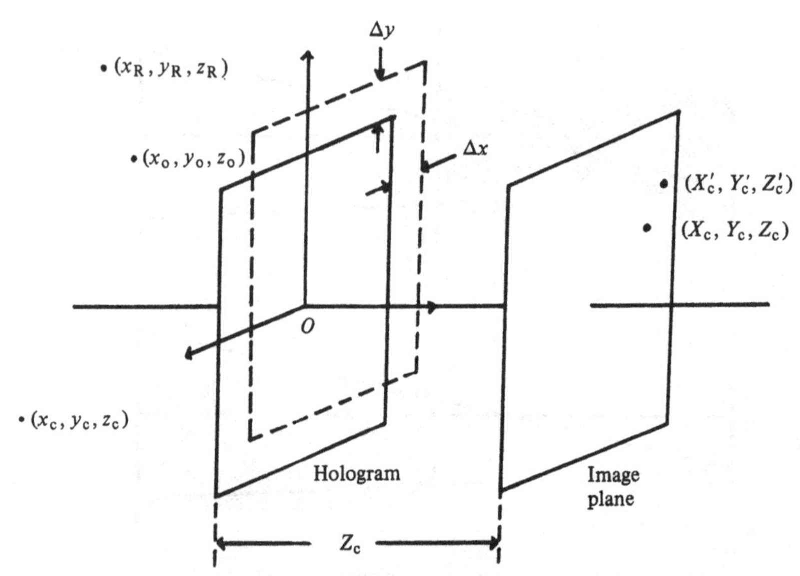

Fig. 6.7 Schematic diagram showing the effect of the hologram movement in its plane by $(\Delta x, \Delta y)$. Coordinates (X_c, Y_c, Z_c) and (X'_c, Y'_c, Z'_c) are the image points of the same object respectively before and after the hologram movement, and with respect to the fixed coordinate system centered at O.

static position. The reconstructed image coordinates are given by Equations (6.1)–(6.3). These coordinates are with respect to the center of the hologram which is also the center of the diffraction pattern formed by the microobject in case of non-diffused illumination.

Suppose the hologram is moved in its plane by $(\Delta x, \Delta y)$. Thus, with respect to the origin of the variable coordinate system situated at the center of the original hologram, the original object and reference source coordinates remain the same. The reconstruction source coordinates shift as

$$x_c^{(H)} = x_c - \Delta x, \tag{6.44}$$

$$y_c^{(H)} = y_c - \Delta y, \tag{6.45}$$

$$z_c^{(H)} = z_c, \tag{6.46}$$

where $x_c^{(H)}$, $y_c^{(H)}$ and $z_c^{(H)}$ are coordinates of the reconstruction source with respect to the variable coordinate center. Suppose Equations (6.44)–(6.46) are substituted into Equations (6.1)–(6.3) keeping in mind that the object and the reference source coordinates remain unchanged with respect to the moving coordinate system. Then we obtain the image coordinates with respect to the moving coordinate system. If the image coordinates with respect to the moving system are $(X_c^{(H)}, Y_c^{(H)}, Z_c^{(H)})$ then with respect to the original system fixed in space:

$$X'_c = X_c^{(H)} + \Delta x, \tag{6.47}$$

$$Y'_c = Y_c^{(H)} + \Delta y, \tag{6.48}$$

$$Z'_c = Z_c^{(H)}, \tag{6.49}$$

where (X'_c, Y'_c, Z'_c) represent the image point after the hologram movement. By performing the necessary algebra we obtain:

$$X'_c = X_c + \frac{M_c \Delta x}{m_o}, \tag{6.50}$$

$$Y'_c = Y_c + \frac{M_c \Delta y}{m_o}, \tag{6.51}$$

$$Z'_c = Z_c, \tag{6.52}$$

where M_c in the transverse magnification given by Equation (6.14) and m_o is the magnification from the recording configuration alone as given by Equation (3.4). A fixed observation point in the image space corresponds to different points in the object space as the hologram moves. The situation is represented in Figure 6.8. The same point in image space (X_c, Y_c, Z_c) with respect to the original fixed coordinate system will become $(X_c - \Delta x, Y_c - \Delta y, Z_c)$ with respect to the moving system. Also, with respect to the moving system, the reconstruction source point is given by Equations (6.44)–(6.46). Substituting these into Equations (6.11)–(6.13) would give object point coordinates $(x_o^{(H)}, y_o^{(H)}, z_o^{(H)})$ with respect to the moving system. The coordinates (x'_o, y'_o, z'_o) with respect to the fixed system will then be given by:

$$x'_o = x_o^{(H)} + \Delta x, \tag{6.53}$$

$$y'_o = y_o^{(H)} + \Delta y, \tag{6.54}$$

$$z'_o = z_o^{(H)}. \tag{6.55}$$

By performing the necessary algebra as stated above, we obtain:

$$x_o = x_o - \Delta x/m_o, \tag{6.56}$$

$$y'_o = y_o - \Delta y/m_o, \tag{6.57}$$

$$z'_o = z_o, \tag{6.58}$$

where Equations (6.3), (6.14) and (6.15) have been used leading to the relationship:

$$M_c/m_o = 1 - Z_c/z_c. \tag{6.59}$$

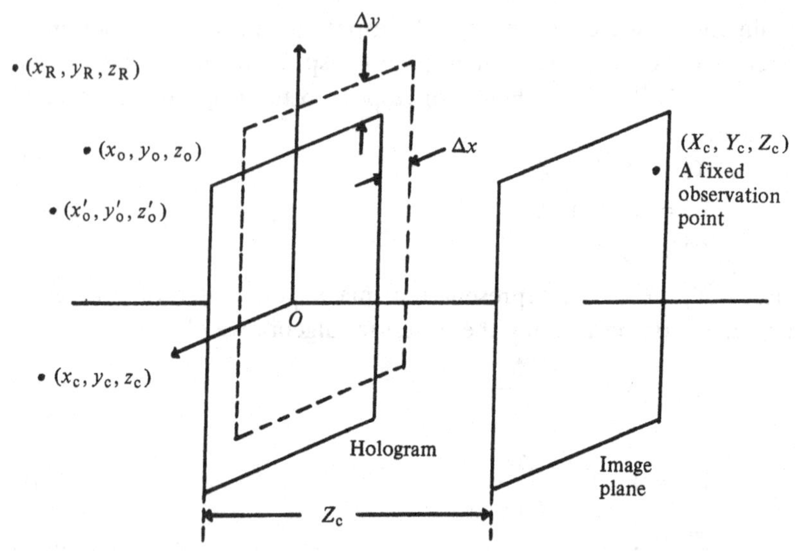

Fig. 6.8 Diagram to determine the new object point so that the $(\Delta x, \Delta y)$ movement of the hologram will give the image at the same point in space. All coordinates are shown with respect to the fixed coordinate system, i.e. centered at the original position of the hologram center at O.

Equations (6.56)–(6.58) conclude that the hologram movement $(\Delta x, \Delta y)$ is equivalent to the corresponding shift $(-\Delta x/m_o, -\Delta y/m_o)$ in the object space.

Similarly, Equations (6.50)–(6.52) mean that the image of the same object point is shifted by $(M_c\Delta x/m_o, M_c\Delta y/m_o)$ in the image plane by the same hologram movement. The object space movement, hologram movement and the image movement are therefore in the ratios of 1, m_o and M respectively.

Equations (6.50) and (6.51) yield that for the in-plane hologram movement $(\Delta x, \Delta y)$, the entire pattern in the image space is moved by $(M_c\Delta x/m_o, M_c\Delta y/m_o)$. However, the relative positions within the image plane remain unchanged. This leads to the transverse magnification being the same as that in the static hologram situation. This can easily be found by differentiating Equation (6.50) or (6.51) as

$$M_c = \frac{dX_c}{dx_o} = \frac{dX'_c}{dx_o} = \frac{dY_c}{dy_o} = \frac{dY'_c}{dy_o}, \tag{6.60}$$

where M_c is given by Equation (6.14).

6.3.3 *Volume magnification*

In Section 6.3.1 we saw the effect of the hologram's longitudinal position on the transverse magnification. The local transverse magnification can be obtained by a few different approaches as discussed. The local longitudinal magnification can be obtained by differentiating one of the relevant expressions, for example Equation (6.3). The local volume magnification can then be determined. This local volume magnification generally changes with the hologram's longitudinal position. If a substantial image depth is to be considered for density calculations the net volume magnification is not straightforward. It is this kind of volume magnification, as has been described by Vikram and Billet,[15] that we discuss in this section.

As seen in Figure 6.4, the hologram is moved between the longitudinal positions $Z_c^{(1)}$ and $Z_c^{(2)}$ to cover the volume of the image. Suppose a fixed cross-sectional area A_c in the conjugate image space is analyzed. If the final detector plane (such as the monitor of the closed circuit television system) does not correspond to the sufficient image space cross-section, the hologram can be moved in-plane for frame-by-frame analysis. We have seen in Section 6.3.2 that this kind of movement does not alter the transverse magnification. The object-space cross-sectional area $A_o = \Delta x_o \Delta y_o$, corresponding to an image-space cross-sectional area $A_c = \Delta X_c \Delta Y_c$, can be determined from Equations (6.11) and (6.12) as:

$$A_o = \frac{A_c}{[1 - nZ_c(1/nz_c + 1/z_R)]^2}. \qquad (6.61)$$

With the main condition of the present mode of the analysis, i.e. Equation (6.27), Equation (6.61) becomes

$$A_o = \frac{A_c z_R^2 (Z_c + d)^2}{(-dnZ_c + dz_R - nZ_c^2)^2}. \qquad (6.62)$$

Now, Equations (6.13) and (6.27) can be combined to yield:

$$z_o = -\frac{nz_R Z_c(Z_c + d)}{-dnZ_c + dz_R - nZ_c^2}. \qquad (6.63)$$

Differentiating Equation (6.63) we obtain

$$\Delta z_o = -\frac{dnz_R^2(d + 2Z_c)\Delta Z_c}{(-dnZ_c + dz_R - nZ_c^2)^2}. \qquad (6.64)$$

From Equation (6.62) and (6.64) image-space cross-sectional area A_c and depth ΔZ_c correspond to the object-space volume as:

$$\Delta V_o = A_o \Delta z_o$$
$$= \frac{A_c d z_R^4}{n^3} \frac{(d + 2Z_c)(d + Z_c^2)\Delta Z_c}{[(Z_c + d/2)^2 - d(z_R/n + d/4)]^4} \quad (6.65)$$

In the limiting case of a collimated reconstruction beam ($d \to \pm\infty$), Equation (6.65) becomes Equation (6.21) at $z_c \to \pm\infty$. By integrating RHS of Equation (6.65) we get the total object volume corresponding to longitudinal positions $Z_c^{(1)}$ and $Z_c^{(2)}$ as:

$$V_o = \int_{Z_c^{(1)}}^{Z_c^{(2)}} dV_o = \left| \frac{A_c d z_R^4}{n^3} \right.$$
$$\left. \times \int_{Z_c^{(1)}}^{Z_c^{(2)}} \frac{(d + 2Z_c)(d + Z_c)^2 dZ_c}{[(Z_c + d/2)^2 - d(z_R/n + d/4)]^4} \right|. \quad (6.66)$$

Equation (6.66) can be written in the form:

$$V_o = |(A_c d z_R^4/n^3)[(d^2/2)\psi_1 + 2d\psi_2 + 2\psi_3]|_{Z_c^{(1)}}^{Z_c^{(2)}}|, \quad (6.67)$$

where

$$\psi_1 = \int \frac{p dZ_c}{(p^2 + q)^4}, \quad (6.68)$$

$$\psi_2 = \int \frac{p^2 dZ_c}{(p^2 + q)^4}, \quad (6.69)$$

and

$$\psi_3 = \int \frac{p^3 dZ_c}{(p^2 + q)^4}, \quad (6.70)$$

For simplicity, we have defined:

$$p = Z_c + d/2 \quad (6.71)$$

and

$$q = (z_R/n + d/4)d \quad (6.72)$$

The solution of ψ_1 is simple resulting in

$$\psi_1 = -\frac{1}{6(p^3 + q)^3}. \quad (6.73)$$

ψ_2 can be solved first by integration by parts using p as one function and $p(p^2 + q)^{-4}$ as the other. The remaining integral can then be solved using the results 2.171.3 and 2.172 of Gradshteyn and Ryzhik[17] to obtain, for non-zero q:

$$\psi_2 = -\frac{p}{6(p^2 + q)^3} + \frac{p}{24q(p^2 + q)^2} + \frac{p}{16q^2(p^2 + q)}$$
$$+ \frac{\theta}{16q^2|q|^{1/2}}, \quad (6.74)$$

where

$$\theta = \begin{cases} -\mathrm{Arth}(p/|q|^{1/2}) & \text{for negative } q \\ \mathrm{arctg}(p/|q|^{1/2}) & \text{for positive } q. \end{cases} \tag{6.75}$$

'Arth' and 'tg' represent inverse hyperbolic and hyperbolic functions respectively. ψ_3 can be solved by the integration by parts using p^2 as one function and $p(p^2+q)^{-4}$ as the other to get the form of ψ_1. After the necessary algebra we obtain:

$$\psi_3 = -\frac{p^2}{6(p^2+q)^3} - \frac{1}{12(p^2+q)^2}. \tag{6.76}$$

Thus, Equation (6.67) along with the Equations (6.73), (6.75) and (6.76) give the object–image space volume relationship. These expressions are for non-zero and finite values of q. However, for some particular cases, zero or infinite values of q are possible resulting in the following particular solutions.

Collimated reference beam For the collimated readout beam case ($z_c = \pm\infty$), the analysis of Section 6.2 is already valid due to constant z_c. However, the collimated reference beam during the recording ($q = z_R = \pm\infty$) will make the general results of Equation (6.67) invalid. In this particular case, Equation (6.66) becomes

$$\begin{aligned} V_o &= \left| \frac{nA_c}{d^3} \int_{Z_\xi^{(1)}}^{Z_\xi^{(2)}} (d + 2Z_c)(d + Z_c)^2 \, dZ_c \right| \\ &= \left| \frac{nA_c}{d^3} \left(\frac{Z_c^4}{2} + \frac{5dZ_c^3}{3} + 2d^2 Z_c^2 + d^3 Z_c \right) \Big|_{Z_\xi^{(1)}}^{Z_\xi^{(2)}} \right|. \end{aligned} \tag{6.77}$$

$q = 0$ case z_R and d both are negative in the common case of divergent beams. This eliminates the possibility of a zero value of q given by Equation (6.72). However, if one of these beams is convergent, then it is possible to have a zero value of q when

$$d = -4z_R/n. \tag{6.78}$$

In this particular situation Equation (6.66) becomes

$$\begin{aligned} V_o &= \left| \frac{2A_c d z_R^4}{n^3} \int_{Z_\xi^{(1)}}^{Z_\xi^{(2)}} \frac{(d + Z_c)^2 \, dZ_c}{(Z_c + d/2)^7} \right| \\ &= \left| \frac{A_c d z_R^4}{n^3} \left(\frac{1}{2p^4} + \frac{2d}{5p^5} + \frac{d^2}{12p^6} \right) \Big|_{Z_\xi^{(1)}}^{Z_\xi^{(2)}} \right|. \end{aligned} \tag{6.79}$$

6.4 Effect of different media

The general discussion so far in this chapter assumes the space to be air (practically unity refractive index). However, one or more of the object, reference and reconstruction spaces can be some other medium entirely or in part. This situation generally arises for the object beam and also for the reference beam in the in-line case. This is for example due to test-section windows, experiments in water tunnels, bubble chamber holography, etc. The situation is represented in Figure 6.9. *P* could be the object or reference source point. Between the point *P* and the recording plane, there are plane parallel surfaces perpendicular to the propagation direction. The point *P* can also act as the reconstruction source point and the recording plane as the hologram during the readout process. However, reconstruction generally involves air only.

There are a total of *N* surfaces including the recording plane. The light originating from point *P* at an actual physical height h_1 from the optical axis refracts at different surfaces. The wave finally reaching the recording plane after covering the total physical longitudinal distance $(d_1 + d_2 + \ldots + d_N)$ would appear to originate at a point, but not necessarily at *P*. The apparent source point can be found from the common paraxial ray tracing equations of first-order optics[18,19] passing through plane parallel surfaces:

$$z' = n_a z / n_b \tag{6.80}$$

and

$$\frac{h'}{h} = \frac{n_b z'}{n_a z}, \tag{6.81}$$

where n_a and n_b are the refractive indices before and after the surface respectively. The transfer between the *j*th and the $(j+1)$th surfaces is governed by:

$$z_{j+1} = z'_j - d_{j+1}. \tag{6.82}$$

Here z and z' are the longitudinal source distances from the surface before and after refraction respectively, h and h' are the corresponding heights from the optical axis and d_{j+1} is the thickness (distance along the optical axis) between the *j*th and the $(j+1)$th surfaces.

Starting from $z = -d_1$ and $h = h_1$ at the first surface, Equations (6.80)–(6.82) can be used to obtain

$$z_N = -\left(n_N \sum_{j=1}^{N-1} d_j/n_j\right) - d_N, \tag{6.83}$$

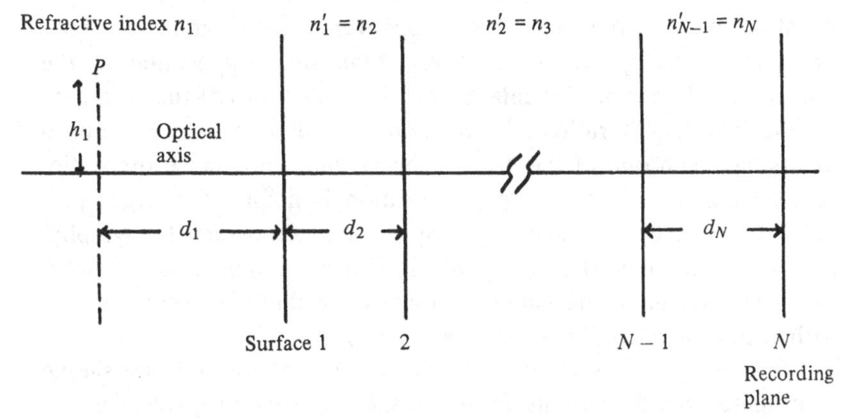

Fig. 6.9 Schematic diagram showing different media between the object or the reference source point and the recording plane. The diagram could also illustrate the situation between the reconstruction source point and the hologram during the readout process.

and the final apparent source height h_N as

$$h_N = h_1. \tag{6.84}$$

The recording plane is generally in air ($n_N = 1$) so that Equation (6.83) can be written as

$$z_N = -\sum_{j=1}^{N} d_j/n_j. \tag{6.85}$$

Therefore, basically each longitudinal distance should be divided by the corresponding refractive index before adding the distance.[20,21] The transverse coordinate remains the same as the actual physical one given by Equation (6.84). To conclude, once the longitudinal coordinates are modified as stated above, the equations of Sections 6.1, 6.2 and 6.3 are valid.

Sometimes the surfaces are not plane parallel, such as in cylindrical tubes, tunnels, etc. These situations will give different apparent points leading to aberrations. Corrections must generally be made for successful reconstructions. Some of these situations are considered in detail in Section 7.6.

6.5 Effect of hologram magnification

The preliminary relationships described in Equations (6.1)–(6.3) are for the case when the hologram is recorded as such and used for the reconstruction. This is indeed the most common situation encountered in practice. However, the hologram is magnified in various special applications. One such application is the removal of aberrations by enlargement (or reduction) of the hologram by the wavelength ratio n.[5,6] Relaxing the film resolution requirements has been one of the uses of the magnification since the sixties (see Section 5.2.1). Even demagnification is helpful for larger particles (see Section 5.2.4). In x-ray or far ultraviolet holography, pre-magnification is desirable even if film resolution is not a problem.[5] This is due to the limited resolution capability at reconstruction with visible wavelengths (see Section 4.8.1).

The analysis of such holograms can be performed by simple coordinate transformations in the results already available for the $m_H = 1$ case. The image–object point relationships[6] (see also Chapter 2) in case of hologram magnification m_H can be written in the forms:

$$X_c = -\frac{(x_c/n)(m_H^2 z_o/z_c) - m_H x_o + x_R(m_H z_o/z_R)}{1 - m_H^2 z_o/nz_c - z_o/z_R}, \quad (6.86)$$

$$Y_c = -\frac{(y_c/n)(m_H^2 z_o/z_c) - m_H y_o + y_R(m_H z_o/z_R)}{1 - m_H^2 z_o/nz_c - z_o/z_R}, \quad (6.87)$$

and

$$Z_c = -\frac{m_H^2 z_o/n}{1 - m_H^2 z_o/nz_c - z_o/z_R}. \quad (6.88)$$

A careful look at Equations (6.86)–(6.88) and comparing them with the $m_H = 1$ case given by Equations (6.1)–(6.3) yields that the hologram magnification can be described by the following coordinate transformations:

$$\left.\begin{array}{lll} x_o & \rightarrow & m_H x_o, \\ x_R & \rightarrow & m_H x_R, \\ y_o & \rightarrow & m_H y_o, \\ y_R & \rightarrow & m_H y_R, \\ z_o & \rightarrow & m_H^2 z_o, \\ z_R & \rightarrow & m_H^2 z_R, \end{array}\right\} \quad (6.89)$$

where the other quantities remain the same. Consequently there are the following transformations also implied:

$$\left.\begin{array}{lll} dx_o dy_o & \rightarrow & m_H^2 dx_o dy_o \\ dz_o & \rightarrow & m_H^2 dz_o. \end{array}\right\} \quad (6.90)$$

Since m_H is a constant, the transformations are easy to apply at every stage in the image analysis. For example in the collimated beam case,

M_c given by Equation (6.19) becomes $m_H M_c$. The longitudinal magnification $\Delta Z_c/\Delta z_o$ from Equation (6.20) would become m_H^2 times the value when m_H is unity.

It is interesting to note that another coordinate transformation is also possible:

$$\left.\begin{array}{lll} x_o & \rightarrow & m_H x_o, \\ x_R & \rightarrow & m_H x_R, \\ y_o & \rightarrow & m_H y_o, \\ y_R & \rightarrow & m_H y_R, \\ n & \rightarrow & n/m_H^2. \end{array}\right\} \tag{6.91}$$

The other quantities remain the same. With this transformation, the relationships of this chapter (the $m_H = 1$ case) remain valid just by multiplying the transverse magnification by m_H (or the area magnification by m_H). The rest is taken care of by the wavelength ratio transformation $n \rightarrow n/m_H^2$.

7

Aberrations and their control

The paraxial approximation or first-order optics considered so far in the analysis for determining the image coordinates and magnification (Chapter 6) is sufficient for many applications. As the minimum required effective hologram aperture increases for smaller objects, the aberrations must be considered for accurate imaging and analysis. This chapter is devoted to the general background development for third-order aberrations dealing with particle field holography. The in-line method is considered in more detail due to its widespread use. The reference source, the object and the hologram center points form a straight line in the in-line method, leading to interesting aspects of aberrations and their control. Cylindrical test-sections with curved windows are also encountered in practical far-field holography. Cylindrical ampules, water tunnels, etc. are, in effect, cylindrical lenses resulting in unwanted astigmatism. Ways to neutralize the cylindrical lens are therefore also discussed in the chapter.

7.1 Primary aberrations

For determining the coordinates of the image point (Chapter 2), the object, reference and reconstruction beams were assumed to be spherical over the hologram aperture. This approximation neglected higher order terms in the phase expressions of the wavefronts. Since particle field holography deals with very small objects, the aberration allowances are also small. In the presence of aberrations, diffraction-limited resolution may not be obtained.[1]

From the discussion of Chapter 2, the ideal image point will form a spherical phase term over the hologram aperture with its center at the image point. For a finite hologram aperture, this is only an approximation and the third-order term is

$$\Phi^{(3)} = \frac{2\pi}{\lambda_c}\Big[-\frac{1}{8Z_c^3}(x^4 + y^4 + 2x^2y^2 - 4x^3X_c - 4y^3Y_c$$
$$- 4xy^2X_c - 4x^2yY_c + 6x^2X_c^2 + 6y^2Y_c^2 + 2x^2Y_c^2$$
$$+ 2y^2X_c + 8xyX_cY_c - 4xX_c^3 - 4yY_c^3$$
$$+ 4xX_cY_c^2 - 4yX_c^2Y_c)\Big] \tag{7.1}$$

for the conjugate image. For the primary image, subscript c should be replaced by p. Similarly, there are the third-order terms of $\Phi_p = \phi_c + \phi_o - \phi_R$ and $\Phi_c = \phi_c + \phi_R - \phi_o$. Third-order aberrations of the image are the phase differences $\Phi^{(3)}$ and Φ_c or Φ_p depending on the image under consideration. As discussed in Chapter 2, the third-order terms of ϕ_c, ϕ_o, and ϕ_R and hence those of Φ_c and Φ_p are known. Due to these third-order terms, the aberrations of the object and the reference beams will alter the hologram fringe pattern itself.[2] Upon reconstruction, the aberrations of the reconstruction beam will also come into the picture. The cumulative effect will finally be seen in the image. By performing the necessary algebra, the aberration can be separated into the five types:[3]

$$\Delta\Phi = \frac{2\pi}{\lambda_c}[-\tfrac{1}{8}\rho^4 S \qquad \text{Spherical}$$
$$+ \tfrac{1}{2}\rho^3(C_x\cos\theta + C_y\sin\theta) \qquad \text{Coma}$$
$$- \tfrac{1}{2}\rho^2(A_x\cos^2\theta + A_y\sin^2\theta + 2A_x A_y\cos\theta\sin\theta) \qquad \text{Astigmatism}$$
$$- \tfrac{1}{4}\rho^2 F \qquad \text{Field curvature}$$
$$+ \tfrac{1}{2}\rho(D_x\cos\theta + D_y\sin\theta)] \qquad \text{Distortion} \tag{7.2}$$

where ρ and θ are the polar coordinates defined by

$$\rho^2 = x^2 + y^2, \qquad x = \rho\cos\theta, \qquad y = \rho\sin\theta. \tag{7.3}$$

For the conjugate wavefront, the various aberration coefficients are given by:[3]

$$S = \frac{n}{m_H^4}\left[\left(\frac{n^2}{m_H^2} - 1\right)\left(\frac{1}{z_o^3} - \frac{1}{z_R^3}\right) - \frac{3n}{z_c}\left(\frac{1}{z_o^2} + \frac{1}{z_R^2}\right)\right.$$
$$\left. + 3\left(\frac{m_H^2}{z_c^2} - \frac{n}{m_H^2 z_o z_R}\right)\left(\frac{1}{z_o} - \frac{1}{z_R}\right) + \frac{6n}{z_o z_R z_c}\right], \tag{7.4}$$

$$C_x = \frac{n}{m_H z_c^2}\left(\frac{x_o}{z_o} - \frac{x_R}{z_R}\right)$$

$$
\begin{aligned}
&- \frac{n}{m_H^3 z_o^2}\left[\frac{x_o}{z_o}\left(1 - \frac{n^2}{m_H^2}\right) + \frac{n x_c}{m_H z_c} + \frac{n^2 x_R}{m_H^2 z_R}\right] \\
&+ \frac{n}{m_H^3 z_R^2}\left[\frac{x_R}{z_R}\left(1 - \frac{n^2}{m_H^2}\right) - \frac{n x_c}{m_H z_c} + \frac{n^2 x_o}{m_H^2 z_o}\right] \\
&+ \frac{2n}{m_H^2}\left(\frac{x_c}{z_c} - \frac{n x_o}{m_H z_o} + \frac{n x_R}{m_H z_R}\right) \\
&\times \left(\frac{1}{z_o z_c} - \frac{1}{z_c z_R} + \frac{n}{m_H^2 z_o z_R}\right),
\end{aligned} \tag{7.5}
$$

$$
\begin{aligned}
A_x = {} & \frac{n x_c^2}{m_H^2 z_c^2}\left(\frac{1}{z_o} - \frac{1}{z_R}\right) \\
&- \frac{n x_o^2}{m_H^2 z_o^2}\left[\frac{1}{z_o}\left(1 - \frac{n^2}{m_H^2}\right) - \frac{n}{z_c} + \frac{n^2}{m_H^2 z_R}\right] \\
&+ \frac{n x_R^2}{m_H^2 z_R^2}\left[\frac{1}{z_R}\left(1 - \frac{n^2}{m_H^2}\right) - \frac{n}{z_c} + \frac{n^2}{m_H^2 z_o}\right] \\
&+ \frac{2n}{m_H}\left(\frac{1}{z_c} - \frac{n}{m_H^2 z_o} + \frac{n}{m_H^2 z_R}\right) \\
&\times \left(\frac{x_o x_c}{z_o z_c} - \frac{x_c x_R}{z_c z_R} + \frac{n x_o x_R}{m_H z_o z_R}\right),
\end{aligned} \tag{7.6}
$$

$$
F = A_x + A_y, \tag{7.7}
$$

and

$$
\begin{aligned}
D_x = {} & \frac{n}{m_H}\left[\left(\frac{n^2}{m_H^2} - 1\right)\left(\frac{x_o^3}{z_o^3} - \frac{x_R^3}{z_R^3} + \frac{x_o y_o^2}{z_o^3}\right)\right. \\
&+ \frac{3x_o}{z_o}\left(\frac{x_c}{z_o} + \frac{n x_R}{m_H z_R}\right)^2 - \frac{n(3x_o^2 + y_o^2)}{m_H z_o^2}\left(\frac{x_c}{z_c} + \frac{n x_R}{m_H z_R}\right) \\
&\left. - \frac{3x_c x_R}{z_c z_R}\left(\frac{x_c}{z_c} + \frac{n x_R}{m_H z_R}\right)\right].
\end{aligned} \tag{7.8}
$$

Deriving Equation (7.8), y_R and y_c are set equal to zero without any loss of generality.[3] These aberration coefficients[3,4] are well accepted in holography and will be used here for further discussion in connection with particle field holography. The coefficients listed above are for the conjugate wavefront Φ_c. For the primary image described by Φ_p, the signs of z_o and z_R in Equations (7.4)–(7.8) should be changed. Only x-coefficients of off-axis aberrations are listed, y-coefficients can be similarly written.

7.2 Tolerance limits

As seen in Equation (7.2), zero aberration is generally not possible. For sufficiently small aberrations, the normalized intensity at a point in the image region is[5]

$$i \sim 1 - (\Delta\phi)^2, \tag{7.9}$$

where ϕ represents the optical phase, i.e. $2\pi/\lambda_c$ times the path. $\Delta\phi$ is the variation from the ideal value. The system is considered well corrected or practically aberration free when $i \geqslant 0.8$ at the diffraction focus.[5] Thus we obtain the condition

$$|\Delta\phi| \leqslant (0.2)^{1/2} \tag{7.10}$$

The condition given by Equation (7.10) is practically equivalent to Maréchal's condition.[5] The goal now is to satisfy the condition of Equation (7.10).

7.3 Hologram center and related parameters

The microobjects in particle field holography are generally back illuminated. The illumination is generally by non-diffused light except sometimes in the off-axis method (Chapter 11). Thus, the line joining the illumination source point and the microobject meets the recording plane at a point which is the center of the object diffraction pattern. This is physically the center of the effective hologram aperture and must be considered so for aberration calculations. This section deals with the details of finding physical parameters with respect to the effective center of the hologram.

7.3.1 *In-line method*

The situation is shown in Figure 7.1. In the conventional sense, the illumination source S is on the optical axis. Normal SO to the recording plane passes through the center O of the conventional origin. However, a microobject at the point P forms the center of the diffraction pattern on the recording plane P'. As discussed in Chapter 4, if the coordinates of the point P are (ξ_o, η_o, z_o), then the coordinates of P' will be $(m_o\xi_o, m_o\eta_o)$ where m_o is the magnification from the recording configuration as given by Equation (3.4). Since P' is the center of the effective hologram aperture, the different parameters in the aberration calculation should be with respect to this

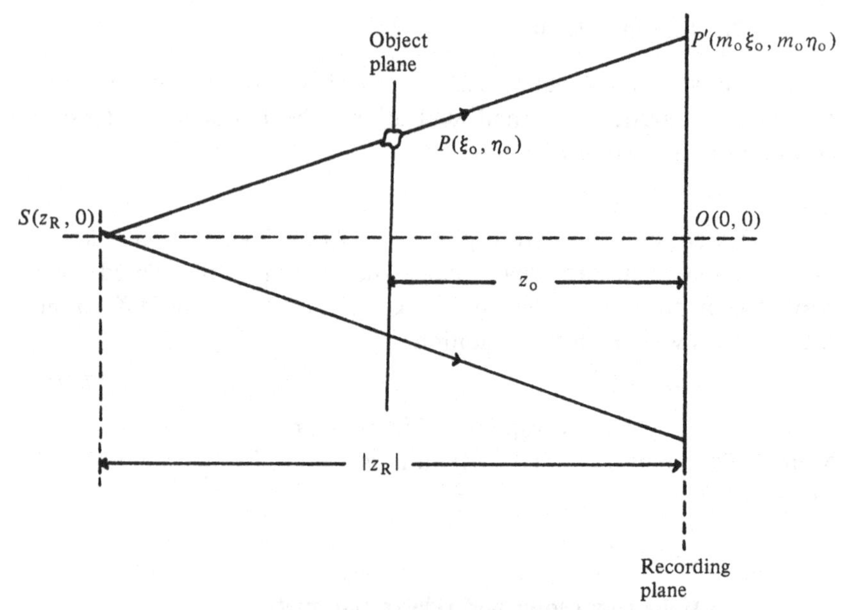

Fig. 7.1 Diagram for determining different recording parameters in the in-line method with respect to the coordinate center at P'.

center. With respect to P' we have the new coordinates as:

$$\left.\begin{aligned} x'_o &= -(m_o - 1)\xi_o, \\ y'_o &= -(m_o - 1)\eta_o, \\ z'_o &= z_o, \\ x'_R &= -m_o\xi_o, \\ y'_R &= -m_o\eta_o, \\ z'_R &= z_R. \end{aligned}\right\} \qquad (7.11)$$

For a collimated beam (we assume the usual case of the propagation direction perpendicular to the recording plane), the effective hologram center and the object point are both at the same distance from the conventional optical axis (fixed). The situation is represented in Figure 7.2. The different recording parameters are:

$$\left.\begin{aligned} x'_o = y'_o = x'_R/z'_R &= y'_R/z'_R = 0, \\ z'_o &= z_o, \\ z'_R &= -\infty. \end{aligned}\right\} \qquad (7.12)$$

The values of x'_o and y'_o represent their central positions. That does not mean zero object size. For aberration calculations, the variations

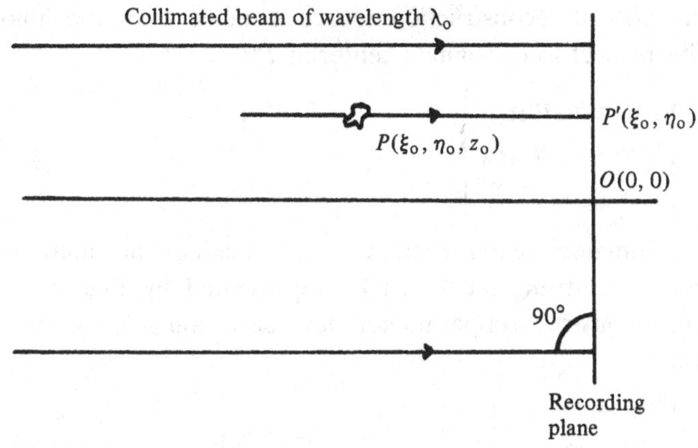

Fig. 7.2 In-line hologram recording with a collimated beam. The direction of propagation is perpendicular to the recording plane. P and P' represent the object and hologram center points respectively.

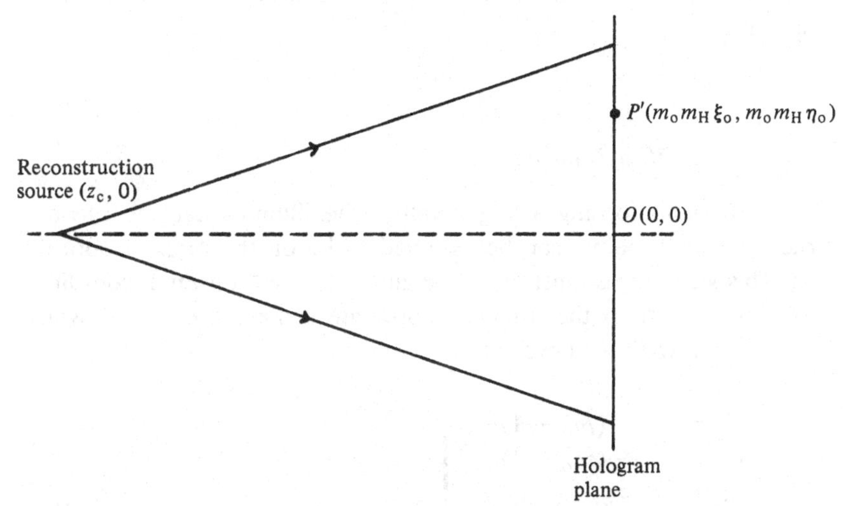

Fig. 7.3 Diagram for determining the different reconstruction parameters with respect to the effective hologram center at P'.

in x'_o and y'_o due to the finite object size must be considered even if the effect can be neglected.

Suppose the hologram is magnified m_H times (say by using lenses as discussed in Section 5.2). During the reconstruction, the coordinate of the point P' will therefore be $(m_o m_H \xi_o, m_o m_H \eta_o)$ with respect to the original optical axis. This situation is shown in Figure 7.3. If z_c is the conventional reconstruction source distance from the hologram, then with respect to coordinate center at P':

$$\left.\begin{aligned} x'_c &= -m_o m_H \xi_o, \\ y'_c &= -m_o m_H \eta_o, \\ z'_c &= z_c. \end{aligned}\right\} \tag{7.13}$$

For a collimated reconstruction beam incident normally on the hologram, the arrangement can be represented by Figure 7.4. The sufficient reconstruction parameters for aberration calculations can be written as:

$$\left.\begin{aligned} x'_c/z'_c &= 0, \\ y'_c/z'_c &= 0, \\ z'_c &= -\infty. \end{aligned}\right\} \tag{7.14}$$

Object and reference beam parameters are given by Equation (7.11) or Equation (7.12) for spherical or collimated recording beams respectively.

7.3.2 *Off-axis method*

Without loosing any generality, the illumination and reconstruction source points can be assumed to be on the original optical axis. This situation is illustrated in Figure 7.5 where different coordinates with respect to the original coordinate center at O are shown. With respect to P' we have

$$\left.\begin{aligned} x'_o &= -(m_o - 1)\xi_o, \\ y'_o &= -(m_o - 1)\eta_o, \\ z'_o &= z_o, \\ x'_R &= -m_o\xi_o + x_R, \\ y'_R &= -m_o\eta_o + y_R, \\ z'_R &= z_R. \end{aligned}\right\} \tag{7.15}$$

If the object illuminating beam is collimated (assuming propagation normal to the recording plane as is the usual practical case), then

$$\left.\begin{aligned} x'_o &= y'_o = 0, \\ z'_o &= z_o. \end{aligned}\right\} \tag{7.16}$$

Similarly, for the collimated reference beam case

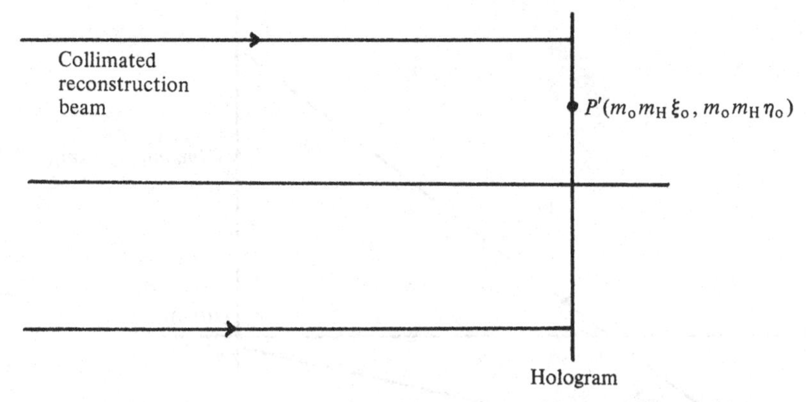

Fig. 7.4 The reconstruction arrangement with the collimated beam incident normally on the hologram.

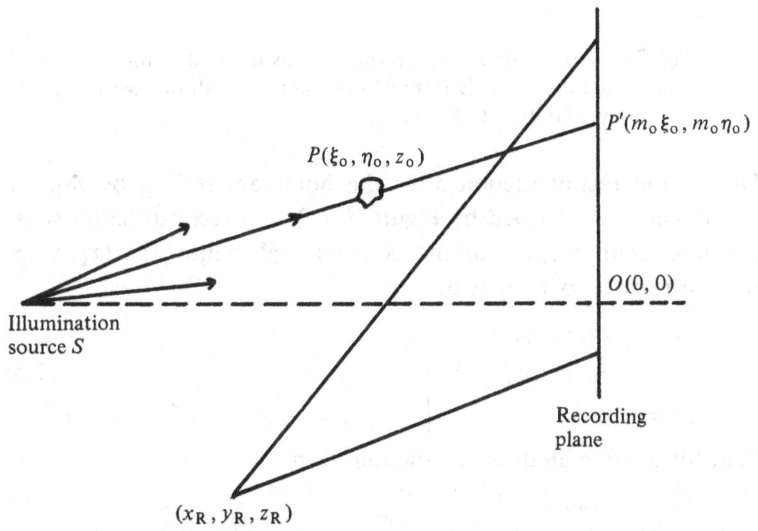

Fig. 7.5 Diagram for determining various coordinates with respect to the diffraction pattern center at P' in off-axis particle field holography. The recording source is assumed to be on the original optical axis. The coordinates shown are with respect to the original center at O.

$$\left.\begin{aligned} x'_R/z'_R &= \tan\theta_{x,R}, \\ y'_R/z'_R &= \tan\theta_{y,R}, \\ z_c &= -\infty, \end{aligned}\right\} \tag{7.17}$$

where $\theta_{x,R}$ and $\theta_{y,R}$ are the angles that the reference beam makes with the positive z-direction in (x, z) and (y, z) planes respectively.

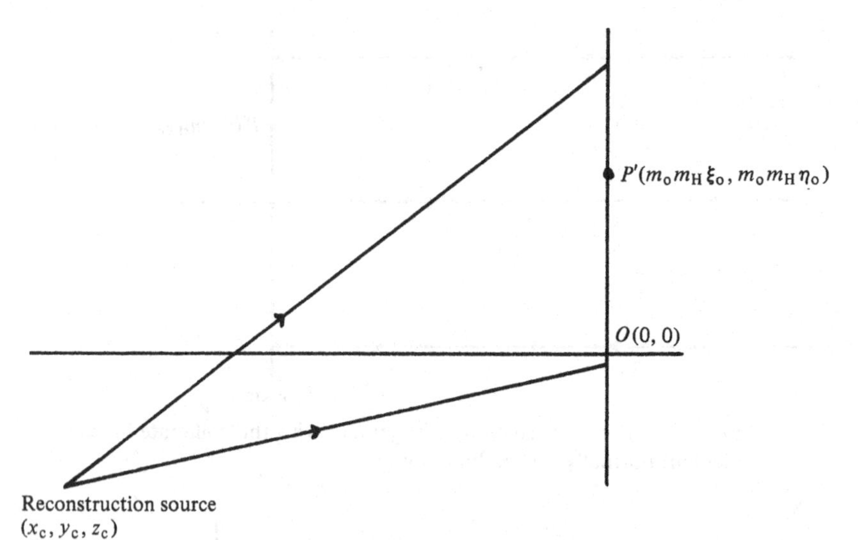

Fig. 7.6 Coordinate system in the off-axis method at the reconstruction stage. Different coordinates are shown with respect to the original center at O.

During the reconstruction after the hologram scaling by m_H, the situation can be described by Figure 7.6. If the reconstruction source coordinates with respect to the conventional center are (x_c, y_c, z_c) then with respect to P' they are

$$\left.\begin{aligned} x'_c &= -m_o m_H \xi_o + x_c, \\ y'_c &= -m_o m_H \eta_o + y_c, \\ z'_c &= z_c. \end{aligned}\right\} \tag{7.18}$$

Again, for a collimated reconstruction beam:

$$\left.\begin{aligned} x'_c/z'_c &= \tan\theta_{x,c}, \\ y'_c/z'_c &= \tan\theta_{y,c}, \\ z'_c &= -\infty, \end{aligned}\right\} \tag{7.19}$$

where $\theta_{x,c}$ and $\theta_{y,c}$ are the angles that the reconstruction beam makes with the positive z-axis in (x, z) and (y, z) planes respectively.

7.3.3 *Diffused illumination*

When off-axis holography is performed with diffused illumination (see Chapter 11), the paraxial center of the object beam cross-section at the hologram plane can be considered as the hologram center. As seen in Figure 7.7, the object illumination beam

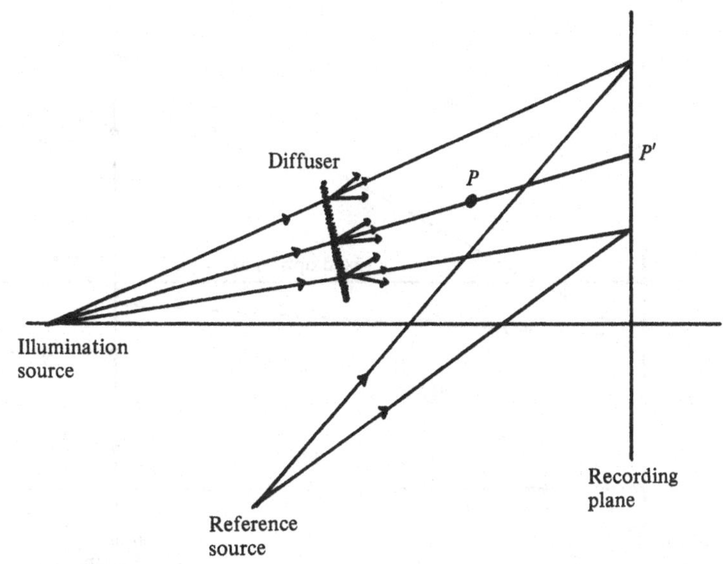

Fig. 7.7 Diagram describing off-axis holography with diffuse illumination.

encounters a diffuser before reaching the microobjects. The halo cross-section after the diffuser is centered at the axis of propagation. Therefore, the effective hologram center is basically the same as that dicussed in Section 7.3.2 without a diffuser. We assumed here the commonly occurring situation that the microobjects are around the axis of the cone of propagation.

7.3.4 *Hologram movement during reconstruction*

The discussions of Sections 7.3.1 and 7.3.2 are valid only when the hologram is fixed and the image is searched for in space by moving the detector or the camera. Another approach is to move the hologram on the $x-y-z$ stage with a fixed image plane (see Section 6.3). For a non-collimated reconstruction beam this means a variable distance z_c or z_c' given by Equation (6.27). This situation is illustrated in Figure 7.8. The in-plane hologram movement $(\Delta x, \Delta y)$ yields the shift in the reconstruction source coordinates given by Equations (6.44)–(6.46). Thus, with respect to the effective hologram center:

$$\left.\begin{aligned} x_c' &= -m_o m_H \xi_o + x_c - \Delta x, \\ y_c' &= -m_o m_H \eta_o + y_c - \Delta y, \\ z_c' &= Z_c + d. \end{aligned}\right\} \qquad (7.20)$$

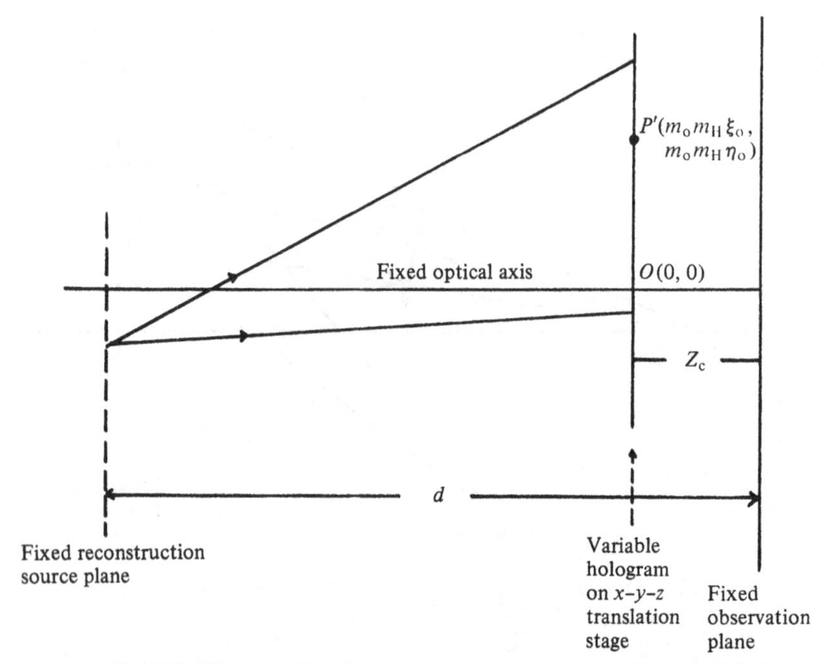

Fig. 7.8 Diagram showing the reconstruction arrangement with the hologram on the moving stage. d is positive or negative if the reconstruction source is at RHS or LHS respectively of the observation plane.

For a collimated reconstruction beam, the hologram movement does not alter any parameter with respect to the effective hologram. The reconstruction parameters will then be described by Equation (7.19).

A practical and common situation is in-line reconstruction with spherical beams (such as that shown in Figure 7.3) and observation of the images near the fixed optical axis. This means the effective hologram center is brought to the fixed optical axis for the image analysis. The reconstruction parametes are therefore

$$\left.\begin{aligned} x'_c &= y'_c \approx 0, \\ z'_c &= Z_c + d. \end{aligned}\right\} \tag{7.21}$$

7.4 Limitations of collimated beams without scaling

Let us consider the commonly used case of collimated recording and reconstruction beams in the in-line method. The situation can be described by Figures 7.2 and 7.4 with $m_H = 1$ and $n \neq 1$. The case corresponds to the common situation of different recording and reconstruction wavelengths. The aberration-limited resolution[6] of this standard procedure is discussed in this section.

The various aberration coefficients in this case, from Equations (7.4)–(7.8) become

$$S = n(n^2 - 1)/z_o^3, \tag{7.22}$$

$$C_x = Sx_o; \quad C_y = Sy_o, \tag{7.23}$$

$$A_x = Sx_o^2; \quad A_y = Sy_o^2, \tag{7.24}$$

$$F = S(x_o^2 + y_o^2) \tag{7.25}$$

and

$$D_x = Sx_o(x_o^2 + y_o^2); \qquad D_y = Sy_o(x_o^2 + y_o^2). \tag{7.26}$$

Although the microobjects are on the optical axis, their sizes are not zero. Therefore the maximum values of x_o and y_o correspond to these sizes. The maximum x_o or y_o is half of the maximum object extent (such as the diameter d for spherical objects).

From Equation (4.36), the required aperture of the hologram is given by

$$\rho = (1 + m)Nd, \tag{7.27}$$

where m is the number of side lobes recorded, N is the number of far-fields, and d is the object diameter.

Thus, at the maximum ρ, the maximum value of x_o/ρ or y_o/ρ is $1/[(1 + m)2N]$; $2(1 + m)N$ is generally a number much greater than unity. Thus, Equations (7.22)–(7.26) in conjunction of Equation (7.2) then show that the spherical aberration is most important and the others can be neglected. However, for small $2(1 + m)N$, coma may also be considered.

In fact, the simplified Equations (7.22)–(7.26) are obtained due to $z_c = z_R = -\infty$. Also, $x_R/z_R = y_R/z_R = x_c/z_c = y_c/z_c = 0$ due to normal recording and reconstruction in the in-line method. Imprecise mounting of the hologram during the recording and/or reconstruction will have finite non-zero angles. This will generally increase the aberrations as many terms in Equations (7.4)–(7.8) may not be then neglected. This aspect is observed experimentally by Staselko and Kosnikovskii.[1]

Let us consider the role of spherical aberration alone. Equations (7.2), (7.10), (7.22) and (7.27) yield, for negligible spherical aberration:

$$\left|\frac{\pi(n^2 - 1)(1 + m)^4 N\lambda_o^2}{4d^2}\right| \lesssim (0.2)^{1/2}, \tag{7.28}$$

where $N = \lambda_o z_o/d^2$ is the number of far-fields at the recording time and $n = \lambda_c/\lambda_o$. Equation (7.28) gives the limit on the object size as

$$\frac{d}{\lambda_o} \geqslant \frac{\pi^{1/2}|n^2 - 1|^{1/2}(1+m)^2 N^{1/2}}{2(0.2)^{1/4}}. \tag{7.29}$$

Equations (7.29) sets the limit on size of microobject that can be practically aberration-free. Using the same recording and reconstruction wavelengths ($n = 1$) is ideal. In practice, however, different wavelengths are generally used. For example, use of a ruby laser at $\lambda_o = 0.6943\ \mu m$ for the recording and HeNe at $\lambda_c = 0.6328\ \mu m$ for the reconstruction is very common. The minimum aberration-free value of d for this situation is plotted in Figure 7.9. For $m = 3$ and $N = 4$, we then require a minimum object diameter of 12 μm. For common applications allowing lower m and/or N resulting in reduced value of $(1 + m)^2 N^{1/2}$, there is no problem.[7,8]

In the example, the situation changes considerably when m or N increases. For $N = 100$, we obtain $d \geqslant 60$ – a considerably higher object size. Similarly, on changing m from 3 to 5 the minimum required object diameter changes from 12 μm to 33 μm. Special techniques can be used to record a hologram at high N while still successfully storing a large number of side lobes (see Chapter 8). In those situations, the spherical aberration must be taken into account. For smaller microobjects, special recording and reconstruction geometries must then be adopted. Some approaches are discussed in Section 7.5.

7.5 Control of aberrations

As seen in Section 7.4, aberration seriously limits the resolution. The discussions of Sections 7.1–7.3 provide the background required for the knowledge and control of third-order aberrations. In the general case of the off-axis method with spherical beams, complete elimination of the aberrations may not be possible. However, for smaller microobjects, where aberrations become important, care is needed. A careful look at Equations (7.4)–(7.8) reveals that the third-order aberrations can be made to disappear by using plane reference and reconstruction beams ($z_c = z_R = -\infty$) of equal but opposite offset angles ($x_c/z_c = -x_R/z_R$, $y_c/z_c = -y_R/z_R$) and by scaling the hologram in the ratio of the wavelengths ($n = m_H$). In the lensless Fourier transform configuration[9–11] where $z_o = z_R$, the third-order spherical aberration is zero. As such, these aspects are needed for aberration control.

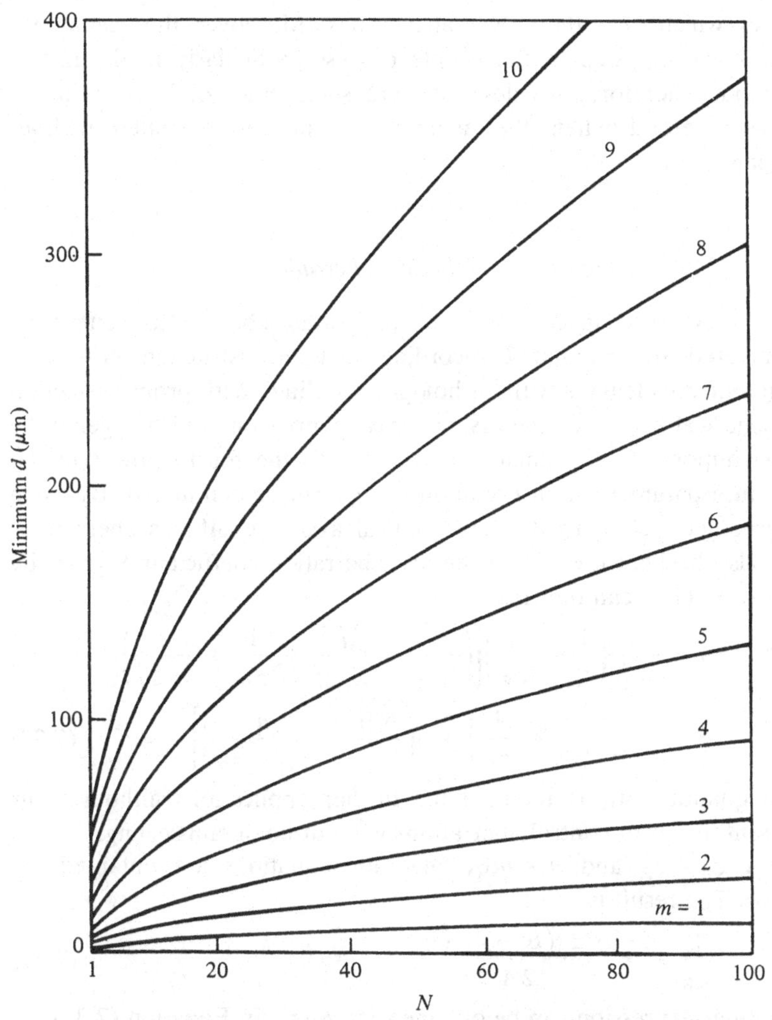

Fig. 7.9 Variation of the minimum aberration-free object diameter against the number N of far-field distances for several side lobes (m) recorded. Recording and reconstruction wavelengths considered are 0.6943 μm (ruby) and 0.6328 μm (HeNe) respectively. (Ref. 6.)

The $n \neq 1$ case is often encountered, as when a ruby laser is used for the recording and a HeNe laser for reconstruction. Using a closer wavelength, such as the krypton–argon red line, for the reconstruction is better.[12] Very special arrangements to get continuous-wave light at the ruby wavelength, such as a dye laser pumped by an argon-ion laser, are possible.[13,14] Such lasers are not common. Theoretically, scaling is then a must.[3,4] This scaling is provided by lenses which give their own aberrations. Moreover the scaling requires another step in the simple process particularly in the in-line method. Therefore, we describe here some cases of interest in the in-line method when the aberrations can be controlled without scaling.

7.5.1 *Elimination of spherical aberration*

As seen in Section 7.4, third-order aberrations cannot be eliminated by collimated recording and reconstruction beams of different wavelengths without hologram scaling. With proper selection of spherical beam parameters, spherical aberration, which is generally most important[15] for small objects (due to the fourth power in the aperture parameter ρ in Equation (7.2)), can be eliminated. By using microobjects close to the fixed optical axis, the off-axis aberrations can also be kept low. The spherical aberration coefficient S given by Equation (7.4) can be written as

$$S = \frac{n}{m_H^4}\left(\frac{1}{z_o} - \frac{1}{z_R}\right)\left[\left(\frac{n^2}{m_H^2} - 1\right)\left(\frac{1}{z_o^2} + \frac{1}{z_R^2} + \frac{1}{z_o z_R}\right) - \frac{3n}{z_c}\left(\frac{1}{z_o} - \frac{1}{z_R}\right) + 3\left(\frac{m_H^2}{z_c^2} - \frac{n}{m_H^2 z_o z_R}\right)\right]. \quad (7.30)$$

The quantity m_H is retained for further applications although our present aim is to control aberrations without hologram scaling.

For $z_o \neq z_R$ and $n \neq m_H$, quadratic equations are obtained for $S = 0$. The result is

$$\frac{z_o}{z_R} = \frac{-B \pm (B - 4AC)^{1/2}}{2A}. \quad (7.31)$$

A similar expression can be obtained for z_o/z_c. In Equation (7.31),

$$A = \frac{n^2}{m_H^2} - 1,$$

$$B = \frac{n^2}{m_H^2} - 1 + \frac{3nz_o}{z_c} - \frac{3n}{m_H^2},$$

$$C = \frac{n^2}{m_H^2} - 1 - \frac{3nz_o}{z_c} + \frac{3m_H^2 z_o^2}{z_c^2}. \quad (7.32)$$

From knowledge of the ruby laser wavelength ($\lambda_o = 0.6943\ \mu$m) and the HeNe laser wavelength ($\lambda_c = 0.6328\ \mu$m), we can calculate $n = \lambda_c/\lambda_o$. If there is no hologram scaling ($m_H = 1$) and there is a

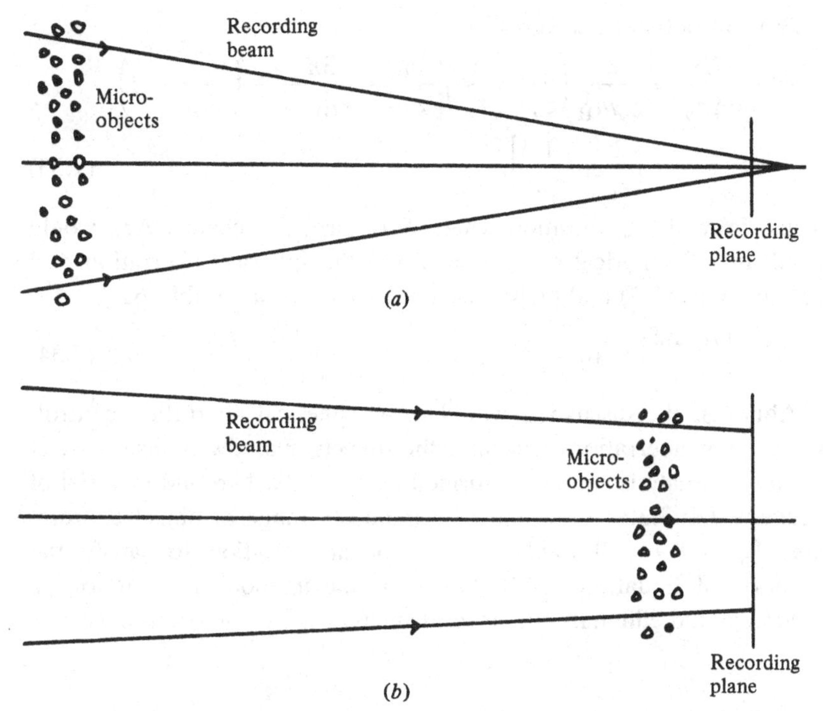

Fig. 7.10 Schematic diagram representing convergent beam in-line holographic recording so that the collimated reconstruction beam gives zero spherical aberration. The case is $\lambda_o = 0.6943\ \mu m$, $\lambda_c = 0.6328\ \mu m$ and $m_H = 1$ as discussed in the text: (*a*) $z_o/z_R \approx -17.091$. (*b*) $z_o/z_R \approx 0.05851$.

collimated reconstruction beam ($z_c = -\infty$) Equation (7.31) gives $z_o/z_R = -17.091$ or -0.05851. Both these recordings are schematically shown in Figure 7.10. The case corresponding to Figure 7.10(*a*) represents $z_o/z_R = -17.091$. This is a rather highly convergent beam for recording. The transverse magnification of the conjugate image $M_c = (1 - z_o/z_R)^{-1}$ yields a demagnification of about 18. This case can be used for large objects at a distance to keep the image–hologram distance conveniently small. A similar purpose can be achieved by using lenses (see Section 5.2.4) but presently spherical aberration is simultaneously eliminated.

The case $z_o/z_R = 0.05851$ is represented by Figure 7.10(*b*). Here, a slightly convergent beam is used for the recording. The magnification M_c is about 0.945 and this case is useful for common applications.

The practical range for the object distances can be found from the variation of S against z_o. By differentiating S given by Equation

(7.30), we obtain (near $S = 0$):

$$\frac{\Delta S}{\Delta z_o} \simeq \frac{n}{z_o^2 m_H^4}\left(\frac{1}{z_o} - \frac{1}{z_R}\right)\left[\frac{3n}{z_c} + \frac{3n}{m_H^2 z_R} - \left(\frac{n^2}{m_H^2} - 1\right) \times \left(\frac{2}{z_o} + \frac{1}{z_R}\right)\right]. \qquad (7.33)$$

Thus, from the z_o position where S is zero, the change Δz_o would result in ΔS spherical aberration. From the spherical aberration part in Equations (7.2) and (7.10), we must have, for allowable Δz_o,

$$\left|\frac{\pi \rho^4 \Delta S}{4\lambda_c}\right| \leqslant (0.2)^{1/2}. \qquad (7.34)$$

Although the discussion would allow spherical aberration control, the off-axis aberrations will limit the objects near the optical axis. A numerical analysis can be performed using the background material of Sections 7.1–7.3. This would determine the range of object coordinates ξ_o and η_o allowable in a particular situation to satisfy the criterion of Equation (7.10). However, the method of the following section would eliminate the need of such a cumbersome approach.

7.5.2 *Plane wave recording and spherical wave reconstruction*[16]

Plane wave recording with the usual beam propagating normally to the recording plane is described by Figure 7.2. As seen in Figure 7.2 or described by Equation (7.12), each object point is on the optical axis of the effective hologram. The small deviation from the axis is only due to the finite dimensions of the microobjects.

At the reconstruction stage, the hologram is generally moved on a translation stage and the images are observed near the optical axis again. With this mode of recording and reconstruction, the object P, the effective hologram center P' and the image point are near the optical axis during the reconstruction. The situation is illustrated in Figure 7.11. At any time, a frame on the monitor is studied for the image analysis. The entire frame generally corresponds to a few millimeters at a net transverse magnification of a few hundred. For the image on the outer side of the frame, the effective hologram center P' will be slightly away from the optical axis. A rigorous aberration calculation can be performed using Equation (7.13) for the slightly-off-effective-hologram-center case. However, in a practical sense, it is very easy to bring the image to the center of the monitor

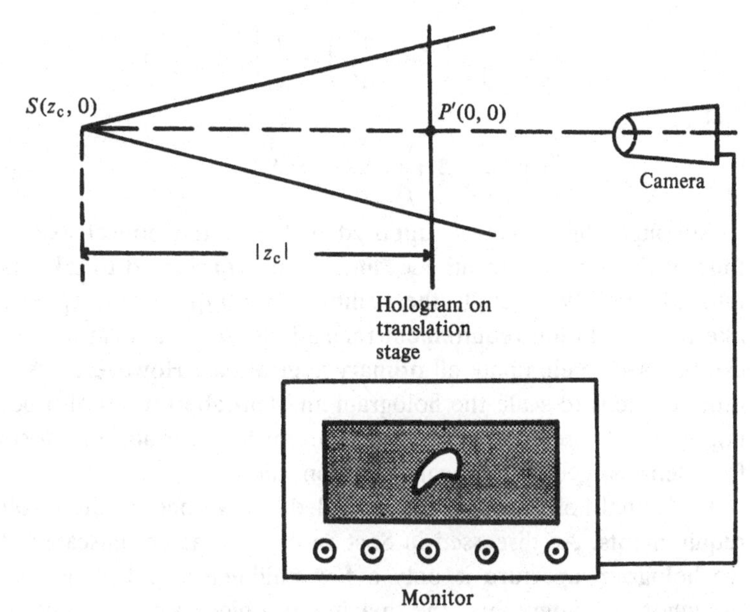

Fig. 7.11 Schematic diagram for spherical beam reconstruction of an in-line hologram recorded with a collimated beam. The hologram is moved in its plane so that the image, the effective hologram center P', and the reconstruction source S all are on the optical axis. By proper selection of z_c, primary aberrations can be completely eliminated with this mode. (Ref. 16.)

by hologram movement. In this situation:

$$\left.\begin{aligned} x'_c &= 0, \\ y'_c &= 0, \\ z'_c &= z_c. \end{aligned}\right\} \tag{7.35}$$

Using these parameters and the recording parameters given by Equation (7.12), the aberration coefficients given by Equations (7.4)–(7.8) become:

$$S = \frac{n}{m_H^4}\left[\left(\frac{n^2}{m_H^2} - 1\right)\frac{1}{z_o^3} - \frac{3n}{z_c z_o^2} + \frac{3m_H^2}{z_c^2 z_o}\right], \tag{7.36}$$

$$C_x = \frac{n x_o}{m_H z_c^2 z_o} - \frac{n x_o}{m_H^3 z_o^3}\left(1 - \frac{n^2}{m_H^2}\right) - \frac{2n^2 x_o}{m_H^3 z_c z_o^2}, \tag{7.37}$$

$$A_x = -\frac{n x_o^2}{m_H^2 z_o^2}\left[\frac{1}{z_o}\left(1 - \frac{n^2}{m_H^2}\right) + \frac{n}{z_c}\right], \tag{7.38}$$

$$F = -\frac{n}{m_H^2 z_o^2}\left[\frac{1}{z_o}\left(1 - \frac{n^2}{m_H^2}\right) + \frac{n}{z_c}\right](x_o^2 + y_o^2) \qquad (7.39)$$

and

$$D_x = \frac{n}{m_H}\left(\frac{n^2}{m_H^2} - 1\right)\left(\frac{x_o^3 + x_o y_o^3}{z_o^3}\right). \qquad (7.40)$$

Although the object is supposed to be on the optical axis at the time of the recording, finite x_o and y_o are considered to take its size into account. We include the scaling factor $m_H \neq n$ for applications like film resolution requirement relaxations. As stated earlier $m_H = n$ can be used to eliminate all primary aberrations. However, this would require a lens to scale the hologram and lens aberrations also become important. Therefore $m_H = 1$ is important here for aberration control from lensless geometrical considerations alone.

The actual hologram aperture needed is governed by the resolution requirements. As discussed in Section 4.4, the actual unscaled effective hologram aperture is only a few millimeters. Also, in common Fraunhofer holography, the maximum object size is only a few hundred micrometers. The maximum extent of x_o and y_o is the nominal object radius $d/2$. Equations (7.2) and (7.10) then yield the result that the object size related terms given by Equations (7.37)–(7.40) can generally be neglected. This is an important result. The reconstruction beam parameter z_c can then be adjusted to result in zero primary spherical aberration given by Equation (7.36). Thus, in principle, all primary aberrations can be practically eliminated.[16]

Equating the RHS of Equation (7.36) to zero, we obtain:

$$\frac{2z_o}{z_c} = \frac{n}{m_H^2} \pm \left[\left(\frac{n}{m_H^2}\right)^2 - \frac{4}{3m_H^2}\left(\frac{n^2}{m_H^2} - 1\right)\right]^{1/2}, \qquad (7.41)$$

which, for the current case of no hologram scaling, i.e. for $m_H = 1$ becomes

$$\frac{2z_o}{z_c} = n \pm [n^2 - \tfrac{4}{3}(n^2 - 1)]^{1/2}. \qquad (7.42)$$

When this relationship is satisfied, the spherical aberration will be zero. Thus all the primary aberrations can be eliminated practically. Let us consider a common situation. Ruby wavelength recording ($\lambda_o = 0.6943\ \mu m$), HeNe wavelength reconstruction ($\lambda_c = 0.6328\ \mu m$) and hence known $n = \lambda_c/\lambda_o$ yields z_o/z_c given by Equation (7.42) as

$$z_o/z_c = -0.05820 \qquad (7.43)$$

or

$$z_o/z_c = +0.9696. \qquad (7.44)$$

Equation (7.43) demands a slightly convergent beam reconstruction. The transverse magnification $M_c = 1/(1 - z_o/nz_c)$ of the conjugate image is about 0.94. This situation can be used for common applications where the image is magnified by a closed circuit television system. From Equations (6.3) and (6.14), $Z_c = -z_o M_c/n$. Since z_o is negative, Z_c is positive and hence the image is real.

Equation (7.44) describes the situation of near equal z_o and z_c. That means the reconstruction beam source is in the region of the original object space. The transverse magnification M_c in this case comes out to be about -15.66. The image is highly magnified but virtual. As mentioned in Section 7.1, the aberration coefficients of the primary image can be obtained by changing the signs of z_o and z_R. The primary image magnification $M_p = 1/(1 + z_o/nz_c)$ will again remain -15.66 and the image will remain virtual. Thus, with the condition corresponding to Equation (7.44), the aberration-free image is magnified and virtual.

The conclusion is that near unity magnification real and highly magnified virtual images with zero primary aberrations can be obtained. The proper choice of z_c is related to z_o. One way is first to perform an experiment to see an image in focus. Equation (6.28) can then be used to determine z_o from knowledge of the positions of the reconstruction source hologram and image planes. Once z_o is known, a proper z_c can be selected satisfying Equation (7.42) to obtain the aberration-free image. The conjugate image plane will then be governed by Equation (6.3) which, in this situation, becomes

$$Z_c = -\frac{2z_o}{2n \mp [n^2 - 4(n^2 - 1)/3]^{1/2}}. \tag{7.45}$$

When aberration-free analysis of the size is very important for a few microobjects, the above mentioned procedure can be applied. However, for routine applications which analyze a large number of microobjects, this would be very time consuming. Therefore, it is important to study the effect of the object distance variation on the primary spherical aberration. Also when the longitudinal object space is covered by the longitudinal hologram movement during the reconstruction, z_c is not a constant. For $m_H = 1$, Equation (7.36) can be differentiated with respect to z_o and z_c to obtain:

$$\begin{aligned}\Delta S = \frac{3n}{z_o^4} \{&[-(n^2 - 1) + 2n(z_o/z_c) - (z_o/z_c)^2]\Delta z_o \\ &+ [n(z_o/z_c)^2 - 2(z_o/z_c)^3]\Delta z_c\},\end{aligned} \tag{7.46}$$

where ΔS is the change in the primary spherical aberration coefficient

S with small changes Δz_o in the object distance and Δz_c in the reconstruction source distance. If the condition given by Equation (7.42) is followed, then ΔS will yield the deviation of the coefficient S from zero. In that case, Equation (7.34) can be used for allowable Δz_o and Δz_c.

To explain Equation (7.46), let us first use Equation (4.27) to relate the hologram aperture radius $\rho = r'$, the object diameter $d = 2a$, and the number m of the side lobes to be recorded by

$$\rho/z_o = (1 + m)\lambda_o/d. \tag{7.47}$$

Equations (7.34), (7.46) and (7.47) can be combined to obtain:

$$\frac{3n\pi}{4}\left[\frac{(1+m)\lambda_o}{d}\right]^4 \left|\left[1 - n^2 + 2n\left(\frac{z_o}{z_c}\right) - \left(\frac{z_o}{z_c}\right)^2\right]\frac{\Delta z_o}{\lambda_c} + \left[n\left(\frac{z_o}{z_c}\right)^2 - 2\left(\frac{z_o}{z_c}\right)^3\right]\frac{\Delta z_c}{\lambda_c}\right| \leqslant (0.2)^{1/2}. \tag{7.48}$$

With a properly determined value of z_o/z_c for zero spherical aberration, such as one given by Equation (7.43), Equation (7.48) can easily be used to determine the allowable Δz_o and Δz_c. With the numerical situation of Equation (7.43), i.e. ruby laser recording and HeNe laser reconstruction, Equation (7.48) becomes:

$$\left[\frac{(1+m)\lambda_o}{d}\right]^4 (0.29\Delta z_o + 0.017\Delta z_c) \leqslant \lambda_c. \tag{7.49}$$

The conclusion can be immediately drawn that the effect of hologram movement or uncertainty in the source distance Δz_c is far less serious than object distance uncertainty Δz_o.

Let us consider a minimum of three side lobes ($m = 3$) and small objects, say $d/\lambda_o = 10$ (remembering that λ_o here is 0.6943 μm). The LHS of equation (7.49) becomes approximately $0.0074(\Delta z_o/\lambda_c) + 0.00044(\Delta z_c/\lambda_c)$. With $\lambda_c = 0.6328$ μm, whereas a Δz_c movement of about 1.5 mm is allowed for zero Δz_o, a Δz_o movement of only about 0.1 mm is allowed for zero Δz_c. Thus, for small objects, exact setting of z_c for each z_o seems necessary.

The situation changes for larger objects. For $d/\lambda_o = 50$ (i.e. about a 35 μm diameter object), the LHS of Equation (7.49) becomes approximately $0.00001(\Delta z_o/\lambda_c) + 0.0000007(\Delta z_c/\lambda_c)$. Now about a 6 cm change in z_o is allowed with a large permissible value of Δz_c.

Thus, except for very small microobjects, exact settings of z_c for each z_o can be avoided and analysis by the mode of hologram movement can be performed. One intermediate position in the image space can be set satisfying Equation (7.42) or particularly Equation

(7.43) with the present example. Then the hologram can be moved to cover a volume satisfying Equation (7.49), in particular, or Equation (7.48) in general.

7.6 Cylindrical test-sections

The discussions on aberrations so far in this chapter have dealt with microobjects in open air or enclosed by flat parallel surfaces. The third-order aberrations considered are basically due to the finite size of the hologram. There is another class of aberrations which is introduced even if first-order optics is considered. The microobjects under study can be in a complex shaped container or in the test-section. In practical situations such as antiobiotic ampules or water and wind tunnels, the containers are often cylindrical. In the in-line method, the light beams become cylindrical before reaching the recording plane. In this section, this situation and the ways to neutralize the effect are discussed.

7.6.1 *Medical ampules*

The presence of particulate contamination in injectable liquids can be hazardous. In view of the importance of detecting this, Zeiss *et al.*[17] and Crane *et al.*[18] demonstrated the use of in-line far-field holography as a reliable inspection tool. The cylindrical ampules are generally a few centimeters in size. Immersing them in a liquid of controlled refractive index to neutralize the cylindrical power is therefore a practical solution.[17,18] A typical cross-section of the ampule is shown in Figure 7.12: n' is the refractive index of the ampule (glass), n'' is the index of liquid inside (say antiobiotic), R is the radius of curvature of the outer surface of the ampule, and d is the wall thickness. n_s is the refractive index of the surrounding medium. For air, n_s is 1.

Suppose a ray parallel to the optical axis is incident on the ampule. The ray path in the cross-section faces curved surfaces. Starting with the source point at infinity, the paraxial ray can be traced[19] to obtain the apparent distance l' at the first inner surface of the ampule as

$$\frac{n''}{l'} = \frac{n'n''(R-d) - n_s(n'R - n''d)}{[n'R - (n' - n_s)d](R-d)}. \tag{7.50}$$

Ray tracings without the paraxial approximation can also be performed.[17,18] Remember that in the perpendicular direction, the ray does not face curvatures and remains parallel to the optical axis. The

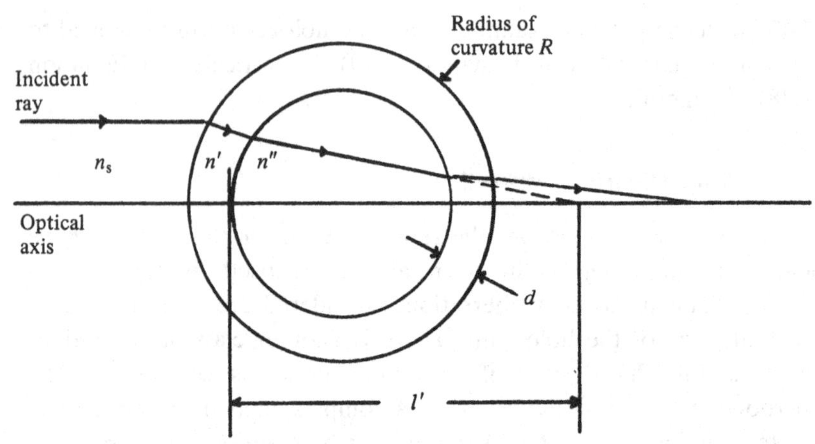

Fig. 7.12 Cross-section of a typical cylindrical ampule. The refractive index of the surrounding medium is n_s (= 1 for air). A schematic ray path represents the cylindrical lens effect showing that the collimated beam becomes cylindrical.

resulting astigmatism effect is shown in Figure 7.13. 40 μm pollen grains suspended in the ampule render the reconstruction unusable.

Now suppose the ampule is immersed in a fluid of refractive index n_s such that

$$n_s = \frac{n'n''(R - d)}{n'R - n''d}. \tag{7.51}$$

In this situation, Equation (7.50) shows that the ray remains parallel to the optical axis ($l' = -\infty$) even inside the ampule. Due to the symmetry of the system, the emergent ray from the ampule will also remain parallel to the optical axis. This is the basis of the index-matching approach.[17,18] Several holograms of the particles in the index-matched situation using relay lens have been recorded.[17,18] The astigmatism introduced by the object beam and the lens combination was investigated by Crane *et al.*[18] and found to be practically insignificant. Figure 7.14 shows some reconstructions from recordings performed before and after the proper index matching. The image astigmatism is practically eliminated. Reconstructed images of ampule contaminants are also satisfactory when the index-matching technique is applied.[17,18]

7.6.2 *Cylindrical tunnels*

Conventional experimental test tunnels, such as water and glycerine tunnels were designed decades ago when the flow was

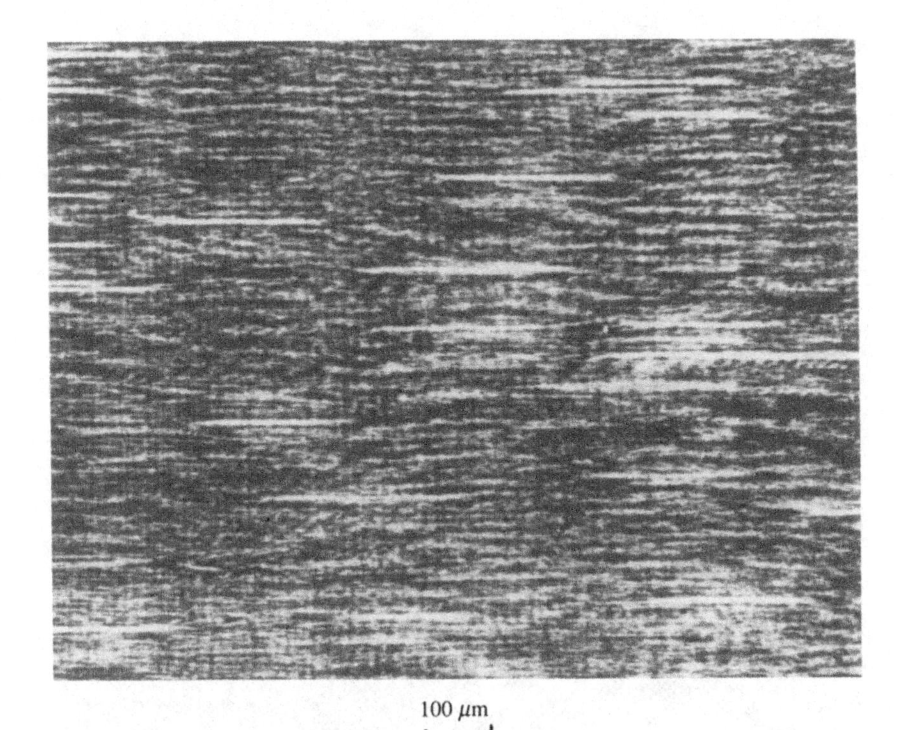

Fig. 7.13 Reconstruction of 40 μm pollen grains in a cylindrical ampule. The unwanted astigmatism effect is clearly seen. (Crane *et al.*[18].)

measured by non-optical techniques such as hot film anemometry. Since optical methods are now more common[20], it is necessary to have transparent (generally plastic or glass) windows on the test-section. In many situations the inner surface of the window must be cylindrical due to symmetry requirements. If the outer surface is kept flat, the resulting astigmatism generally spoils the micro-imaging process.

Since these tunnels are big, the index matching approach is not convenient. The outer surface of the window can be made suitably cylindrical to neturalize the effect. Additional cylindrical lenses can also be introduced to do the same job. The method of outer window curvature modification was proposed by Vikram and Billet.[21] As shown in Figure 7.15, suppose the inner radius of curvature of the cylindrical tunnel is R and d is the thickness of each window. The aim is to find the outer radius of R' such that the net cylindrical power becomes zero. Let us assume the refrative index of the window

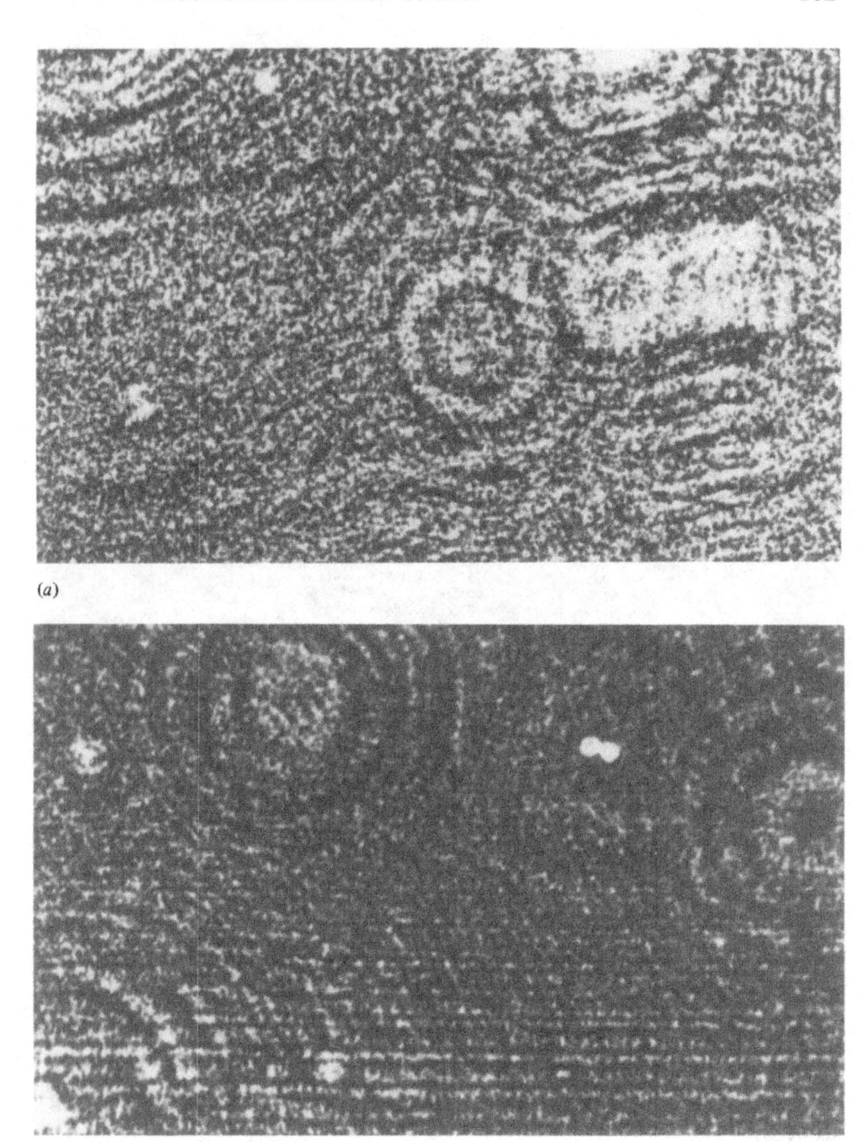

Fig. 7.14 (*a*) Reconstructed images of 20 μm pollen grains recorded with the refractive index mismatched by 0.01. (*b*) The reconstruction with proper index matching before recording. (Crane *et al.*[18].)

material is n_1 and that of the tunnel medium is n_2. From the general thick lens formula,[19] the power P of half of the system is

$$P = \frac{n_1 - 1}{R'} + \frac{n_2 - n_1}{R} - \frac{(n_1 - 1)(n_2 - n_1)d}{n_1 R R'}. \tag{7.52}$$

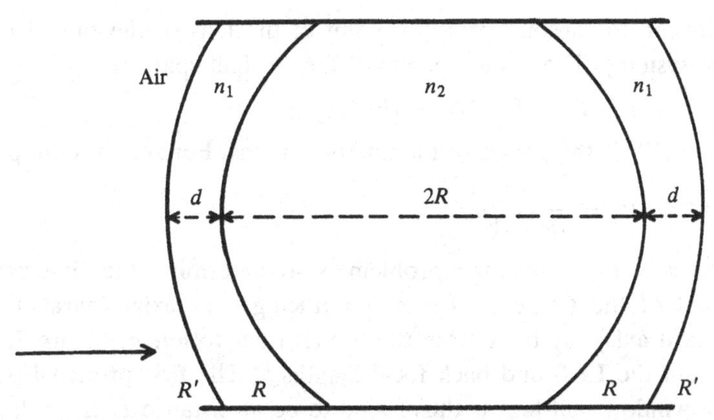

Fig. 7.15 Cross-section of a cylindrical tunnel. A general radius of curvature R' of the outer surface is considered to solve for net zero power. (Ref. 21.)

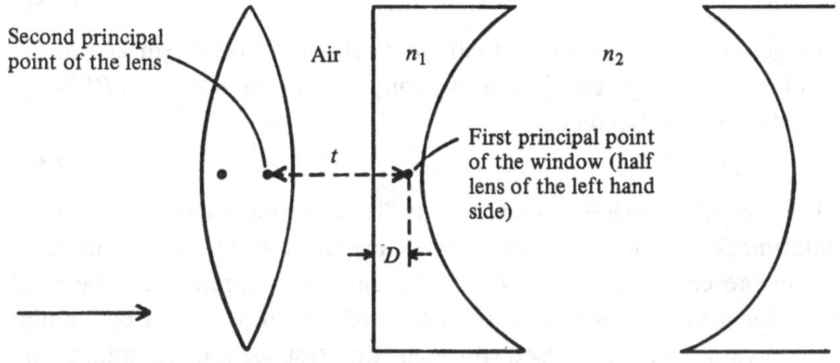

Fig. 7.16 Schematic diagram showing the use of external cylindrical lens to neutralize window curvature effects. (Ref. 21.)

The power becomes zero when

$$R'(P = 0) = \frac{(n_1 - 1)[n_1 R - (n_2 - n_1)d]}{(n_1 - n_2)n_1}. \tag{7.53}$$

The curvature modification of the outer surface will yield zero cylindrical power the way index matching did in Section 7.6.1. It may not be convenient to modify the outer surface of the existing window. An additional cylindrical lens can be used to neutralize the power in those situations.[21] The procedure is represented in Figure 7.16. The half space of the tunnel test-section, described by the power relationship of Equation (7.52), can be treated as a cylindrical lens. An additional parallel cylindrical lens (each side) kept outside the window is required to neutralize the power. Suppose t is the distance between the second principle point of the lens of the tunnel (half of the system). The system power of the net half space is[19]

$$P_{\text{system}} = P + P' - tPP', \tag{7.54}$$

where P' is the power of the additional lens. For zero system power:

$$P' = \frac{P}{tP - 1}. \tag{7.55}$$

Basically the remaining problem is to determine the first principal point of the tunnel half-lens. By tracing a paraxial (parallel to the optical axis) ray back from the tunnel right to left in Figure 7.16, we obtain the LHS and back focal lengths.[19] The first principal point of the window half-lens is thus found to be at distance D from the outer window surface toward the center of the tunnel, given by

$$D = \frac{d(n_1 - n_2)/Rn_1}{(n_1 - n_2)/R + (1 - n_1)/R' + d(1 - n_1)(n_1 - n_2)/RR'n_1}. \tag{7.56}$$

The general case of radius of curvature R' of the outer surface of the window is considered. For a typically flat outer surface ($R' = \infty$), Equation (7.56) becomes

$$D = d/n_1. \tag{7.57}$$

The second principle point of the lens to be used can also be determined.[19] For a thin lens, the principal point can be assumed to be on the center of the lens. A thin lens approximation can be used even for a thick lens to obtain an approximate solution. A collimated light beam can be passed through the test-section to adjust the separation t to obtain the collimated output beam.

If a lens of the window material can be fabricated, knowledge of the principal points is not necessary. By keeping one surface of the additional lens in contact with the outer window of the tunnel, the relationship (7.53) is sufficient. An interesting case exists when the first surface of the additional lens is plane. In this case the second principal point lies on the second surface itself.

Thus, it is not difficult to choose an additional cylindrical lens for the purpose of neutralizing the power. One can obtain a nearly appropriate lens from stock and adjust the distance t for fine matching.

Sometimes, it is convenient to replace the conventionally metallic test-section by commercially available cylindrical tubes of transparent materials such as glass or plexiglass. The cylindrical power can be neutralized by using additional cylindrical power outside the tube.

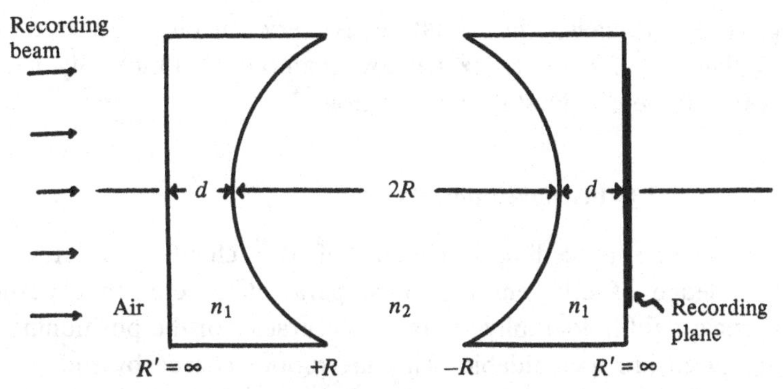

Fig. 7.17 The recording arrangement to determine the tolerance limits of $\Delta n = n_2 - n_1$ when the refractive indices n_1 and n_2 are very close.

Using the approach described in this section, the parameters of the external system can be obtained. Different design parameters in this connection are discussed by Vikram and Billet.[22]

With the cylindrical inner and flat outer surfaces of the window, the astigmatism effects can be small if the refractive indices of the window material and the tunnel medium are very close. This aspect is discussed in detail by Vikram and Billet.[23] In a practical sense, it may not be possible to have a perfect match. However for some cases, such as glycerine tunnels, a vartiey of plastic glasses are available with close refractive indices. Collimated beam in-line Frauhofer holography with the recording plane in contact with the outer window is shown in Figure 7.17. For a very small refractive index difference, $\Delta n (= n_2 - n_1)$, the differences between cylindrical and flat transverse and longitudinal magnifications of the conjugate image are[23]

$$\Delta M_c \approx -\frac{2(l + d)\Delta n}{\bar{n} R} \tag{7.58}$$

and

$$\Delta M_{\text{long}} \approx -\frac{4(l + d)\Delta n}{\bar{n} R}. \tag{7.59}$$

respectively when $\bar{n} \approx n_1 \approx n_2$. For a very small Δn, the differences can be neglected. We know the allowable movement of the micro-object during the exposure is generally as high as one tenth of its diameter. This knowledge can be used to set the maximum allowable ΔM_c as high as 0.1.

With available plastic or glass materials, obtaining ΔM_c under the tolerance limits is difficult with water tunnels.[23] However, with glycerine tunnels, the situation is very encouraging due to the availability of close index window materials.[24] Practically negligible ΔM_c can be obtained in this situation.[23]

7.7 Other developments

The preceding discussion of this chapter assumes precise knowledge of different positions, parameters, etc. In experimental particle field holography, certain inaccuracies of the positioning of the hologram are unavoidable. The aberrations caused by such misalignments have been experimentally observed.[1] More recently, Banyasz, Kiss and Varga[25] presented a detailed analysis of such aberrations. Naon, Bjelkhagen, Burnstein and Voyvodic[26] described the aberrations for the modified in-line holographic system of the Bubble Chamber at the Fermi National Accelerator Laboratory. Phase errors can also be subtracted optically, as discussed in Section 11.6. An application of such an approach in lens-assisted in-line holography is discussed in Ref. 27.

Aberration theory dealing with holography is still under study. The theory in the non-paraxial region is of particular interest.[28]

8

Hologram fringe-contrast and its enhancement

As it is clear now, the in-line Fraunhofer hologram is recorded in the form of a high frequency interference pattern. The contrast of these fringes becomes low away from the center of the diffraction pattern due to the falling envelope of the pattern. Even at the center of the diffraction pattern, the contrast becomes poor if the object is very far from the recording plane. Depending on the contrast recording capability of the recording medium, this aspect sets the maximum object distance, as discussed in Section 4.3. A very low contrast pattern is mixed in noise ultimately limiting the effective size of the hologram and hence the resolution. Therefore, this chapter is devoted to the contrast aspect in detail. Approaches to enhance the contrast for better hologram recordability are also discussed. These are based on utilizing a divergent recording beam, a Gaussian recording beam, detuned interference filters, and non-blocked zero frequency filtering.

8.1 Fringe-contrast in Fraunhofer holography

To describe the contrast, let us consider an opaque two-dimensional object cross-section. The normalized irradiance I at the recording plane can then be obtained (see Section 4.1) as:

$$I = 1 - \frac{2m_o}{\lambda_o|z_o|}\sin\left[\frac{\pi r'^2}{\lambda_o m_o|z_o|} + \phi\left(\frac{x'}{\lambda_o|z_o|}, \frac{y'}{\lambda_o|z_o|}\right)\right]$$
$$\times \tilde{A}_o\left(\frac{x'}{\lambda_o|z_o|}, \frac{y'}{\lambda_o|z_o|}\right) + \left(\frac{m_o}{\lambda_o z_o}\right)^2 \tilde{A}_o^2\left(\frac{x'}{\lambda_o|z_o|}, \frac{y'}{\lambda_o|z_o|}\right). \quad (8.1)$$

For the corresponding one-dimensional objects (such as a long thin wire) the distribution is

$$I = 1 - 2\left(\frac{m_o}{\lambda_o|z_o|}\right)^{1/2} \cos\left[\frac{\pi x'^2}{m_o\lambda_o|z_o|} + \phi\left(\frac{x'}{\lambda_o|z_o|}\right) - \frac{\pi}{4}\right]$$
$$\times \widetilde{A}_o\left(\frac{x'}{\lambda_o|z_o|}\right) + \frac{m_o}{\lambda_o|z_o|}\widetilde{A}_o^2\left(\frac{x'}{\lambda_o|z_o|}\right). \tag{8.2}$$

The high frequency sine or cosine terms demand a film resolution as discussed in Section 4.2. Let us consider the contrast here in more detail. For the two-dimensional case given by Equation (8.1), the Fourier transform term $\widetilde{A}(x'/\lambda_o|z_o|, y'/\lambda_o|z_o|)$ is rather slowly varying resulting in many fine fringes given by the sine term within a broad diffraction ring. Thus, in a region (x', y'), the maximum and the minimum intensities are (assuming a positive value of m_o)

$$I_{\max} = 1 + \frac{2m_o}{\lambda_o|z_o|}\widetilde{A}_o\left(\frac{x'}{\lambda_o|z_o|}, \frac{y'}{\lambda_o|z_o|}\right)$$
$$+ \left(\frac{m_o}{\lambda_o z_o}\right)^2 \widetilde{A}_o^2\left(\frac{x'}{\lambda_o|z_o|}, \frac{y'}{\lambda_o|z_o|}\right) \tag{8.3}$$

and

$$I_{\min} = 1 - \frac{2m_o}{\lambda_o|z_o|}\widetilde{A}_o\left(\frac{x'}{\lambda_o|z_o|}, \frac{y'}{\lambda_o|z_o|}\right) + \left(\frac{m_o}{\lambda_o z_o}\right)^2$$
$$\times \widetilde{A}_o^2\left(\frac{x'}{\lambda_o|z_o|}, \frac{y'}{\lambda_o|z_o|}\right) \tag{8.4}$$

respectively. The visibility of the fringes is thus

$$V(x', y') = \frac{I_{\max} - I_{\min}}{I_{\max} + I_{\min}}$$
$$= \frac{\dfrac{2m_o}{\lambda_o|z_o|}\widetilde{A}_o\left(\dfrac{x'}{\lambda_o|z_o|}, \dfrac{y'}{\lambda_o|z_o|}\right)}{1 + \left(\dfrac{m_o}{\lambda_o z_o}\right)^2 \widetilde{A}_o^2\left(\dfrac{x'}{\lambda_o|z_o|}, \dfrac{y'}{\lambda_o|z_o|}\right)} \tag{8.5}$$

For one-dimensional objects, Equation (8.2) can be similarly used to obtain the visibility:

$$V(x') = \frac{2(m_o/\lambda_o|z_o|)^{1/2}\widetilde{A}_o(x'/\lambda_o|z_o|)}{1 + (m_o/\lambda_o|z_o|)\widetilde{A}_o^2(x'/\lambda_o|z_o|)}. \tag{8.6}$$

At a large distance z_o, the contrast tends to become zero. The result is limited object distance range[1–5] as discussed in Section 4.3. Depending upon the hologram aperture requirements, the envelope of the visibility should be at least a certain minimum $V_{\min}$ governed by the contrast response of the recording medium.[6]

8.2 Object shape and fringe visibility

Equations (8.5) and (8.6) show that the nature of the visibility is different for one- and two-dimensional objects. Even for a given dimension (one or two), the Fourier transform $\tilde{A}_o$ depends on the object shape (square, circular, etc.). This also results in different visibilities for different shapes even if the nominal size is the same. This kind of object shape related visibility and hence the resolution is discussed by Dunn and Thompson.[7] Detailed analysis is also provided by Cartwright, Dunn and Thompson[8] and Dunn and Thompson.[9] The visibility has been discussed in detail for a wire (one-dimensional), rectangular cross-section objects, and objects with circular cross-sections.[9] Of these, one-dimensional and two-dimensional circular cross-section objects are discussed here in detail due to their practical significance. The one-dimensional case represents objects like long thin fibers whereas the circular cross-section case represents bubbles, spray droplets, rain drops, pollen grains, etc.

The Fourier transform $\tilde{A}$ for a circular cross-section object of radius a is

$$\tilde{A}\left(\frac{x'}{\lambda_o|z_o|}, \frac{y'}{\lambda_o|z_o|}\right) = \pi a^2 \left[\frac{2J_1\left(\frac{2\pi a r'}{\lambda_o|z_o|}\right)}{\frac{2\pi a r'}{\lambda_o|z_o|}}\right], \tag{8.7}$$

where $r'^2 = x'^2 + y'^2$. For a one-dimensional object of width $2a$, the Fourier transform is

$$\tilde{A}\left(\frac{x'}{\lambda_o|z_o|}\right) = 2a \operatorname{sinc}\left(\frac{2\pi a x'}{\lambda_o|z_o|}\right), \tag{8.8}$$

where $\operatorname{sinc}(q) = \operatorname{sinc}(q)/q$ has been assumed.

Substituting Equation (8.7) into Equation (8.5), we obtain, for circular cross-section objects,

$$V(r') = \frac{8\pi N m_o \left|\frac{2J_1(2\pi a r'/\lambda_o|z_o|)}{2\pi a r'/\lambda_o|z_o|}\right|}{16N^2 + \pi^2 m_o^2 \left[\frac{2J_1(2\pi a r'/\lambda_o|z_o|)}{2\pi a r'/\lambda_o|z_o|}\right]^2}, \tag{8.9}$$

where $N = \lambda_o|z_o|/(2a)^2$ is the number of far-fields.

Similarly, for one-dimensional opaque objects, equations (8.6) and (8.8) yield:

$$V(x') = \frac{2m_o^{1/2}N^{1/2}|\operatorname{sinc}(2\pi a x'/\lambda_o|z_o|)|}{N + m_o \operatorname{sinc}^2(2\pi a x'/\lambda_o|z_o|)}. \tag{8.10}$$

Equations (8.9) and (8.10) provide sufficient background for visibilities and their comparison. It is noticeable that the diffraction pattern related modulation parameters $|2J_1(2\pi ar'/\lambda_o|z_o|)/(2\pi ar'/\lambda_o|z_o|)|$ and $|\mathrm{sinc}\,(2\pi ax'/\lambda_o|z_o|)|$ vary between maxima and zero. The visibility at the maxima of these modulation parameters gives the envelope of the visibility pattern in the recording plane.

8.2.1 *Visibility at the hologram center*

This situation is useful in the study of the role of increasing object distance (related by N) on the visibility. By setting $r' \approx 0$ in Equation (8.9) and $x' \approx 0$ in Equation (8.10) we obtain the visibility at the hologram center as:

$$V(0)|_{\text{2D-circular}} = \frac{8\pi N m_o}{16N^2 + \pi^2 m_o^2}, \tag{8.11}$$

for circular cross-section objects and

$$V(0)|_{\text{1D}} = \frac{2m_o^{1/2} N^{1/2}}{N + m_o}, \tag{8.12}$$

for line objects. For $m_o = 1$, the expressions are derived and discussed by Dunn and Thompson.[9] In Figure 8.1, these have been plotted for the collimated beam recording ($m_o = 1$) case. It is interesting to note that the visibility is generally very high for the one-dimensional case as compared to that for a two-dimensional circular object with the same diameter. In the limiting case of large N, when the hologram of the circular object may not be recorded, the one-dimensional object (of the same diameter) might still be recordable. For $m_o = 1$ and large N, Equations (8.11) and (8.12) yield:

$$\frac{V(0)|_{\text{1D}}}{V(0)|_{\text{2D circular}}} \approx \frac{4N^{1/2}}{\pi}. \tag{8.13}$$

For $N = 100$, the ratio is about thirteen! In other words, if the low contrast response of the recording emulsion allows N far-fields in the case of two-dimensional circular case, the same emulsion will allow more than N^2 far-fields for line objects!

8.2.2 *Visibility at outer positions in the hologram*

The analysis of Section 8.2.1 discussed the fringe visibility effects in the hologram center. This way to study the role of increasing N on the visibility and to see the dramatic effects caused

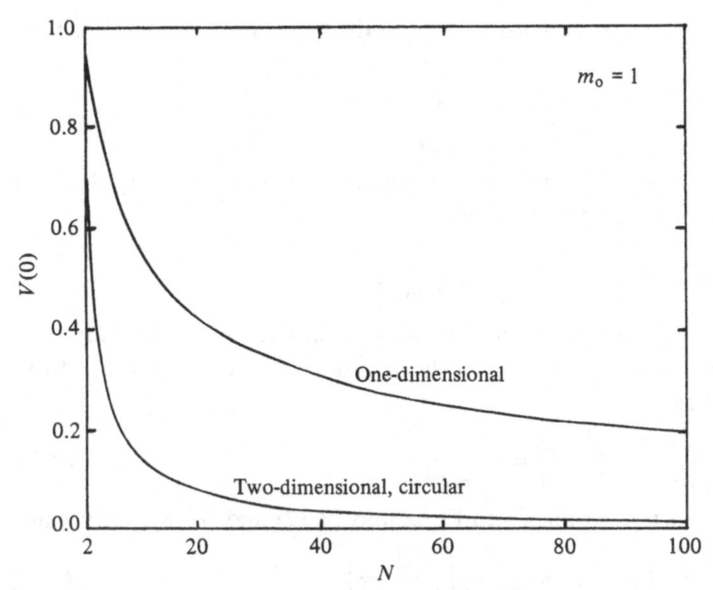

Fig. 8.1 Variation of fringe visibility at the center of the hologram against the number N of far-field distances. The collimated recording beam case ($m_o = 1$) is considered here.

by the object shape. Since a few side lobes are to be recorded in a far-field hologram, the visibility at the outer edge of the effective hologram is also important. The visibility will be zero at zeros of $\tilde{A}_o(x'/\lambda_o|z_o|, y'/\lambda_o|z_o|)$ or $\tilde{A}_o(x'/\lambda_o|z_o|)$, resulting in no hologram formation in those regions. However, the hologram is formed in the regions where $\tilde{A}_o$ is sufficiently large. The envelope of the visibility function is then an important single parameter for the visibility study in the entire hologram. The study is basically associated with low contrast response of the recording medium. Therefore, the case of large N and r' is specifically important. In those situations, Equation (8.9) can be approximated to

$$V(r') \approx \frac{\pi m_o}{2N}\left|\frac{2J_1(2\pi ar'/\lambda_o|z_o|)}{2\pi ar'/\lambda_o|z_o|}\right|. \tag{8.14}$$

Similarly, Equation (8.10) becomes

$$V(x') \approx \frac{2m_o^{1/2}}{N^{1/2}}\left|\operatorname{sinc}\left(\frac{2\pi ax'}{\lambda_o|z_o|}\right)\right|. \tag{8.15}$$

For non-zero values of q, $J_1(q)$ has a crude approximated form:

$$J_1(q) \sim \left(\frac{2}{\pi q}\right)^{1/2}\cos\left(q - \frac{3\pi}{4}\right). \tag{8.16}$$

Equations (8.14) and (8.16) give the envelope of the visibility function as

$$V(r')\big|_{\text{envelope}} \approx \frac{m_o}{2\pi N}\left|\frac{\lambda_o z_o}{ar'}\right|^{3/2}. \tag{8.17}$$

Similarly, the envelope of the one-dimensional visibility function given by Equation (8.15) becomes

$$V(x')\big|_{\text{envelope}} \approx \frac{m_o^{1/2}}{\pi N^{1/2}}\left|\frac{\lambda_o z_o}{ax'}\right|. \tag{8.18}$$

At this stage, let us recall Equation (4.27) to relate the hologram aperture radius r' (or x') and the number m of the side lobes to be recorded as

$$\left|\frac{\lambda_o z_o}{ar'}\right| = \left|\frac{\lambda_o z_o}{ax'}\right| \approx \frac{2}{1+m}. \tag{8.19}$$

Substituting these values into Equations (8.17) and (8.18), we obtain

$$V(r')\big|_{\text{envelope}} \approx \frac{m_o}{2\pi N}\left(\frac{2}{1+m)}\right)^{3/2} \tag{8.20}$$

and

$$V(x')\big|_{\text{envelope}} \approx \frac{m_o^{1/2}}{\pi N^{1/2}}\left(\frac{2}{1+m}\right). \tag{8.21}$$

These envelopes of visibility are plotted in Figures 8.2 and 8.3. The visibility for the circular object case is generally far less than that for the one-dimensional case. Suppose a given number (m) of the side lobes are to be recorded and the envelope of the visibility function is a certain minimum value governed by the recording medium. Then, if N is the maximum number allowed for the circular object, the number N' for one-dimensional objects can be determined using Equations (8.20) and (8.21):

$$N' = 2(1+m)N^2/m_o. \tag{8.22}$$

For m_o not very large, N' should be generally very large compared to N. Similarly, for a fixed object distance (fixed N) and fixed visibility storage capability, the allowed number m of the side lobes in the circular object case can be related to the number m' in the one-dimensional object case by

$$m' = \frac{4N^{1/2}}{m_o^{1/2}}\left(\frac{1+m}{2}\right)^{3/2} - 1. \tag{8.23}$$

Again, m' should generally be large compared to m. At this stage it is clear that a line object of diameter $2a$ will have a significantly more effective hologram aperture compared to that for a spherical object of

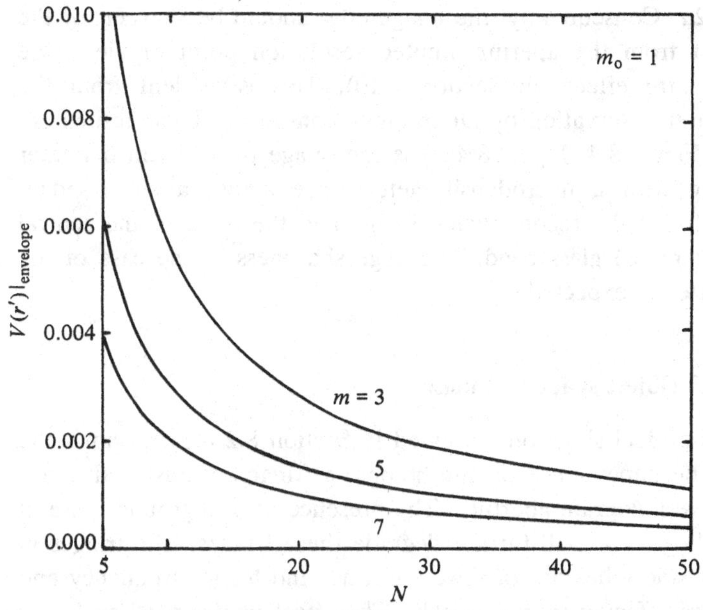

Fig. 8.2 Envelope of visibility for circular cross-section objects as a function of N for a few values of m and $m_o = 1$.

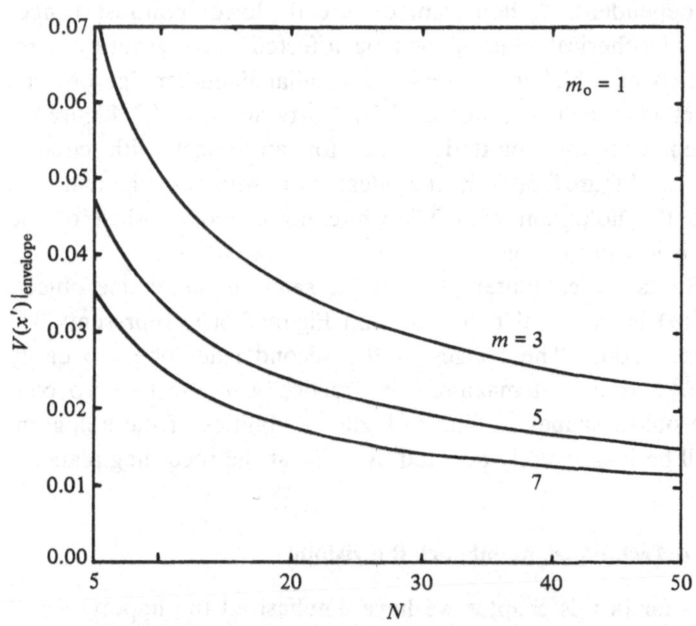

Fig. 8.3 Envelope of visibility for one-dimensional objects as a function of N for a few values of m and the collimated beam recording ($m_o = 1$).

diameter $2a$. Consequently, the image edge should be sharper for the line object from the aperture-limited resolution point of view (see finite aperture effects in Section 5.10). This is evident from the experimental observation by Dunn and Thompson.[9] Their results are shown in Figure 8.4. Figure 8.4(*a*) is the image of a 75 μm diameter wire along with a microdensitometer trace across a single edge. Figure 8.4(*b*) is the reconstructed image and the trace of the central (75 μm diameter) glass bead. The edge sharpness in the case of the wire is better as expected.

8.3 Object shape and noise

The ideal situation discussed in Section 8.2 clearly shows the object shape dependence on the hologram fringe contrast and hence the effective hologram aperture. The presence of background noise at the recording plane will further degrade these fringes. The fringes in the higher side lobes are of lower contrast and higher frequency and they will be affected more severely. The effect of the noise will thus be limiting the effective hologram aperture and hence resolution.

For a given relative noise, the effect on the hologram recordability is shape dependent. A hologram of already lower contrast fringes (like one of spherical objects) will be affected more severely compared to that of a higher contrast one (similar diameter line object). This aspect has been demonstrated by Cartwright *et al.*[8] Figure 8.5 shows their computer plotted results for an object with circular cross-section. Figure 8.5(*a*) is the ideal case whereas Figure 8.5(*b*) represents the hologram with 3% white noise added. Most of the hologram is lost in the noise.

Figure 8.6 is the computer plot[8] of the same diameter line object. Figure 8.6(*a*) is the ideal hologram and Figure 8.6(*b*) represents 3% white noise added. The fringes in the second side lobe are easily detectable. The central maximum is practically unaffected. To conclude, the object shapes leading to higher visibilities of the hologram fringes will be less severely affected by noise at the recording plane.

8.4 Techniques to enhance the visibility

So far in this chapter we have emphasized the importance of the visibility of the high frequency interference fringes. We have also discussed the dependence of the visibility on the object shape. However, for a given shape, size, etc. of the object, the visibility puts a limit on the object–hologram distance, effective hologram aperture, and hence the resolution capability. Visibility enhancement is therefore very useful to have flexibility in experimental parameters such as object–hologram distance, recording emulsion contrast response, etc. In this section, approaches to enhance the contrast are discussed.

8.4.1 *Divergent recording beam*

The collimated recording beam case ($m_o = 1$) yields a certain experimental object–recording plane distance limit. A well-accepted upper limit is 100 far-fields for two-dimensional object cross-sections (see Section 4.3). For a 10 μm diameter object and ruby laser recording ($\lambda_o = 0.6943$ μm), 100 far-fields is only about 1.5 cm. This distance is very small for many practical situations. The test-section chamber window is often itself this thick.

Using divergent recording beams ($m_o > 1$)[10] increases the fringe visibility. In the limiting low contrast range, Equation (8.14) gives a contrast increase of m_o times for two-dimensional objects compared to the collimated beam case. From the envelope relationship given by Equation (8.17), a given low contrast response of the emulsion and object distance (fixed N), we obtain

$$r' = m_o^{2/3} r'_{\text{coll}}, \tag{8.24}$$

where r'_{coll} is the effective aperture radius for the collimated beam case of $m_o = 1$: r' is the effective hologram aperture when a divergent beam ($m_o > 1$) is used for the recording. Thus, the effective hologram aperture is increased by the factor $m_o^{2/3}$. For a fixed number m of the side lobes to be recorded, Equation (8.20) yields

$$N' = m_o N, \tag{8.25}$$

where N and N' are the maximum number of far-fields allowed for collimated and divergent beam cases respectively.

For line objects, Equation (8.18) can be used, for fixed distance parameter N:

$$x' = m_o^{1/2} x'_{\text{coll}}. \tag{8.26}$$

Therefore the effective hologram aperture can be increased by $m_o^{1/2}$. For a fixed number of side lobes to be recorded, Equation (8.21) results in the relationship given by Equation (8.25), i.e. the number of far-fields can again be increased by m_o times.

Obviously, a divergent beam very easily increases the allowed object–hologram distance. Witherow[11] observed that particle–hologram separation at the recording stage can be well over 2000 far-

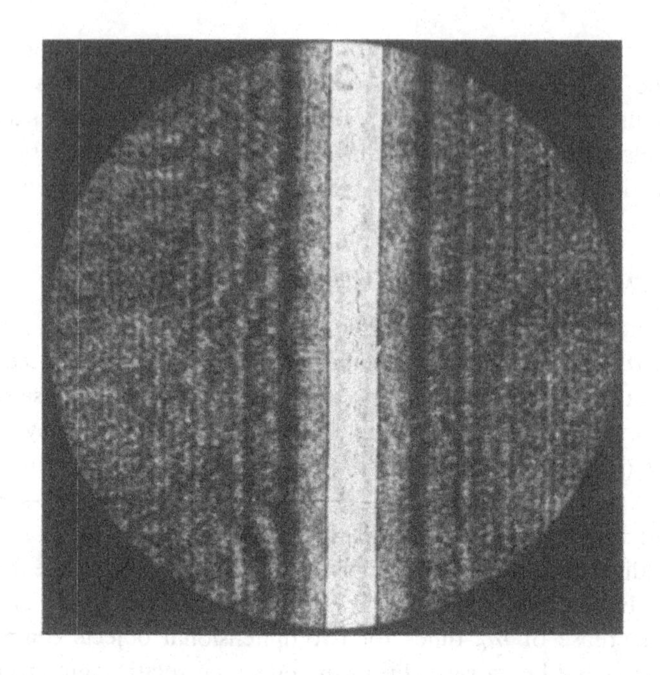

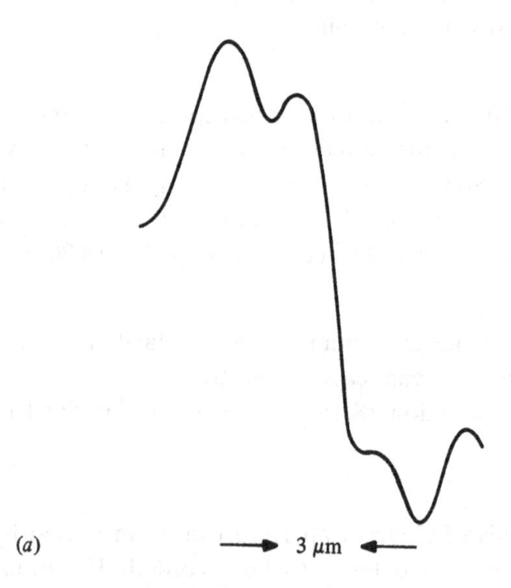

(a)

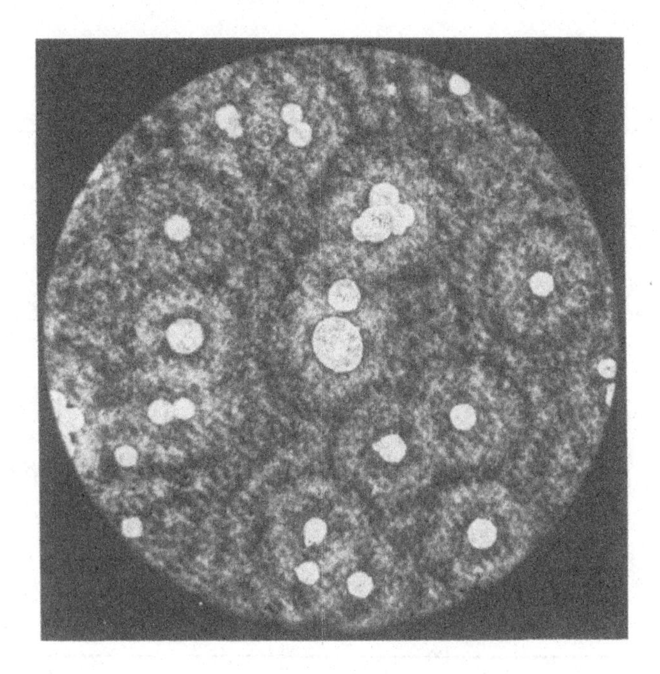

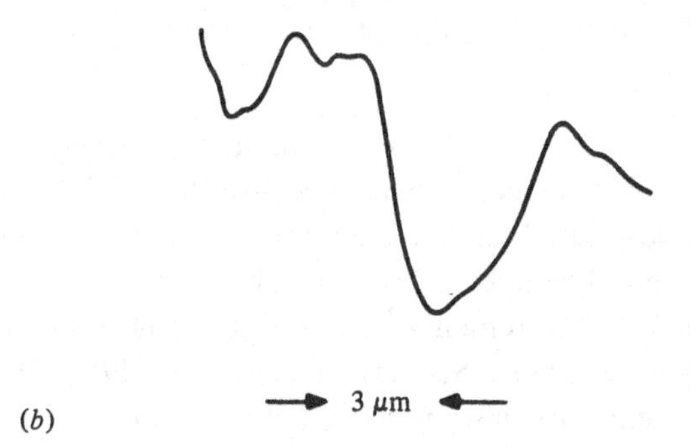

(*b*)

Fig. 8.4 (*a*) Photograph of a reconstructed image and the micro-densitometer trace across a single edge. The object was a 75 μm diameter wire. (*b*) The image and the trace across one edge of the central (75 μm) bead for the objects consisting of glass beads. (Dunn and Thompson[9].)

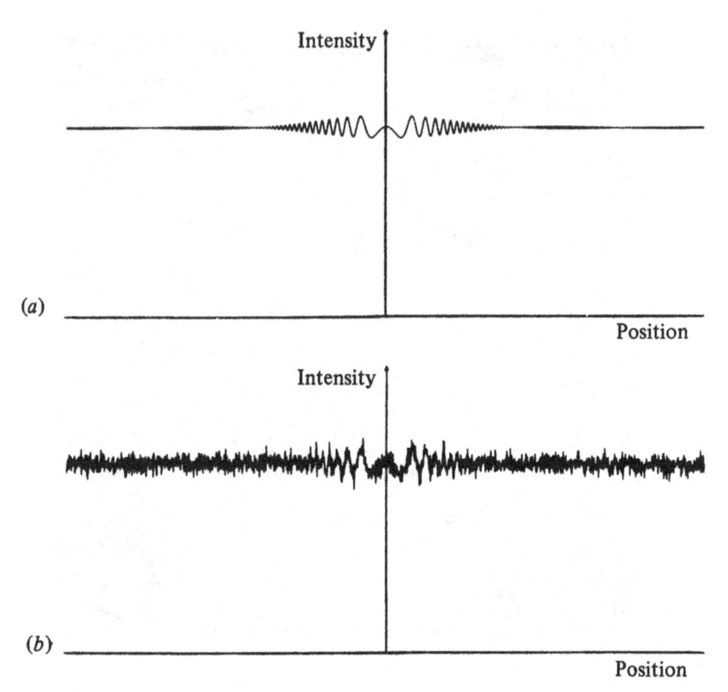

Fig. 8.5 Computer simulated intensity distribution in the recording plane for a circular cross-section object: (*a*) an ideal distribution; (*b*) the distribution when 3% white noise is introduced. (Cartwright *et al.*[8].)

fields when using the divergent beam configuration! Experimentally too, a divergent beam is convenient because the light from the microscope objective–spatial filter combination can be used as such. In fact, the divergent beam case contains the added advantage of relaxed film resolution requirements.[10] (See Equations (8.1) and (8.2) for the high frequency terms. See also equation (4.29).) Temporal coherence requirements are also relaxed (see Section 4.5.2) by m_o times.

8.4.2 *Gaussian recording beam*

For a collimated beam ($z_R = -\infty$), the light amplitude over the beam cross-section is generally assumed to be a constant, i.e. 'B' is constant according to Equation (3.1). This assumption is based on the fact that only a small cross-section of the expanded laser beam is used. If a significant portion is used then the radiation from TEM_{00} mode is Gaussian. The amplitude at the recording plane can then be written as

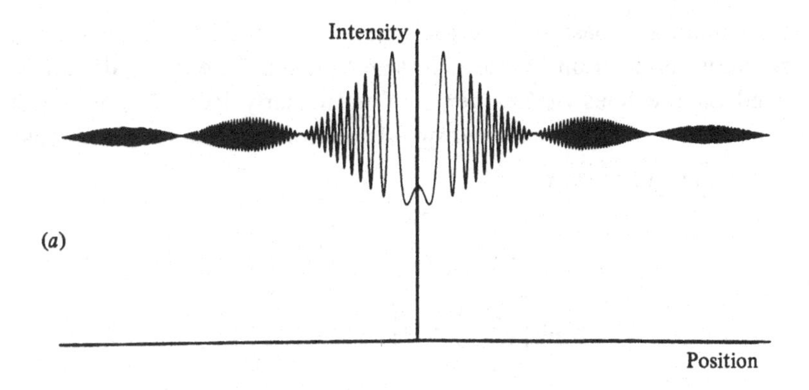

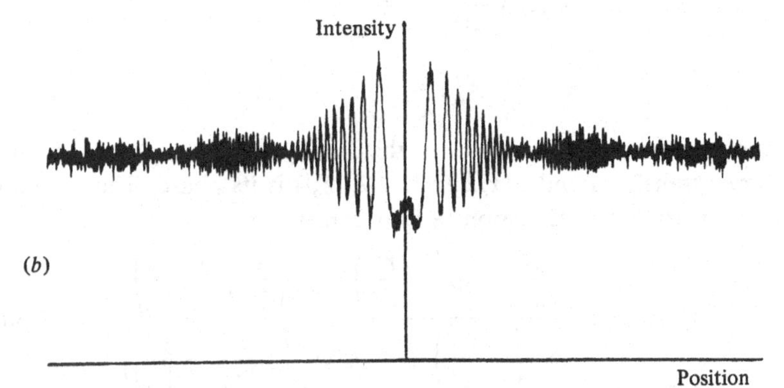

Fig. 8.6 Computer simulated intensity distribution in the recording plane corresponding to a long thin wire of the width equal to the object diameter in Figure 8.5(*a*): (*a*) the ideal noise-free case; (*b*) with 3% white noise added. (Cartwright *et al.*[8].)

$$E_o = B \exp\left(-\frac{\xi^2 + \eta^2}{\omega^2}\right), \tag{8.27}$$

where B is now the maximum amplitude at the center ($\xi = \eta = 0$) and ω is the radius on the half width of the beam. The Huygens–Fresnel principle can be used to determine the field distribution at the recording plane (x, y) at a distance z_o as (see Section 3.1)

$$\begin{aligned} \psi(x, y) = {} & \frac{\mathrm{i}B}{\lambda_o z_o} \exp(-\mathrm{i}k_o z_o) \\ & \times \iint_{-\infty}^{+\infty} [1 - A(\xi, \eta)] \exp\left(-\frac{\xi^2 + \eta^2}{\omega^2}\right) \\ & \times \exp\left\{-\frac{\mathrm{i}k_o}{2z_o}[(x-\xi)^2 + (y-\eta)^2]\right\} \mathrm{d}\xi\,\mathrm{d}\eta \end{aligned} \tag{8.28}$$

This equation is basically the same as Equation (3.2) except for the amplitude modulation by the Gaussian function. The integrals can be solved on the lines of Chapter 3 or particularly Ref. 12. For object points near the optical axis and two-dimensional object cross-sections:

$$\begin{aligned} I(x, y) &= |\psi(x, y)|^2 \\ &\approx B^2\left\{ \exp\left(-\frac{2r^2}{\omega^2}\right) - \frac{2}{\lambda_o|z_o|}\exp\left(-\frac{r^2}{\omega^2}\right) \right. \\ &\quad \times \sin\left[\frac{\pi r^2}{\lambda_o|z_o|} + \phi\left(\frac{x}{\lambda_o|z_o|}, \frac{y}{\lambda_o|z_o|}\right)\right] \\ &\quad \times \widetilde{A}_o\left(\frac{x}{\lambda_o|z_o|}, \frac{y}{\lambda_o|z_o|}\right) \\ &\quad \left. + \frac{1}{\lambda_o^2 z_o^2}\widetilde{A}_o^2\left(\frac{x}{\lambda_o|z_o|}, \frac{y}{\lambda_o|z_o|}\right)\right\}, \end{aligned} \tag{8.29}$$

where $\widetilde{A}_o(x/\lambda_o|z_o|, y/\lambda_o|z_o|)$ is the real positive amplitude of the Fourier transform and $\phi(x/\lambda_o|z_o|, y/\lambda_o|z_o|)$ is its phase. The visibility factor analogous to Equation (8.5) becomes

$$V(x, y) = \frac{\dfrac{2}{\lambda_o|z_o|}\exp\left(-\dfrac{r^2}{\omega^2}\right)\widetilde{A}_o\left(\dfrac{x}{\lambda_o|z_o|}, \dfrac{y}{\lambda_o|z_o|}\right)}{\exp\left(-\dfrac{2r^2}{\omega^2}\right) + \dfrac{1}{\lambda_o^2 z_o^2}\widetilde{A}_o^2\left(\dfrac{x}{\lambda_o|z_o|}, \dfrac{y}{\lambda_o|z_o|}\right)}. \tag{8.30}$$

For the limiting case of low visibility with a uniform collimated beam, $\widetilde{A}_o(x/\lambda_o|z_o|, y/\lambda_o|z_o|)$ is small, resulting in a visibility of

$$V(x, y)|_{\text{uniform beam}} \approx \frac{2}{\lambda_o|z_o|}\widetilde{A}_o\left(\frac{x}{\lambda_o|z_o|}, \frac{y}{\lambda_o|z_o|}\right). \tag{8.31}$$

In the Gaussian beam case, it becomes

$$V(x, y)|_{\text{Gaussian}} \approx \exp(r^2/\omega^2)V(x, y)|_{\text{uniform beam}}. \tag{8.32}$$

A visibility gain of $\exp(r^2/\omega^2)$ is therefore achieved. Thus, in poor visibility cases, the Gaussian beam helps in gradually increasing the visibility when r increases. For small objects and/or large distances, when a large hologram aperture is used, the situation is therefore helpful.

From the recording point of view, $(r/\omega)_{\max}$ should be such that the exposure falls in a reasonably linear region of the recording medium response. In the case of the Agfa-Gevaert 10E75 emulsion, the response is practically linear in the exposure range of 10–30 erg/cm^2.[13] $(r/\omega)_{\max}$ can therefore be as high as $(0.5 \ln 3)^{1/2}$ or about 0.74 for the linear recording of the bias term $\exp(-2r^2/\omega^2)$.

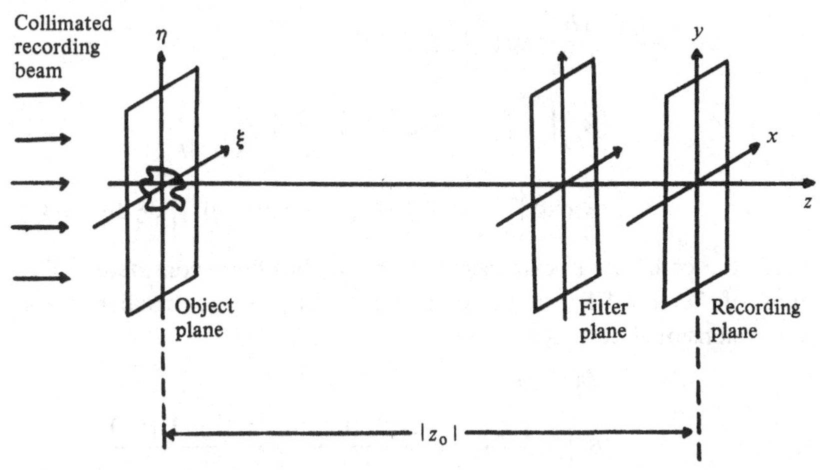

Fig. 8.7 In-line holographic recording with a detuned interference filter as a Fourier processor. The filter can be between the conventional object and recording planes. For simplicity of analysis, the filter is considered to be immediately before the recording plane in the text.

With this r/ω, the visibility modulation $\exp(r^2/\omega^2)$ is about 1.73. For some other emulsion where the maximum allowable r/ω is higher, the gain will be more. From Equations (8.14) and (8.32), the effect of $\exp(r^2/\omega^2)$ for two dimensional objects is the same as the magnification m_o. To conclude, the Gaussian beam nature of the laser light can be used to some advantage in the contrast enhancement problem.

8.4.3 *Interference filters as Fourier processors*

This technique, introduced by Molesini, Bertani and Cetica[14] employs a detuned interference filter between the diffracting object and the recording planes. A detuned interference filter means that it is designed for the peak transmission for a certain wavelength but is being used at a slightly different wavelength. The transmittance of such filters is spatial frequency dependent and hence a certain region of the high frequency fringes can be contrast enhanced. A schematic diagram of the recording arrangement with collimated beams is shown in Figure 8.7. Suppose the transfer function of the filter is $t(x/\lambda_o z_o, y/\lambda_o z_o)$. If the filter is immediately before the recording plane, the conventional recording amplitude of Equation (3.2) is modified to

$$\psi(x, y) = \frac{\mathrm{i}B}{\lambda_o z_o} \exp(-\mathrm{i}k_o z_o)$$
$$\times \iint_{-\infty}^{+\infty} [1 - A(\xi, \eta)] t\left(\frac{x}{\lambda_o|z_o|}, \frac{y}{\lambda_o|z_o|}\right)$$
$$\times \exp\left\{-\frac{\mathrm{i}k_o}{2z_o}[(x-\xi)^2 + (y-\eta)^2]\right\} \mathrm{d}\xi\,\mathrm{d}\eta, \quad (8.33)$$

where the collimated beam case ($z_R = -\infty$) has been considered. The steps of Section 4.1 can be repeated to obtain the irradiance for two-dimensional objects as

$$I(x, y) = |\psi(x, y)|^2$$
$$= B^2\left(|t(0, 0)|^2 - \frac{2t(0, 0)t(x/\lambda_o|z_o|, y/\lambda_o|z_o|)}{\lambda_o|z_o|}\right.$$
$$\times \left\{\sin\left(\frac{\pi r^2}{\lambda_o|z_o|}\right)\right.$$
$$\times \mathrm{Re}\left[\tilde{A}\left(\frac{x}{\lambda_o|z_o|}, \frac{y}{\lambda_o|z_o|}\right) t\left(\frac{x}{\lambda_o|z_o|}, \frac{y}{\lambda_o|z_o|}\right)\right]$$
$$+ \cos\left(\frac{\pi r^2}{\lambda_o|z_o|}\right)$$
$$\left.\times \mathrm{Im}\left[\tilde{A}\left(\frac{x}{\lambda_o|z_o|}, \frac{y}{\lambda_o|z_o|}\right) t\left(\frac{x}{\lambda_o|z_o|}, \frac{y}{\lambda_o|z_o|}\right)\right]\right\}$$
$$\left.+ \left(\frac{1}{\lambda_o z_o}\right)^2 \left|\tilde{A}\left(\frac{x}{\lambda_o|z_o|}, \frac{y}{\lambda_o|z_o|}\right) t\left(\frac{x}{\lambda_o|z_o|}, \frac{y}{\lambda_o|z_o|}\right)\right|^2\right). \quad (8.34)$$

The visibility factor like one given by Equation (8.5) then becomes

$$V(x, y) = \frac{\dfrac{2}{\lambda_o|z_o|}\left|t(0, 0)t\left(\dfrac{x}{\lambda_o|z_o|}, \dfrac{y}{\lambda_o|z_o|}\right)\tilde{A}\left(\dfrac{x}{\lambda_o|z_o|}, \dfrac{y}{\lambda_o|z_o|}\right)\right|}{|t(0, 0)|^2 + \left(\dfrac{1}{\lambda_o z_o}\right)^2\left|\tilde{A}\left(\dfrac{x}{\lambda_o|z_o|}, \dfrac{y}{\lambda_o|z_o|}\right) t\left(\dfrac{x}{\lambda_o|z_o|}, \dfrac{y}{\lambda_o|z_o|}\right)\right|^2}. \quad (8.35)$$

Let us consider the visibility in the center of the hologram. Substituting $x = y = 0$ in Equation (8.35), $V(x, y)$ becomes 1 for the classical situation given by Equation (8.5). Thus, the use of the Fourier processor does not alter the visibility in the central region of the hologram. At other hologram positions, the visibility will be altered depending on the nature of the function $t(x/\lambda_o|z_o|, y/\lambda_o|z_o|)$. As we know, the actual visibility function given by Equation (8.35) will vary and it will be zero at the zeros of $\widetilde{A}(x/\lambda_o|z_o|, y/\lambda_o|z_o|)$, i.e. the diffraction zeros. However, the envelope in the classical method (i.e. without the Fourier processor) is maximum at the center of the hologram or the diffraction pattern and then decreases as r increases. According to Equation (8.17), a circular cross-section object at a given N and a will have the envelope function

$$V(r')|_{\text{envelope}} \propto \left(\frac{r}{\lambda_o|z_o|}\right)^{-3/2}, \tag{8.36}$$

where $r = x^2 + y^2$. The envelope decreases rapidly as the frequency $r/\lambda_o|z_o|$ increases. Ultimately it becomes practically unrecordable, as shown, schematically in Figure 8.8(a). Figure 8.8(b) shows a schematic diagram of the transfer function of a detuned interference filter. Generally, a larger detuning means a larger value of $|F|$ and sharper peaks at $\pm F$. However, the actual $t(r/\lambda_o|z_o|)$ variation depends on a particular situation. Figure 8.8(c) is the envelope of the contrast function when the detuned interference filter is used. As represented, the contrast is enhanced around the frequencies $\pm F$. A large number of interference filters are commercially available. Their practical transfer curves $t(x/\lambda_o|z_o|, y/\lambda_o|z_o|)$ also have a large number of possibilities regarding the visibility enhancement. Filter coatings can be designed and coated on optical surfaces already in use in the system retaining the simplicity of the in-line approach.

8.4.4 *Non-blocked zero-frequency filtering*

The role of this spatial filtering has been discussed in detail by Özkul, Allano and Trinité.[15] The basic principle is illustrated in Figure 8.9. A lens of focal length f is used to relay the object's image. For simplicity, it is assumed that the object is on the optical axis at a distance $2f$ from the lens so that the image is also formed at a distance $2f$ on the other side of the lens. Illumination is provided by a spherical beam converging at the center of the lens. The center of the lens is covered by a filter coating with a real amplitude transmittance α. The extent of the coating is important. If the entire lens is covered, the entire pattern reaching the lens will be transmitted as such but with intensity reduced to α^2 times that without the filter. For partial filter coverage, the transmitted pattern becomes a complex function best described by numerical analysis.[15–17] Özkul *et al.*[15] have discussed several numerical situations for one-dimensional objects (long thin wire). We describe here the application of enhancing the fringe contrast when it is very low.

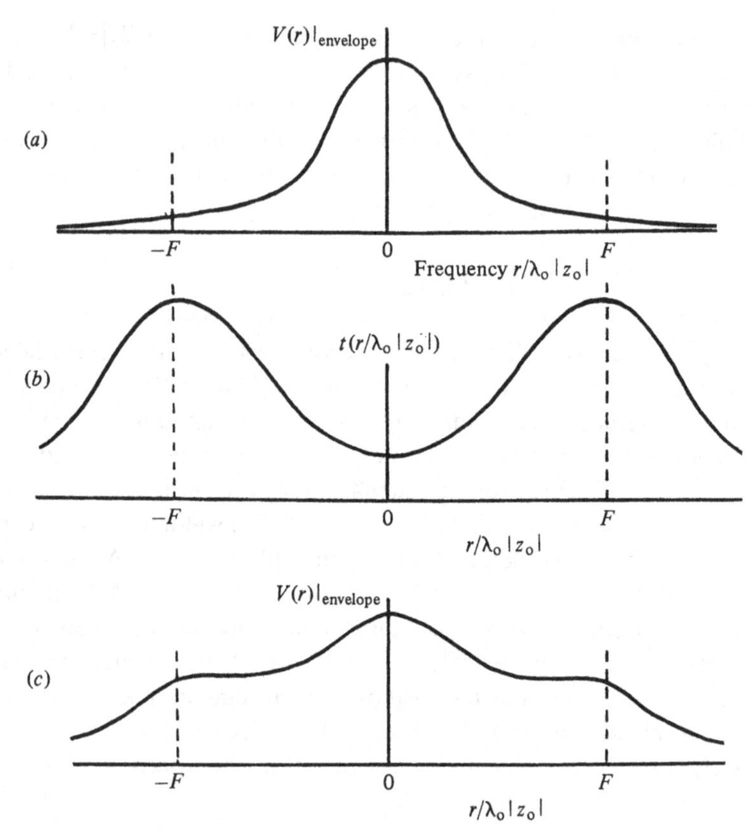

Fig. 8.8 Schematic diagram for showing the effect of the detuned interference filter on the contrast of the in-line Fraunhofer fringes. (*a*) Variation of the envelope of the fringe visibility in a classical hologram. (*b*) Transfer function of a detuned interference filter with peaks at frequencies $\pm F$. (*c*) The envelope of the contrast function when the detuned interference filter is used. The contrast enhancement around the frequencies $\pm F$ is represented.

Thus, let us assume that the object–lens distance is such that the lens aperture contains several side lobes of the diffraction pattern. Also, the focal spot diameter of the converging beam at the lens aperture is small as compared to the central diffraction pattern due to the microobject. If the aperture of the filter is slightly more than the spot (converging beam) diameter, it will practically cover the entire reference or background beam. It is worth mentioning here that the aperture-limited diffraction spot diameter of the converged beam is about $1.22\lambda_0 \cot\theta$. Also, as discussed in Section 4.2, the number m of the side lobes passing through the lens is $(ca/\lambda_0 f) - 1$.

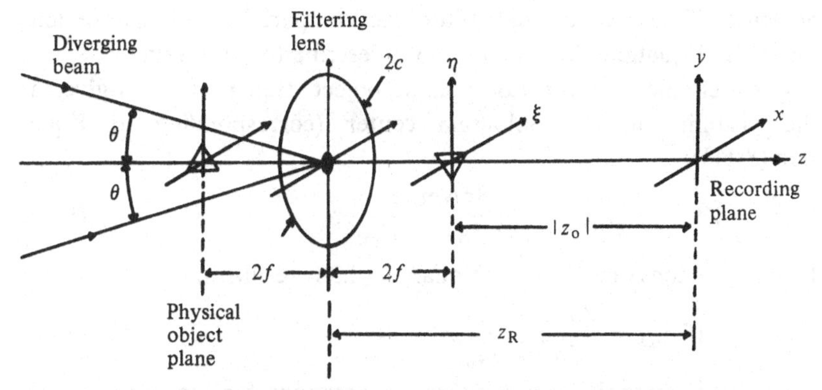

Fig. 8.9 Diagram illustrating the non-blocked zero frequency filtering technique for hologram fringe contrast enhancement. The filtering aperture size is assumed to be small compared to the entire lens aperture or the Fraunhofer diffraction pattern size due to the microobject. The filtering aperture is large compared to the diverging beam spot size.

With these assumptions, the amplitude transmittance α of the filter will control the background intensity without practically altering the diffraction by the microobject. For this situation, Equation (4.14) for the normalized irradiance distribution at the recording plane can be written as

$$I(x, y) \approx \alpha^2 - \frac{2m_o\alpha}{\lambda_o|z_o|}\sin\left[\frac{\pi r^2}{\lambda_o m_o|z_o|} + \phi\left(\frac{x}{\lambda_o|z_o|}, \frac{y}{\lambda_o|z_o|}\right)\right] \times \tilde{A}_o\left(\frac{x}{\lambda_o|z_o|}, \frac{y}{\lambda_o|z_o|}\right) + \left(\frac{m_o}{\lambda_o z_o}\right)^2 \tilde{A}_o^2\left(\frac{x}{\lambda_o|z_o|}, \frac{y}{\lambda_o|z_o|}\right). \tag{8.37}$$

For the one-dimensional case, the normalized irradiance distribution given by Equation (4.16) becomes

$$I(x) \approx \alpha^2 - 2\alpha\left(\frac{m_o}{\lambda_o|z_o|}\right)^{1/2}\cos\left[\frac{\pi x^2}{m_o\lambda_o|z_o|} + \phi\left(\frac{x}{\lambda_o|z_o|}\right) - \frac{\pi}{4}\right] \times A_o\left(\frac{x}{\lambda_o|z_o|}\right) + \frac{m_o}{\lambda_o|z_o|}\tilde{A}_o^2\left(\frac{x}{\lambda_o|z_o|}\right). \tag{8.38}$$

The finite aperture of the filter affects the central part of the diffraction pattern.[15] However, with the assumptions in this section, the effect will be minimum. Also, our aim here is to increase the effective hologram aperture. That situation deals with increasing the visibility of the outer side (higher side lobes) of the hologram aperture. Thus, for a small filter size compared to the entire lens aperture, Equations (8.37) and (8.38) describe the situation.

For a circular cross-section opaque object, Equation (8.37) gives as the visibility in the hologram center (corresponding to Equation (8.11)):

$$V(0)|_{\text{2D circular}} = \frac{8\pi N m_o \alpha}{16N^2\alpha^2 + \pi^2 m_o^2}. \tag{8.39}$$

For one-dimensional objects, Equation (8.38) results in

$$V(0)|_{\text{1D}} = \frac{2m_o^{1/2} N^{1/2} \alpha}{N\alpha^2 + m_o}. \tag{8.40}$$

At the outer hologram positions, expressions for the visibility can simply be derived from Equations (8.37) and (8.38) to get expressions like those given by Equations (8.5) and (8.6) respectively. In the limiting case of small $\tilde{A}$, the envelope function of the visibility is nothing but that given by Equations (8.17) and (8.17) respectively divided by α. In fact, Equations (8.39) and (8.40) give the same conclusion in the limiting case when $16N^2\alpha^2 \geqslant \pi^2 m_o^2$ and $N\alpha^2 \geqslant m_o$ respectively.

With the simple model presented, the contrast of the interference fringes can be increased by $1/\alpha$ as compared to that in the case without the filter ($\alpha = 1$). The magnification m_o from the particular recording configuration must be considered for a comparison with the common collimated recording beam case. For the common in-line Fraunhofer holography $m_o = 1$. In the present system the divergent beam arrangement is virtually provided yielding $m_o > 1$. Thus, a cumulative effect of $\alpha < 1$ and $m_o > 1$ is obtained.

Özkul *et al.*[15] found interesting aspects of the filtering technique like better image edge resolution, improved image focus determination, image contrast, etc.

9

Non-image plane analysis

The general procedure of the analysis is to bring the reconstructed image in focus at the final observation plane. This mode of analysis is very common. However, particularly in the in-line method, there are proposals to perform the analysis at other planes. These are: the recording plane (hologram) itself, the misfocused image plane, and certain transform (Fourier) planes. These approaches have advantages in particular situations. In this chapter, these techniques of non-image plane analysis are discussed. Non-image plane analysis techniques have applications in velocimetry also. Those methods are dicussed in Chapter 10.

9.1 Analysis of the far-field pattern

The pattern at the recording plane of the Fraunhofer hologram itself contains information about the object and its location. The possibility of extracting the desired information from the pattern is as old as Fraunhofer holography itself. Thompson and coworkers[1–5] proposed utilizing the pattern itself for the analysis at the early stages of the development of far-field holography. There are two approaches. One is to study the entire complex pattern to determine the size and the distance of the microobject. This is called the coherent background method due to the presence of the reference beam. The other method, called the spatial filtering method, yields the broad diffraction pattern for further analysis.

9.1.1 *Coherent background method*

The principle of the technique can be described from a typical Fraunhofer diffraction pattern. The schematic diagram is

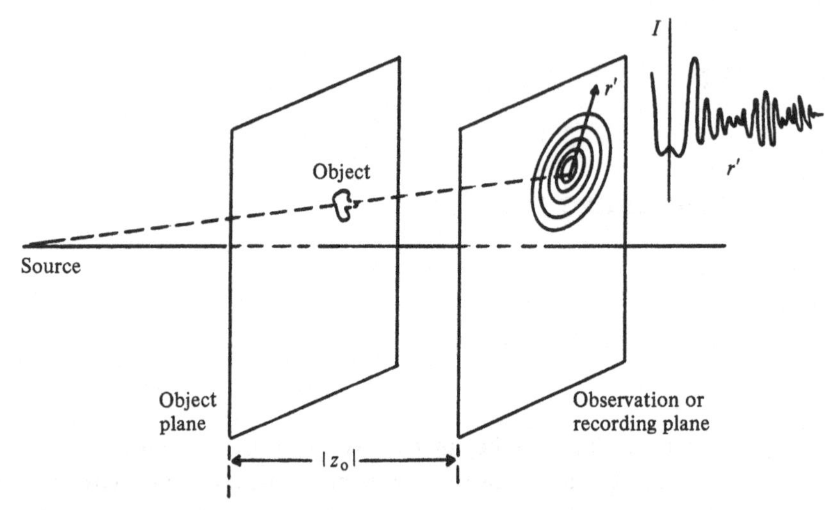

Fig. 9.1 Schematic diagram to illustrate the observation of the diffraction pattern of the object at distance z_o.

shown in Figure 9.1. The normalized irradiance distribution at the recording plane for a two-dimensional object cross-section (see Section 4.1) is

$$I = 1 - \frac{2m_o}{\lambda_o|z_o|}\sin\left[\frac{\pi r'^2}{\lambda_o m_o|z_o|} + \phi\left(\frac{x'}{\lambda_o|z_o|}, \frac{y'}{\lambda_o|z_o|}\right)\right]$$
$$\times \tilde{A}_o\left(\frac{x'}{\lambda_o|z_o|}, \frac{y'}{\lambda_o|z_o|}\right) + \left(\frac{m_o}{\lambda_o|z_o|}\right)^2 \tilde{A}_o^2\left(\frac{x'}{\lambda_o|z_o|}, \frac{y'}{\lambda_o|z_o|}\right). \quad (9.1)$$

As seen in Figure 4.2, typically fine sine fringes are modulated by broad diffraction fringes governed by $\tilde{A}_o(x'/\lambda_o|z_o|, y'/\lambda_o|z_o|)$. Information about the distance (z_o) and the object $[\tilde{A}_o(x'/\lambda_o|z_o|, y'/\lambda_o|z_o|)$ and $\phi(x'/\lambda_o|z_o|, y'/\lambda_o|z_o|)]$ is present. For object distribution with a real Fourier transform, or where $\phi(x'/\lambda_o|z_o|, y'/\lambda_o|z_o|)$ is zero or known, the analysis becomes easy. The fine fringes due to the sine term will yield the object distance z_o and then the broad modulation fringes will give the object size. To illustrate further, let us consider opaque objects with circular cross-section. The pattern then becomes (see Section 4.1).

$$I = 1 - \frac{2\pi m_o a^2}{\lambda_o|z_o|}\sin\left(\frac{\pi r'^2}{\lambda_o m_o|z_o|}\right)\left[\frac{2J_1(2\pi a r'/\lambda_o|z_o|)}{2\pi a r'/\lambda_o|z_o|}\right]$$
$$+ \left(\frac{\pi m_o a^2}{\lambda_o z_o}\right)^2\left[\frac{2J_1(2\pi a r'/\lambda_o|z_o|)}{2\pi a r'/\lambda_o|z_o|}\right]. \quad (9.2)$$

The distribution of Equation (9.2) as well as experimental results

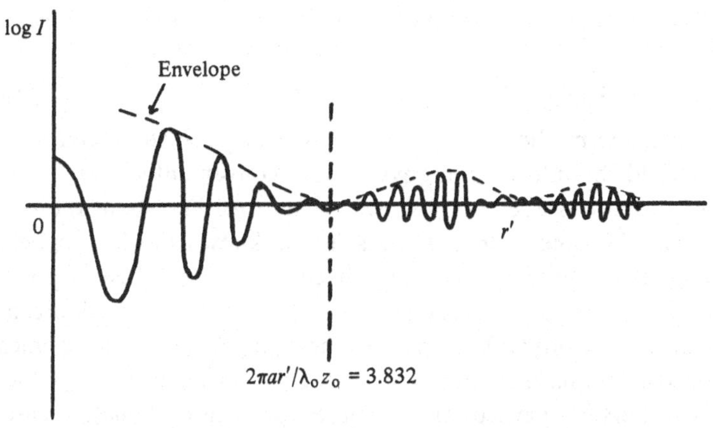

Fig. 9.2 Schematic diagram (not to scale) illustrating the variation of $\log I(r')$ against r'. The dashed vertical line corresponds to the first zero of the Bessel function.

have been plotted for several numerical situations.[1-4] More important is that remarkably good agreement is found between the theoretical curves and microdensitometer traces of photographic records. A typical variation can be schematically represented by Figure 9.2. As seen, the fine interference fringes are modulated by the $2J_1(2\pi ar'/\lambda_o|z_o|)/(2\pi ar'/\lambda_o|z_o|)$ function. In the modulated pattern, the zero ($I = 1$) occurs at the zeros $j_{1,s}$ of the Bessel function, i.e. at $2\pi ar'/\lambda_o|z_o| = 3.832$, 7.016, 10.173, 13.324, etc.

The fine fringes given by the term $\sin(\pi r'^2/\lambda_o m_o|z_o|)$ give the object distance z_o regardless of the object size. Once z_o is so determined, the zeros of the modulated pattern give the object radius 'a' according to

$$a = \lambda_o z_o j_{1,s}/2\pi r', \tag{9.3}$$

where r' is the radius of the ring at the sth zero.

For line objects, the normalized irradiance distribution (see Section 4.1) is

$$I = 1 - 4a\left(\frac{m_o}{\lambda_o|z_o|}\right)^{1/2}\cos\left(\frac{\pi x'^2}{m_o\lambda_o|z_o|} - \frac{\pi}{4}\right)\left[\frac{\sin(2\pi ax'/\lambda_o|z_o|)}{2\pi ax'/\lambda_o|z_o|}\right] + \frac{4m_o a^2}{\lambda_o|z_o|}\left[\frac{\sin(2\pi ax'/\lambda_o|z_o|)}{2\pi ax'/\lambda_o|z_o|}\right]^2. \tag{9.4}$$

In this case, the fine fringes are defined by the cosine term and can be used for determining z_o. Once z_o is so determined, the object radius 'a' can be obtained from the zeros ($I = 1$) of the modulation.

For the sth minima position, occurring at $2\pi ax'/\lambda_o|z_o| = s\pi$, we obtain:

$$a = s\lambda_o|z_o|/2x'. \tag{9.5}$$

As such, the method is not very popular because the reconstructed images yield the information anyway very conveniently. The complex diffraction pattern does not give simple information about the object shape. On the other hand, if the shape is known (such as spherical bubbles), the diffraction pattern can easily give the rest – position and size. In such special situations, the photographic emulsion can be replaced by a photodiode array and the pattern can be electronically processed in a quasireal time manner. This has been done by Özkul and coworkers,[6–8] particularly in connection with automatic measurement of the diameter and position of glass fibers.[7,8] Tsuno and Takahashi[9] used the method for size and position measurements of a wire.

Yonemura[10] proposed the use of another collimated coherent beam incident at the angle at the observation plane. The straight line interference fringes between this beam and the object illumination beam are observed. The intensities between such patterns with and without microobjects can be electronically stored and processed. The difference can yield the size and position of particles. The subtraction suppresses noise and the relatively more difficult data analysis of the conventional photographic recording method can be avoided.

We have discussed the far-field case. However, the irradiance distributions in Fresnel diffraction situations are available in literature.[11–13] Fraunhofer diffraction patterns due to tilted planer objects[14,15] and volumetric bodies[16] have also been described. Several related developments are described in Section 9.4.

9.1.2 *Spatial filtering method*

In this approach,[1,4,5] the coherent background is eliminated by spatial filtering. Thus, a pure diffraction pattern by the microobject is observed. For the filtering, a relay lens arrangement (see Section 5.2.3) can be used with an opaque filter at the common focus. The telescopic system is convenient because it retains constant magnification. If two lenses have a common focal length f, then the magnification is unity. As seen in Figure 9.3, if the object is at a distance u from the first lens, then the image distance v from the

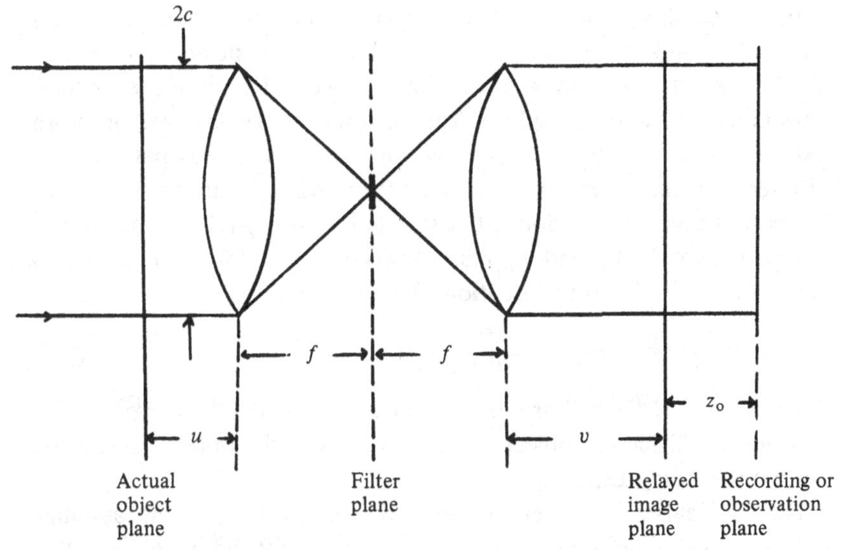

Fig. 9.3 Diagram representing the spatial filtering method of observing the pure diffraction pattern due to the microobject.

second lens will be (see Section 5.2.3)

$$v = 2f - u. \tag{9.6}$$

From this object (relayed image) plane, suppose the observation or the recording plane is at a distance z_o as shown. As such, the situation is like recording a hologram of the object at distance z_o. Now, suppose an opaque filter at the filter plane is introduced to cut off the background light. Then the amplitude of the transmitted field is proportional to the Fourier transform (see Section 4.1). The normalized irradiance I at the observation plane is therefore

$$I = \begin{cases} \tilde{A}_o^2\left(\dfrac{x}{\lambda_o|z_o|}, \dfrac{y}{\lambda_o|z_o|}\right) & \text{2D objects} \\ \tilde{A}_o^2\left(\dfrac{x}{\lambda_o|z_o|}\right) & \text{1D objects} \end{cases} \tag{9.7}$$

For an object with circular cross-section:

$$I = \left[\frac{2J_1(2\pi ar'/\lambda_o|z_o|)}{2\pi ar'/\lambda_o|z_o|}\right]^2. \tag{9.8}$$

This represents the pattern in a dark background with a bright ($I = 1$) central maxima. The analysis can be performed using Equation (9.3). Since the pattern contains combined information about a

and z_o, one should be known for the analysis of the other. The mean value of z_o in a particular experimental arrangement and a thin slice of the object space at a time can be used. For a more precise determination, a beam splitter can be used to record two or more patterns simultaneously.[17] As shown in Figure 9.4, two patterns for different distances $z_o^{(1)} = \overline{ABC}$ and $z_o^{(2)} = \overline{AB'C'}$ can be recorded. Suppose a given diffraction order (a fixed s or $j_{1,s}$) is observed to get the pattern radii r_1 and r_2 respectively corresponding to the paths ABC and $AB'C'$. From Equation (9.8), we have:

$$\frac{r_1}{|z_o^{(1)}|} = \frac{r_2}{|z_o^{(2)}|} = \frac{r_2 - r_1}{|z_o^{(2)}| - |z_o^{(1)}|}. \tag{9.9}$$

Since r_1, r_2, $z_o^{(2)} - z_o^{(1)}$ are experimentally known, $z_o^{(1)}$ or $z_o^{(2)}$ can be determined. Once the object distance is known, the object size can be determined using Equation (9.3).

Ref. 17 deals with observing an unfiltered pattern or the assumption that the fine interference fringes may not be resolved. The situation yields the pattern of Equation (9.7) but with a uniform background. This background will reduce the contrast of the pattern. In these respects, the optical filtering of the background[4,5] as discussed in this section is better.

The arrangement illustrated in Figure 9.3 can be altered. Silverman, Thompson and Ward[4] used a mirror at the filter plane to reflect the entire pattern. A pinhole at the mirror then removes the background field.

The size of the filter is important. The beam cross-section of radius c is focused at the distance f. The diffraction spot diameter at the filter plane is (see Section 5.6.2) about $1.22\lambda_o f/c$. The filter diameter should be several times this diameter effectively to cut off the background. In the arrangement of Silverman *et al.*[4], the filter diameter is about 27 times the diffraction spot diameter.

9.2 Analysis by misfocusing

During the reconstruction, suppose a longitudinally different plane (misfocusing) is observed. Rather than a sharp image, a diffraction like pattern is observed. The pattern contains information about the object. There are two main advantages of this kind of analysis. First, the size of the diffraction pattern is large compared to the focused image size. Then, for smaller images, the measurement accuracy can be enhanced. Secondly, in an automated readout

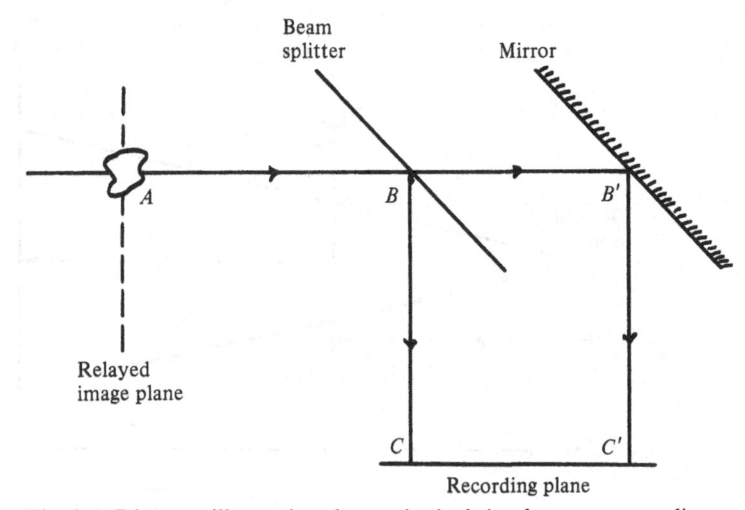

Fig. 9.4 Diagram illustrating the method of simultaneous recording of the diffraction pattern at two or more distances z_o. In the figure, the two paths are ABC and $AB'C'$. The spatial filter arrangement can be inserted between A and B.

system, there is no need to reach the exact image plane. This will reduce the data storage and processing capability requirements. In this section we discuss the form of the pattern at misfocused planes and its role in the analysis.

9.2.1 *The pattern form*

This kind of the non-image plane analysis of the reconstruction has been discussed by Vikram and Billet.[18,19] A typical recording and reconstruction arrangement is shown in Figure 9.5. Figure 9.5(a) represents the recording. The object distribution described by $A(\xi, \eta)$ is illuminated by a light source of wavelength λ_o and the recording is performed at the (x, y) plane. Figure 9.5(b) shows the reconstruction arrangement with wavelength λ_c. The conventional image plane is shown by the dashed line. Our aim here is to study the pattern at a plane a distance Δz away from the conventional image plane. Under the far-field approximation, the reconstructed field at the distance z from the hologram is known (see Section 3.2):

$$\psi(\mu, \nu) \approx -\frac{\mathrm{i}CB'}{\lambda_c z}\exp(\mathrm{i}k_c z)(I_1 + I_2), \tag{9.10}$$

where the object is assumed to be many far-fields away so that the integrals I_3 and I_4 can be neglected. Otherwise the role of those

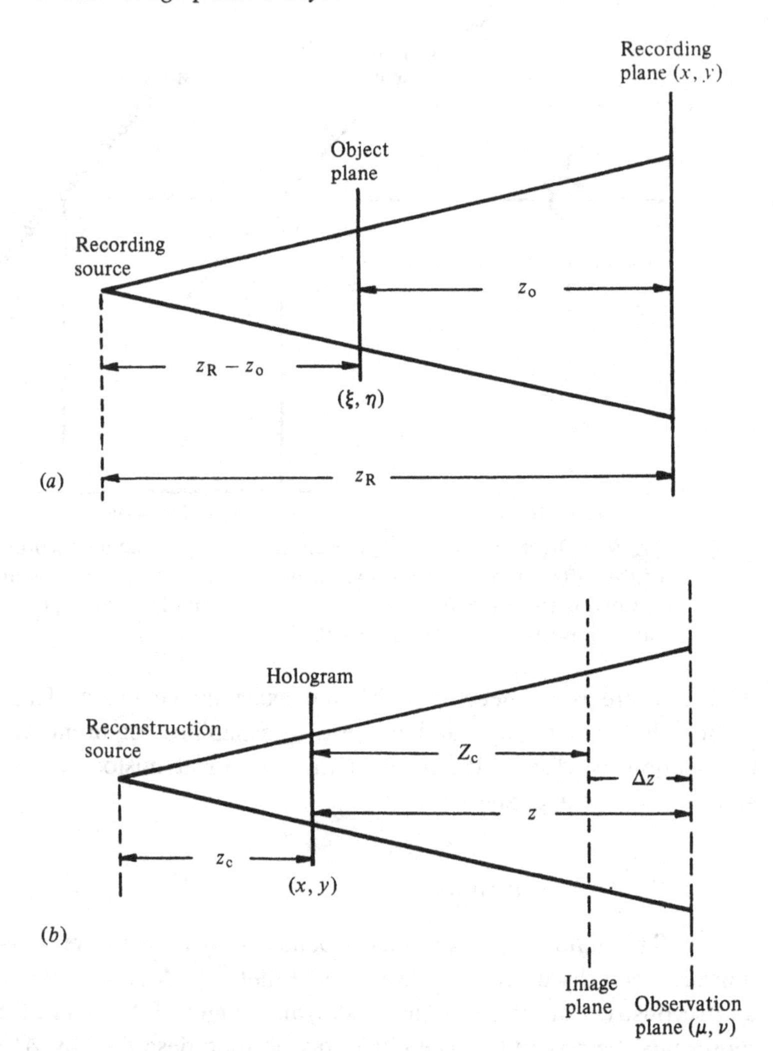

Fig. 9.5 Schematic diagram for recording an in-line far-field hologram and reconstruction at a slightly misfocused plane. (*a*) The recording arrangement with wavelength λ_o. (*b*) The reconstruction with a wavelength λ_c; Δz *is* the amount of misfocusing. (Ref. 19.)

integrals must also be included. I_1 and I_2 are known to be (for the conjugate image)

$$I_1 = \iint_{-\infty}^{+\infty} \exp\left[-\frac{ik_c(x^2 + y^2)}{2z_c}\right] \exp\left\{\frac{ik_c}{2z}\left[(\mu - x)^2 + (\nu - y)^2\right]\right\} dx\, dy \tag{9.11}$$

and

$$I_2 = -\frac{\mathrm{i}\Gamma m_o}{\lambda_o z_o} \iiiint_{-\infty}^{+\infty} A^*(\xi, \eta)$$
$$\times \exp\left\{\frac{\mathrm{i}k_o}{2z_o}[(x-y)^2 + (y-\eta)^2]\right\}$$
$$\times \exp\left[-\frac{\mathrm{i}k_o(x^2+y^2)}{2z_R}\right]\exp\left[\frac{\mathrm{i}k_o(\xi^2+\eta^2)}{2(z_R - z_o)}\right]$$
$$\times \exp\left\{\frac{\mathrm{i}k_c}{2z}[(\mu - x)^2 + (\nu - y)^2]\right\}$$
$$\times \exp\left[-\frac{\mathrm{i}k_c}{2z_c}(x^2+y^2)\right]\mathrm{d}x\,\mathrm{d}y\,\mathrm{d}\xi\,\mathrm{d}\eta. \qquad (9.12)$$

In Chapter 3, the particular case of $z = Z_c$ was considered to get I_1 as the background and I_2 as the focused image. Solution of I_1 is straightforward:

$$I_1 = -\frac{\mathrm{i}\lambda_c z z_c}{z - z_c}\exp\left[\frac{\mathrm{i}k_c R^2}{2(z - z_c)}\right], \qquad (9.13)$$

where $R^2 = \mu^2 + \nu^2$. To solve integral I_2, let us first integrate over x and y. The components can be written in the form of sines and cosines of the square of a function whose integers are known. The method of stationary phase can also be used. The result is

$$I_2 = -\frac{\Gamma m_o \lambda_c z Z_c}{\lambda_o z_o (z - Z_c)} \iint_{-\infty}^{+\infty} A^*(\xi, \eta)$$
$$\times \exp\left\{\frac{\mathrm{i}k_c[(\mu - M_c\xi)^2 + (\nu - M_c\eta)^2]}{2(z - Z_c)}\right\}$$
$$\times \exp\left[-\frac{\mathrm{i}k_c(M_c - M_c^2)(\xi^2 + \eta^2)}{2Z_c}\right]$$
$$\times \exp\left[\frac{\mathrm{i}k_o(\xi^2+\eta^2)}{2(z_R - z_o)}\right]\mathrm{d}\xi\,\mathrm{d}\eta. \qquad (9.14)$$

Equation (9.14) is not valid for the actual image plane, i.e. $z = Z_c$. Now, let us denote

$$\Delta z = z - Z_c, \qquad (9.15)$$

$$P = \frac{2M_c(M_c\xi_o - \mu) + 2\Delta z(M_c^2 - M_c)\xi_o/Z_c + 2n\Delta z\xi_o/(z_R - z_o)}{2\lambda_c \Delta z} \qquad (9.16)$$

and

$$Q = \frac{2M_c(M_c\eta_o - \nu) + 2\Delta z(M_c^2 - M_c)\eta_o/Z_c + 2n\Delta z\eta_o/(z_R - z_o)}{2\lambda_c\Delta z} \tag{9.17}$$

Also, let us have coordinate transformations centered at the object (ξ_o, η_o) and the image $(M_c\xi_o, M_c\eta_o)$ points as:

$$\left.\begin{aligned} \xi' &= \xi - \xi_o, \\ \eta' &= \eta - \eta_o, \\ \mu' &= \mu - M_c\xi_o, \\ \nu' &= \nu - M_c\eta_o. \end{aligned}\right\} \tag{9.18}$$

Now, the far-field conditions for ξ' and η' against the distances $|\Delta z|$, $|z_R - z_o|$ and Z_c result in I_2 given by Equation (9.14) as

$$I_2 = -\frac{\Gamma m_o\lambda_c z Z_c}{\lambda_o z_o\Delta z}\exp\left\{\frac{ik_c}{2\Delta z}\left[(\mu'^2 + \nu'^2) - \frac{(M_c - M_c^2)(\xi_o^2 + \eta_o^2)\Delta z}{Z_c} + \frac{n(\xi_o + \eta_o)\Delta z}{z_R - z_o}\right]\right\}\widetilde{A}^*(P, Q), \tag{9.19}$$

where $A^*(P, Q)$ is the Fourier transform

$$\widetilde{A}^*(P, Q) = \iint_{-\infty}^{+\infty} A^*(\xi', \eta')\exp[-2\pi i(\xi' P + \eta' Q)]\,d\xi' d\eta'. \tag{9.20}$$

For small Δz compared to Z_c and $|z_o - z_R|$, and when the object points near the optical axis so that ξ_o and η_o are not very large, Equations (9.16) and (9.17) become

$$P \approx -\frac{M_c\mu'}{\lambda_c\Delta z} \tag{9.21}$$

$$Q \approx -\frac{M_c\nu'}{\lambda_c\Delta z} \tag{9.22}$$

respectively. Equation (9.19) then becomes:

$$I_2 \approx -\frac{\Gamma m_o\lambda_c z Z_c}{\lambda_o z_o\Delta z}\exp\left[\frac{ik_c R'^2}{2\Delta z} - \phi(P, Q)\right]\widetilde{A}_o^*(P, Q), \tag{9.23}$$

where $R'^2 = \mu'^2 + \nu'^2$; $\widetilde{A}_o(P, Q)$ is the real amplitude and $\phi(P, Q)$ is the phase of the Fourier transform $\widetilde{A}(P, Q)$. Equations (9.10), (9.13) and (9.23) can now be combined to determine the recon-structured irradiance as

$$\begin{aligned} I &= |\psi(\mu, \nu)|^2 \\ &\propto \left(\frac{z_c}{Z_c - z_c}\right)^2 + \left(\frac{\Gamma m_o Z_c}{\lambda_o z_o \Delta z}\right)^2 \tilde{A}_o^2\left(-\frac{M_c \mu'}{\lambda_c \Delta z}, -\frac{M_c \nu'}{\lambda_c \Delta z}\right) \\ &+ 2\left(\frac{z_c}{Z_c - z_c}\right)\left(\frac{\Gamma m_o Z_c}{\lambda_o z_o \Delta z}\right) \sin\left(\frac{k_c R'^2}{2\Delta z} - \phi\right) \\ &\times \tilde{A}_o\left(-\frac{M_c \mu'}{\lambda_c \Delta z}, -\frac{M_c \nu'}{\lambda_c \Delta z}\right). \end{aligned} \quad (9.24)$$

The general behaviour of the function is like a complex diffraction pattern as discussed in Section 9.1. The broad diffraction pattern modulates the high frequency fringes given by the sine function. The envelope minima occur at the zeros of $\tilde{A}_o(P, Q)$. However, if the observation system cannot resolve the fine fringes, then the pattern is $\tilde{A}_o^2(P, Q)$ with a background. All these aspects can be discussed more easily in physical terms for collimated recording and reconstruction beams and objects with circular cross-section.

9.2.2 *Collimated beams and circular objects*

For collimated recording and reconstruction beams, $z_c = -\infty$, $m_o = 1$ and $M_c = 1$. Equation (9.24) then becomes

$$\begin{aligned} I \propto 1 &+ \left(\frac{\Gamma Z_c}{\lambda_o z_o \Delta z}\right)^2 \tilde{A}_o^2\left(-\frac{\mu}{\lambda_c \Delta z}, -\frac{\nu}{\lambda_c \Delta z}\right) - \frac{2\Gamma Z_c}{\lambda_o z_o \Delta z} \\ &\times \sin\left(\frac{k_c R^2}{2\Delta z} - \phi\right) \tilde{A}_o\left(-\frac{\mu'}{\lambda_c \Delta z}, -\frac{\nu'}{\lambda_c \Delta z}\right). \end{aligned} \quad (9.25)$$

In the case of a circular cross-section opaque object of radius a, the circularly symmetric Fourier transform $\tilde{A}_o(-\mu/\lambda_c\Delta z, -\nu/\lambda_c\Delta z)$ is $\pi a^2[2J_1(2\pi aR/\lambda_c\Delta z)/(2\pi aR/\lambda_c\Delta z)]$. For this case, Equation (9.25) becomes

$$\begin{aligned} I \propto 1 &+ \left(\frac{\Gamma Z_c}{\lambda_o z_o \Delta z}\right)^2 (\pi a^2)^2 \left[\frac{2J_1(2\pi aR/\lambda_c\Delta z)}{2\pi aR/\lambda_c\Delta z}\right]^2 \\ &- \frac{2\Gamma Z_c}{\lambda_o z_o \Delta z} (\pi a^2) \left[\frac{2J_1(2\pi aR/\lambda_c\Delta z)}{2\pi aR/\lambda_c\Delta z}\right] \sin\left(\frac{k_c R^2}{2\Delta z}\right)^2. \end{aligned} \quad (9.26)$$

As the number $N = \lambda_c\Delta z/(2a)^2$ of far-fields increases, the number of fine fringes within broad diffraction rings also increases. In an observation system, the envelope zeros occurring at $2\pi aR/\lambda_c\Delta z =$ 3.832, 7.016, 10.173, etc. (as discussed in Section 9.1) can be used for the analysis. If, for the observation system, the spatial frequency of $\sin(k_c R^2/2\Delta z)$ is too fine to resolve, then the intensity averaged over an elementary resoluble area will be seen. In that case, the contribution due to the sine term will be zero, i.e.

$$I \propto 1 + \left(\frac{\Gamma Z_c}{\lambda_o z_o \Delta z}\right)^2 (\pi a^2)^2 \left[\frac{2J_1(2\pi aR/\lambda_c\Delta z)}{2\pi aR/\lambda_c\Delta z}\right]^2 \qquad (9.27)$$

In any case, several interesting possibilities exist. With proper resolution capability, Equation (9.26) can yield the position and size of the microobject in the same way as discussed in Section 9.1.1. Spatial filtering can also be performed (Section 9.1.2) to see only the broad diffraction pattern. During the reconstruction, the contrast can also be enhanced (see Chapter 8).

9.2.3 *Role in accurate measurements*

The size of the diffraction pattern is large compared to the size of the microobject. To illustrate this, let us consider the first diffraction zero in case of the circular cross-section object. This occurs at

$$2\pi aR/\lambda_c\Delta z = 3.832. \qquad (9.28)$$

Writing $\Delta z = (2a)^2 N/\lambda_c$ where N is the far-field number corresponding to the image–observation plane separation, Equation (9.28) gives the radius R of the diffraction ring as

$$R \approx 2.5aN. \qquad (9.29)$$

With N greater than unity, R even at the first diffraction zero ring is large as compared to a. For the higher order diffraction rings, R is still larger. Therefore, with the same absolute error in the measurement of the dimension on the observation screen (such as a television monitor), the measurement accuracy at the non-image planes is much better.[18,20]

The measurement accuracy in the longitudinal position Z_c is also improved.[20] At the focus, the image size $R = a$ is a minimum resulting in

$$\mathrm{d}R/\mathrm{d}z = 0. \qquad (9.30)$$

At non-image planes $|\mathrm{d}R/\mathrm{d}z|$, i.e. $|j_{1,s}\lambda_c/(2\pi a)|$, is greater than zero. Errors $\varepsilon(R)$ in R and $\varepsilon(z)$ in z are thus related by[20]

$$\varepsilon(z) \approx \frac{\varepsilon(R)}{|\mathrm{d}R/\mathrm{d}z|}. \qquad (9.31)$$

Thus, two out-of-focus pattern diameters ($2R$) at two different values of Δz can be measured for more accurate position determination of the microimage.

In Figure 9.6, a typical demonstration is presented. The object is a spherical polystyrene particle of 20 μm diameter on a transparent glass slide. A HeNe laser ($\lambda_o = \lambda_c = 0.6328$ μm) is used to record the hologram on an Agfa 10E75 plate as well as to reconstruct the image. The distance z_o between the object and the recording plane is 6 cm.

Figure 9.6(*a*) shows the reconstruction of the image plane on a television monitor screen with net magnification of 100. The image diameter is about 2 mm as expected. Figure 9.6(*b*) shows the pattern observed at $\Delta z = 900$ μm. The second minimum ring diameter ($2R$) is 128 μm. Since $j_{1,2}$ is 7.01559, the calculated particle diameter is 19.87 μm. Stanton, Caulfield and Stewart[20] have also presented an experimental demonstration with the image of a liquid droplet. Thus, at lower net magnifications and/or for smaller object sizes, the out-of-focus measurements help improve the accuracy. For example, a 10 μm diameter particle at a net magnification of 100 will appear as 1 mm diameter image on the monitor screen. This can be only a few times the television system resolution capability resulting in unclear image boundaries. Out-of-focus analysis eliminates this problem.

9.2.4 *Role in automated analysis*

The common approach in automated hologram reconstruction analysis is to digitize (x, y) planes in the image volume for many z-coordinates. The proper z plane, i.e. the image plane is determined from the image brightness, edge gradients, etc. This exceptionally simple approach is difficult to implement due to severe demands on the computer storage capability and processing time. A typical 10^4 z-steps at many $x-y$ positions for each step and $(525)^2$ points per hologram position are to be digitized and processed.[20] As proposed by Stanton *et al.*[20] and Caulfield[21], the need to search all depth planes can be eliminated by the out-of-focus method. Only a limited number of z-steps can be sufficient. In a given region, the data from two z-steps can be used for further processing. In the spherical particle model, the situation is shown schematically in Figure 9.7. A given diffraction ring order (s) is considered. Corresponding to the defocusings $(\Delta z)_1$ and $(\Delta z)_2$ suppose the ring radii are R_1 and R_2 respectively. From the behaviour of the pattern in the case of circular

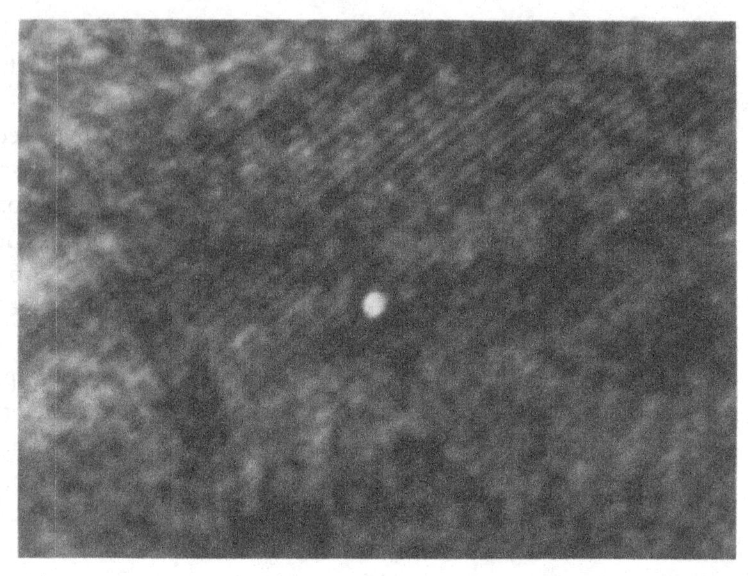

(*a*)

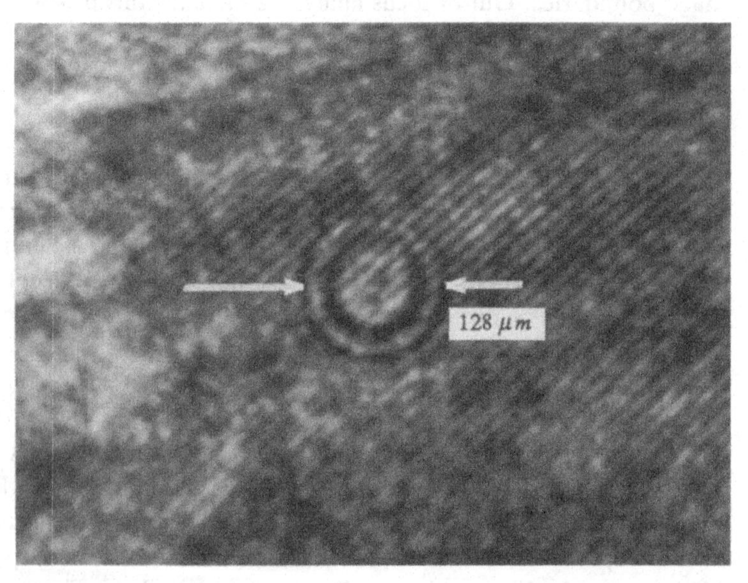

(*b*)

Fig. 9.6 (*a*) Reconstruction of a 20 μm polystyrene sphere at the image plane for $\lambda_o = \lambda_c = 0.6328$ μm, $z_o = 6$ cm. (*b*) Reconstruction at the non-image plane for $\Delta z = 900$ μm. (Ref. 18.)

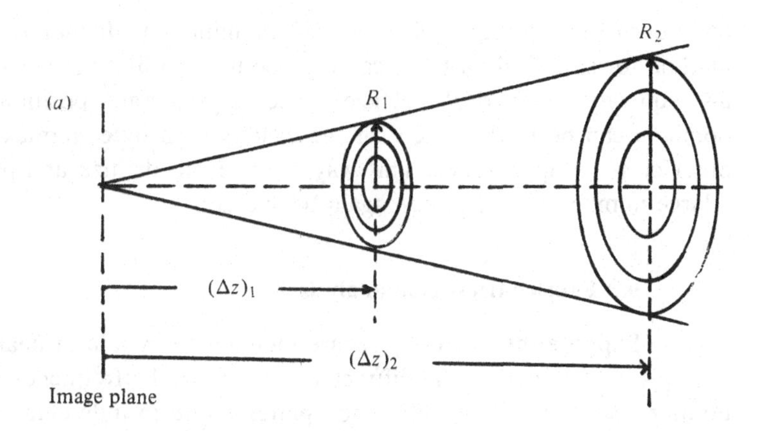

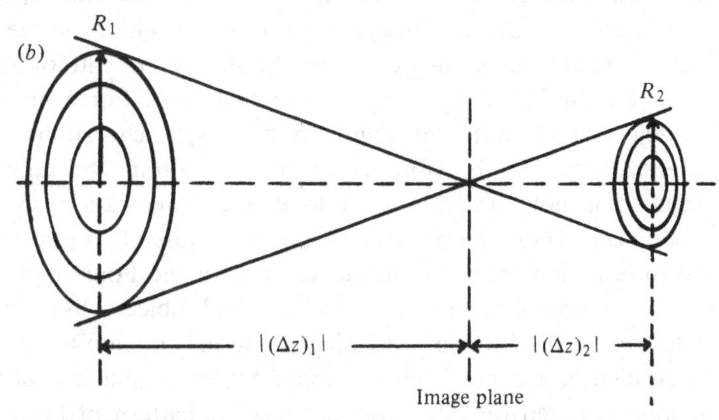

Fig. 9.7 Diagram illustrating the determination of the focus plane position by observing two out-of-focus patterns. (*a*) The situation of observing the same side of the focus plane and (*b*) the situation when the defocused planes are on different sides of the image plane.

cross-section objects,

$$\frac{R_1}{|(\Delta z)_1|} = \frac{R_2}{|(\Delta z)_2|} = \frac{R_2 - R_1}{||(\Delta z)_2| - |(\Delta z)_1||}. \tag{9.32}$$

Or, if the two out-of-focus patterns are on the same side of the focus and $R_2 > R_1$, then

$$\frac{R_1}{|(\Delta z)_1|} = \frac{R_2}{|(\Delta z)_2|} = \frac{R_2 - R_1}{|(\Delta z)_2 - (\Delta z)_1|}. \tag{9.33}$$

Since $|(\Delta z)_2 - (\Delta z)_1|$, R_1 and R_2 are known, $(\Delta z)_1$ or $(\Delta z)_2$ can be precisely determined. From knowledge of the behaviour of the Bessel function, the particle diameter can then be determined.

For objects with arbitrary shapes, the approach may be to encircle the out-of-focus pattern with a circle of minimum diameter.[20] The method for spherical particles can then be used to find the position of an 'equivalent spherical particle'. The approximate position thus obtained can be further used for the exact search by edgetracking or other classical approaches. This way the need to digitize and process a large number of z-steps can again be avoided.

9.3 Fourier transform analysis

Suppose the microobjects are illuminated by a laser beam and a lens is used to collect the diffracted pattern. In the frequency or the Fourier transform plane, diffraction patterns due to individual microobjects are centered at the center of the plane. Thus, their positions in the frequency plane are independent of their positions in the object space. There is some scaling effect on the size of the pattern and that too can be eliminated by using collimated beams. For a number of microobjects with different shapes and sizes, each pattern in the frequency plane is therefore superimposed. From the cumulative pattern, it is not straightforward to extract information about the microobjects. There have been significant efforts to extract object space information from the diffraction data in the Fourier plane.[22–5] For known object shape (such as spherical objects like droplets, bubbles, etc.), the inverse processing of the information to determine the size distribution has been developed by Swithenbank *et al.*[26] The technique is based on measuring the angular pattern of the forward scattered energy and then performing a matrix inversion to determine the size distribution. There are continuing efforts in particle sizing by such Fourier techniques.[27,28]

The same approach can be applied to study the reconstructed microobjects from a hologram. This has been demonstrated by Ewan[29] and Ewan, Swithenbank and Sorusbay[30] for the in-line method and Hess, Trolinger and Wilmot[31] and Hess and Trolinger[32] for the off-axis method. The main advantage is that very fast analysis of the reconstruction in an automated fashion can be done although exact shape information cannot be obtained. However, in practice, the shape is already known in some situations. This is particularly true for spray droplets, bubbles, etc. where the objects are spherical. In those situations, Fraunhofer plane analysis is of particular interest.

For accurate size distribution analysis, the angular distribution in the frequency plane should be measured at as many zones as possible. The inversion can be performed either by a matrix approach[29,30] or

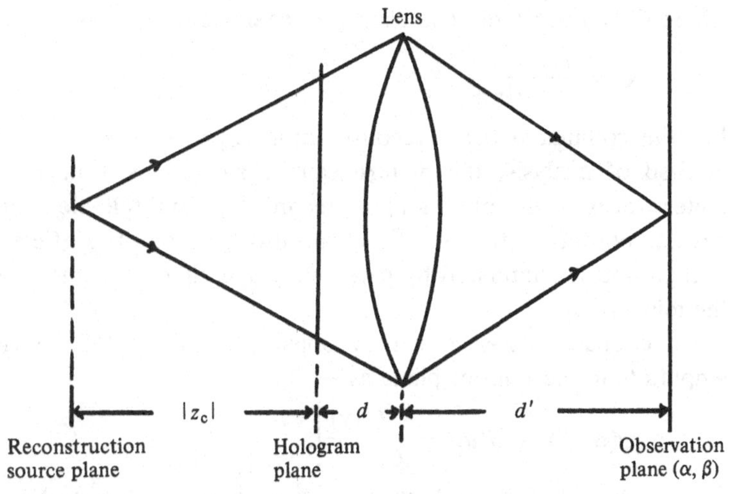

Fig. 9.8 Diagram illustrating the reconstruction geometry of an in-line hologram. If the hologram is formed with a collimated beam then the observation plane is also the Fraunhofer plane for primary and conjugate images. For collimated beam reconstruction also, the observation plane is the back focal plane of the lens.

the integral approach[31,32]. The basic principle is the same for direct Fourier transform analysis from the object scene or from the reconstructed image. A typical arrangement for the reconstruction from an in-line hologram and the observation at the Fourier plane is illustrated in Figure 9.8. Collimated beam formation of the hologram is assumed. In this case primary and conjugate images have their frequency planes common at the observation plane. The observation plane, as shown in Figure 9.8, is basically the image plane for the reconstruction source plane. In this arrangement, the presence of two patterns (corresponding to primary and conjugate images) for each microobject enhances the signal. From the lens formula:

$$\frac{1}{d'} + \frac{1}{d + |z_c|} = \frac{1}{f}, \tag{9.34}$$

where f is the focal length of the transform lens. For a two-dimensional object cross-section described by $A(\xi, \eta)$, the diffracted amplitude in the transform plane is

$$\psi(\alpha, \beta) = G \exp\left[\frac{ik_c(\alpha^2 + \beta^2)}{2d'}\right] \iint_{-\infty}^{+\infty} A(\xi, \eta) \exp\left[ik_c C(\xi\alpha + \eta\beta)\right] d\xi d\eta, \tag{9.35}$$

where G is a constant. The transform coefficient C[29,30] is

$$C = \frac{f + d + |z_c|}{f|z_c|}. \tag{9.36}$$

For the collimated beam reconstruction ($z_c = -\infty$), $C = 1/f$. In the method of analysis, the pattern can be studied by a microdensitometer traces of the photographic record.[26] A mulitple-element semicircular photodetector array[29–32] can also be used. The effect of the background is eliminated by performing a similar experiment without the microobjects.

For circular cross-section microobjects, Equation (9.35) yields the amplitude in the Fourier plane as

$$\psi(\alpha, \beta) = G'a\left[\frac{J_1(k_c C a \gamma)}{C\gamma}\right], \tag{9.37}$$

where $\gamma^2 = \alpha^2 + \beta^2$ and G' is another constant (complex). Writing $C\gamma$ as angle θ, Equation (9.37) becomes

$$\psi(\theta) = G'aJ_1(k_c a\theta)/\theta. \tag{9.38}$$

For a number of microobjects of the same radius a, the amplitude will simply be multiplied by the number of microobjects. However, in practice, there is a distribution which has to be determined. The integral approach[31,32] for the inverse processing to determine the distribution is described here.

Suppose a distribution of particle size $f(a)$ exists. The contribution of particles between radii a and $a + \mathrm{d}a$ to the amplitude at an angle θ is

$$\frac{G'f(a)aJ_1(k_c a\theta)}{\theta},$$

resulting in the total amplitude in the direction θ

$$\psi_t(\theta) = \frac{G'}{\theta}\int_0^\infty f(a)aJ_1(k_c a\theta)\,\mathrm{d}a. \tag{9.39}$$

The distribution $f(a)$ then becomes

$$f(a) = -\frac{2}{k_c a}\int_0^\infty \theta J_1(k_c a\theta)Q(\theta)Y_1(k_c a\theta)\,\mathrm{d}\theta, \tag{9.40}$$

where Y_1 is a Neumann function and

$$Q(\theta) = \frac{\mathrm{d}}{\mathrm{d}\theta}\left(\frac{\pi I k_c^3}{|G'|^2}\right). \tag{9.41}$$

Expression (9.40) is the basis for particle sizing instruments which use angular spectrum measurements. I is the intensity scattered by the particles. Thus, the angular distribution in the Fourier plane yields the particle size distribution $f(a)$.

The experimental results[26-32] are generally quantified using the best curve fit of the well-established Rosin–Rammler distribution. This defines that the weight or the volume fraction V of the microobjects larger than the size X is given by

$$V = \exp[-(X/\bar{X})^{N}]. \tag{9.42}$$

The volume distribution is therefore

$$P(X) = \frac{\mathrm{d}V}{\mathrm{d}X} = \frac{NX^{N-1}}{\bar{X}^{N}} \exp\left[-\left(\frac{X}{\bar{X}}\right)^{N}\right]. \tag{9.43}$$

$P(X)$ is the weight fraction per unit size increment at the size X, N is a measure of the spread of the size distribution and $\bar{X}$ is the characteristic size.

Several experiments have been performed[29-32] to compare the results obtained directly from the sizer against those obtained by holography as an intermediate step. Figure 9.9 shows such comparisons obtained by Ewan[29] for in-line reconstructions. Generally a good agreement is found. However, the holographic measurements are slightly biased towards smaller particles. This could be due to grain-noise from the emulsion, which can be minimized by subtracting the background reading without microobjects. Another cause could be the far-field structure present on the hologram itself. The flare due to the transform of this structure can contribute to irradiance in the frequency plane. Thus, the object–hologram separation should be sufficiently large, i.e. $\geqslant 50$ far-field distances.[29] Comparisons for known size distributions in the off-axis case[31,32] yield similarly good agreements. The mean particle size was measured with $\pm 5\%$ error. The distribution broadening was estimated with $\pm 15\%$ error. The broadening could be due to hologram noises as well as due to inversion scheme software.[31,32]

In any case, the commercial availability of the Fourier transformation instrumentation[29-32] and only a few minutes analysis time per hologram make the approach very attractive. Consequently, there is significant current interest in the Fourier transformation approach.[27,28,33-5]

9.4 Related developments

Optical scintillations produced by raindrops have been used to study the rain parameters.[36-8] Fraunhofer diffraction patterns have

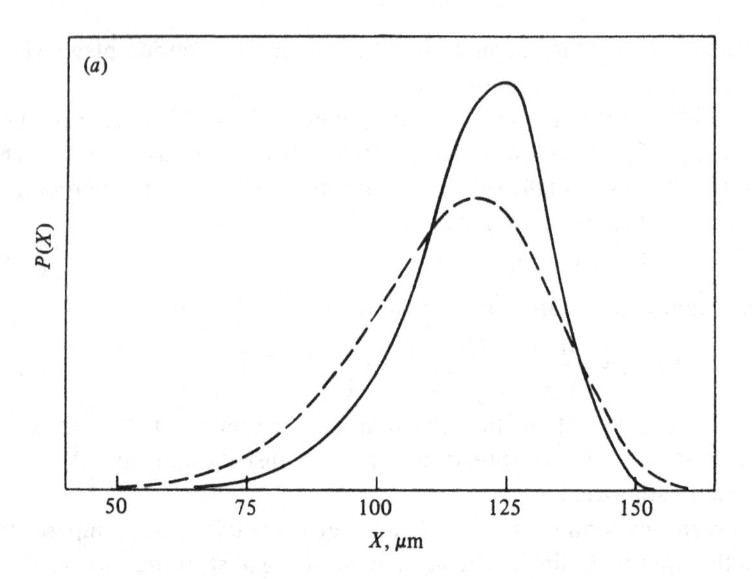

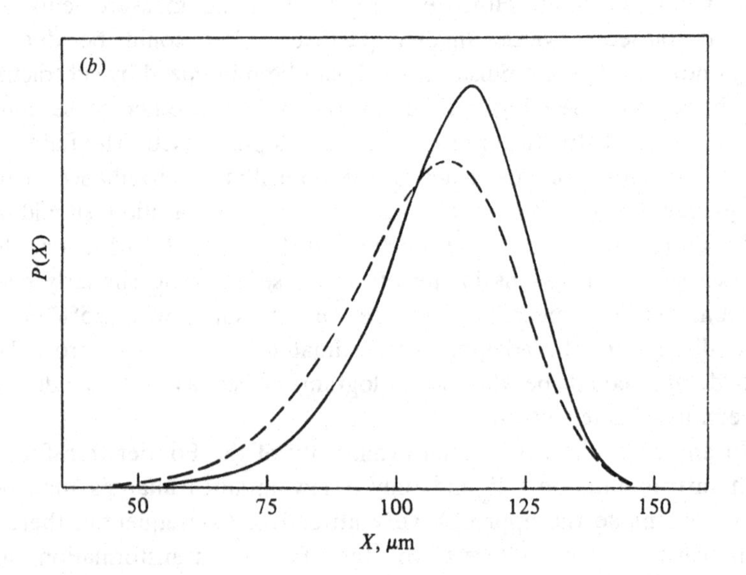

Fig. 9.9

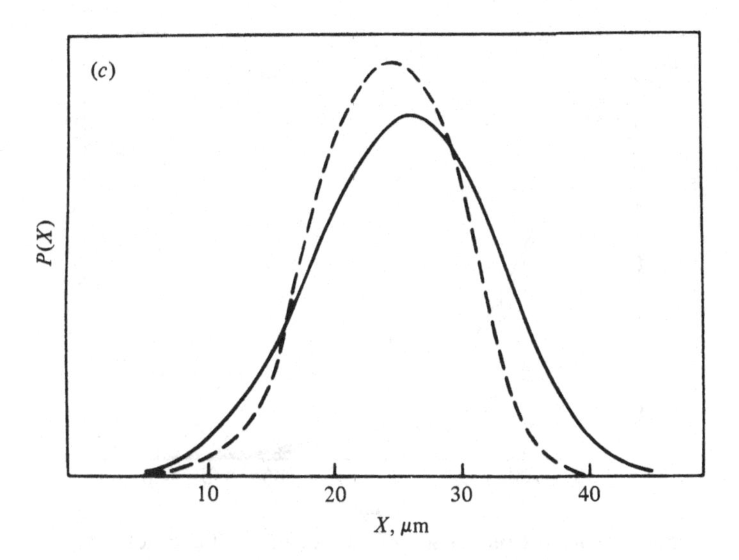

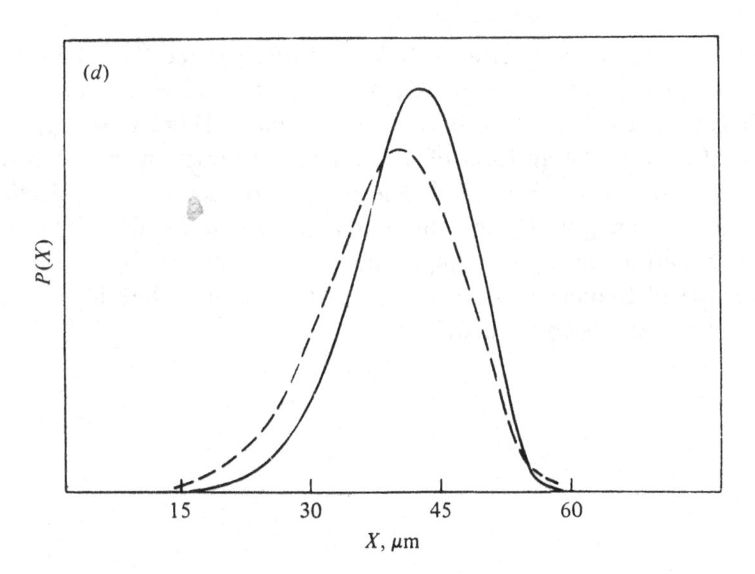

Fig. 9.9 (contd)

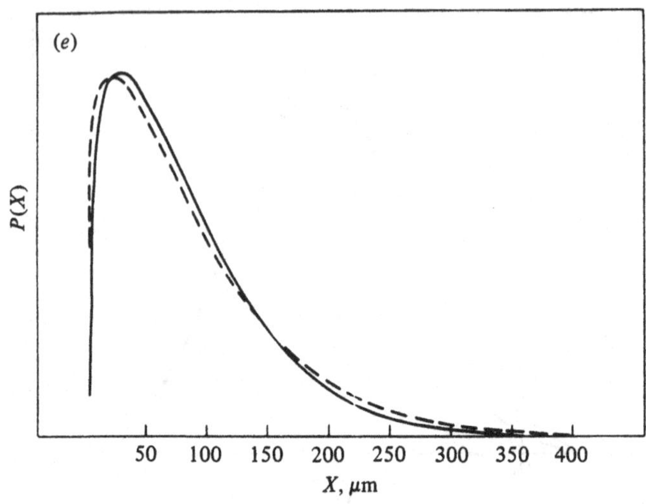

Fig. 9.9 (contd) Comparison of direct (full line) and holographically (dashed line) measured size distributions using the in-line approach: (*a*) metallic particles; (*b*) glass ballotini; (*c*) alumina particles; (*d*) glass ballotini and (*e*) aerosol spray. (Ewan[29].)

also been used to determine isolated surface defect[39] and other[40] parameters. Patterns from thick planar objects,[41] Koch fractals[42] and self-similar structures[43] have also been studied. There is an application of a Hankel transform of a Fraunhofer diffraction spectrum for the particle size distribution.[44] The roles of refraction and reflection in the size analysis by the diffraction pattern have also been studied.[45] Diffraction by an opaque sphere[46] has been studied in detail. The role of frequency doubled Nd : YAG laser in in-line high speed holography has been studied.[47]

10

Velocimetry and high speed holography

Two or more recordings of moving microobjects can be recorded to the same recording medium. The different sets of images provide the displacement and other changes (such as shape and change in size of burning coal particles). Non-image plane analysis techniques (see Chapter 9) have also been applied for velocimetry applications. Different reference or object beam directions for storing different frames provide another way of multiplexing. The techniques of high speed photography have also been extended to holographic applications. In this chapter we will discuss the methods applicable to particle field holography.

10.1 Multiple-exposure holography

The object–image point relationships (see Chapter 6) are known. The concepts of high speed photography can be used there to extract the velocity and other time variant changes with the object field. The advantage of using holography is that microobjects in a volume with changes in three dimensions can be studied. As reported by Boettner and Thompson,[1] a microobject can be allowed to move during the exposure to obtain a streak yielding the path of the movement. Trolinger and coworkers[2,3] introduced the use of multiple exposures to study the dynamics of the particle field. Multiple exposures can be provided, for example by several lasers with synchronized fires, or more conveniently by active or passive *Q*-switches.[3] Trolinger, Belz and Farmer[3] reported several double exposure holograms and reconstructions demonstrating the usefulness of the technique in velocity and other transient phenomena studies. A double-exposure hologram of the moving particle field reconstructs two sets of images. The image movement is related to the object movement and hence the velocity during the recording. Fourney, Matkin and Waggoner[4] verified the role of the double-exposure technique for studying motion in three dimensions. They used water droplets and glass particles for their demonstrations.

Since these initial developments, a considerable number of applications and demonstrations have been reported for size and velocity analysis of microobjects. Webster[5] performed velocity and trajectory measurements for different particle and droplet systems. Boettner and Thompson[1] and LeBaron and Boettner[6] used the technique to study the fall of tremolite and asbestos dust through air. The study was aimed at understanding the mechanism of deposition of inhaled particles in human lungs. Belz and Menzel[7] studied liquid droplets and particles in air flows using the in-line as well as the off-axis approach. Brenden[8] used a miniature ruby laser capable of triple-pulsed holograms for velocity and acceleration measurement of particles. The capability of two pulses for determining the velocity vector has also been demonstrated by Royer and Albe.[9] Besides the common in-line approach,[10] Royer[11,12] used a modified in-line approach (see Section 5.4.1) for velocimetry applications. Uyemura, Yamamoto and Tenjinbayashi[13] further demonstrated the use of double pulsed Q-switched ruby lasers for three-dimensional velocity measurements.

Due to the general usefulness of the approach, an automatic holocinematographic velocimeter can be designed.[14] In the approach, displacements can be obtained from sequential hologram reconstructions by tracking individual microimages.

For velocimetry applications, the mere distinguishing of two or more images is sufficient. Even if sharp image edge details are not available for accurate size analysis, the velocity can still be measured. This fact has been used for velocity determination of sub-micrometer particles.[12] In gas and fluid dynamics studies, these particles are used as tracers.[14–16] For velocimetry, the size analysis is not then needed. Thus, there has been much interest in and application of two or more recordings in the same emulsion. In the commonly used multiple-exposure approach, each pulse duration is small enough to freeze the field. The duration between the pulses yields the velocity and other phenomena under study. The general methodology is to relate two or more reconstructed images to the object location at the time of the exposures.[17] Figure 10.1 shows a doubly-exposed hologram of free falling dust and Figure 10.2 shows the reconstruction of a section clearly distinguishing the two images.

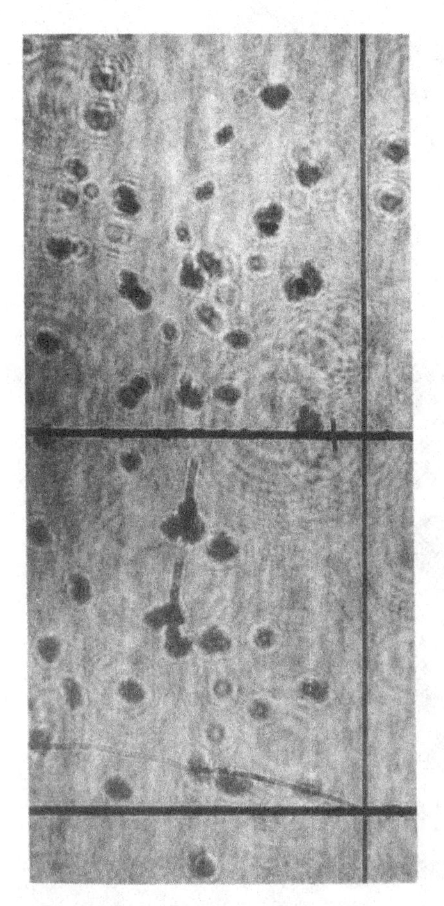

Fig. 10.1 Double-exposure hologram of falling dust particles taken at 10 ms apart. Each pulse duration was 10 μs. obtained by a mechanically chopped argon-ion laser. (Boettner and Thompson[1].)

10.1.1 *Direction analysis*

From typical double- or multiple-exposure reconstructions, only the object displacement (and hence the speed) can be determined. For direction information, one method is to provide two (or more) pulses with different amounts of exposure.[3] The reconstructed microimages will be of different intensities yielding the sense of the direction.

In the off-axis method, the problem can be solved using two reference beams. This has been done by Briones and Wuerker.[18,19] The Pockels cell of the ruby laser is electronically switched so that the second light pulse passes through the second reference path. Both the

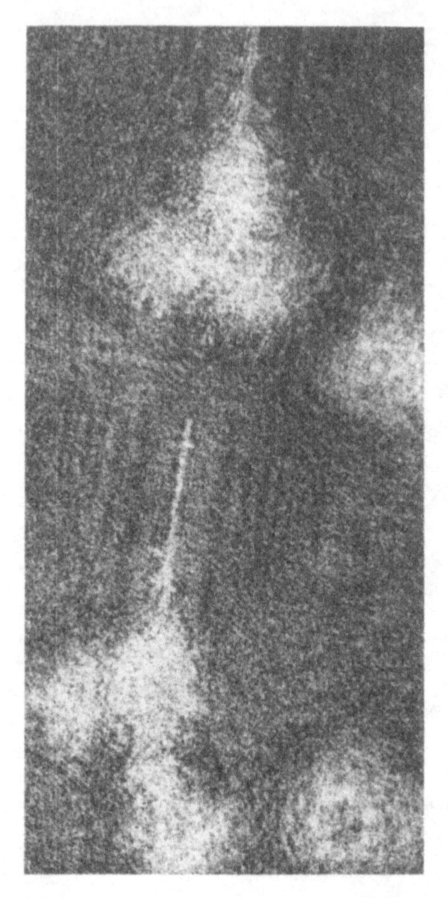

Fig. 10.2 A section of the reconstructed image from the hologram shown in Figure 10.1 (Boettner and Thompson[1].)

holograms are recorded on the same plate but with spatially different reference beams. One reference beam at a time can be used for the reconstruction. Thus, two particle fields can be separately reconstructed with exact information about the order. There is another advantage of the approach. In case of dense microobjects, the measurement is improved. In the single reference double-exposure method, images of two or more microobjects can overlap resulting in confusion.

10.1.2 *Holocinematography and transient phenomena analysis*

The concepts of high speed photography and holography can be combined to study not only the velocity but also the general object

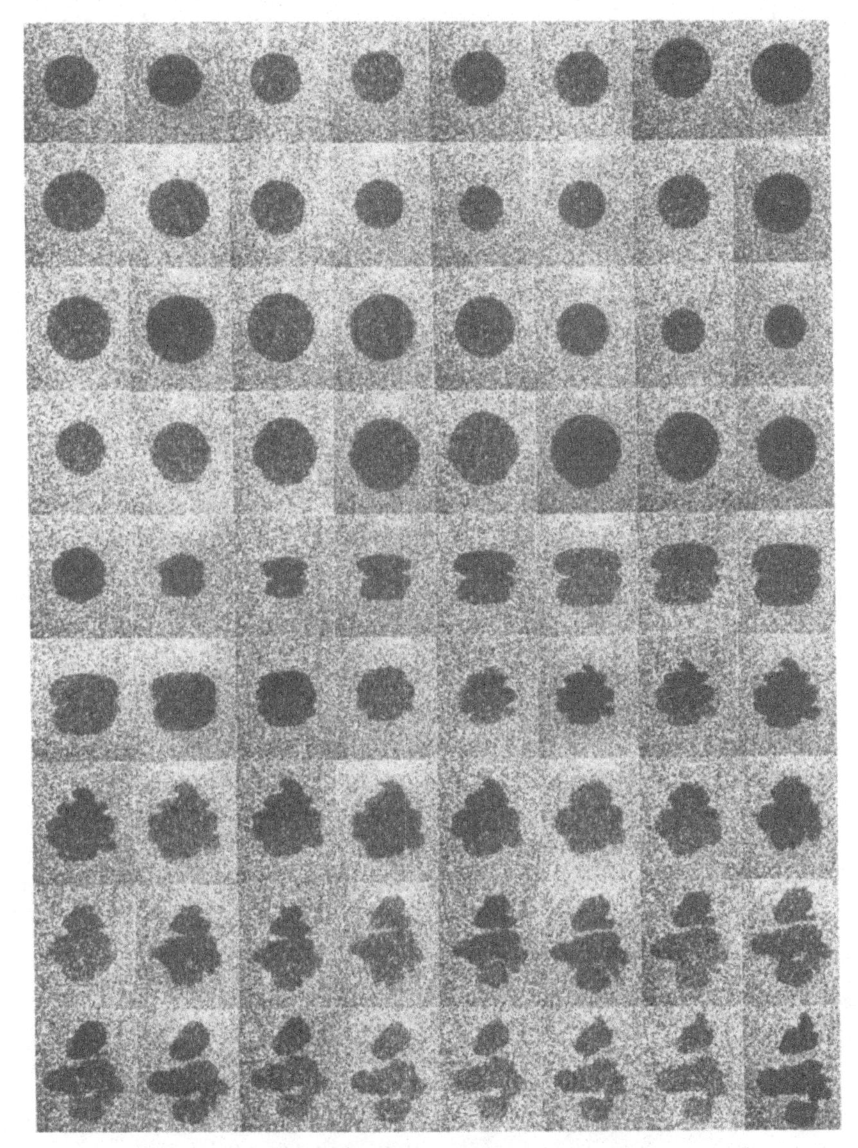

Fig. 10.3 Oscillations and decay of a bubble of about 1 mm diameter in a sound pressure field of increasing amplitude. The sequence of 72 frames was taken at 66.7 kHz. (Hentschel and Lauterborn[31].)

behavior against time. Holographic motion pictures of living planktonic marine organisms have been made by Knox and Brookes[20] and Stewart, Beers and Knox[21] using the in-line method, and by Heflinger, Stewart and Booth[22] using the off-axis approach. High speed cine camera systems utilizing flexible films are used to store frame by frame holograms of the event. Small details such as plankton movement and food manipulation can be studied.

High speed events can likewise be studied with proper pulsing, synchronization, etc. Using triple-exposure in-line far-field holography, Raterink and Van der Meulen[23] studied traveling bubble cavitation in a water tunnel. Not only the bubble velocity but its shape and size changes including collapse, can be studied.

Several techniques for providing multiple-exposure are available. Landry and McCarthy[24] described the role of a multiple cavity laser in this connection.

Extensive work using acoustooptical switchable beam splitters for spatial frequency multiplexing and/or spatial multiplexing has been reported by Hinsch and Bader,[25] Lauterborn and Ebeling,[26,27] Lauterborn,[28] Ebeling and Lauterborn,[29,30] Hentschel and Lauterborn[31] and Lauterborn and Hentschel[32,33]. In the acoustooptical modulation approach, successive light pulses can be spatially separated. Spatial multiplexing is generally provided by rotation of the hologram plate in its plane with a fixed aperture to record different holograms in different regions. These two techniques can also be combined. The applications of these studies deal particularly with bubble studies, laser-induced breakdown, and shock wave formation in liquids. Figure 10.3 shows holographic reconstructions from a sequence of 72 frames. An electrolytically produced bubble of about 1 mm diameter is exposed to a sound field. The bubble first experiences spherical and then asymmetric oscillations, then finally starts to split into smaller bubbles.

Prisms can also be used to develop a high speed holocamera to store ultrafast events by recording different holograms at different locations.[34] Other examples of high speed holography applications are studies of coal particle combustion,[35] reacting sprays,[36] solid propellant combustion,[18,19] break up of bubbles and droplets,[37,38] and bubble chamber studies[39–44]. Ruff, Bernal and Faeth[45] demonstrated the usefulness of copper-vapor lasers for holocinematographic applications.

10.2 Displaced diffraction pattern analysis

As seen in Section 9.1, the complex diffraction is formed at the recording plane. The movement of the pattern in the recording plane will yield the transverse object movement. Murakami and

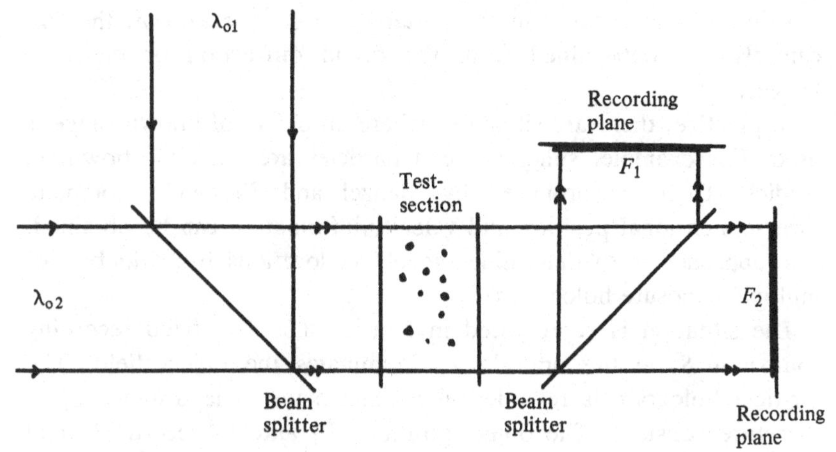

Fig. 10.4 Diagram for recording diffraction patterns of a moving scene at different times with different wavelengths λ_{o1} and λ_{o2}. Relay lenses can also be used in the path. A delayed trigger circuit provides proper pulse separation. Interference filters F_1 and F_2 separate out the colors.

Ishikawa[46] demonstrated this approach for velocity distribution measurement of particles accelerated by a shock wave. They used two dye laser pulses of different wavelengths (0.59 μm and 0.632 μm). Delayed trigger circuits provide the firing time difference for velocimetry. A schematic diagram of the recording arrangement is shown in Figure 10.4. The first beam splitter combines the two light beams for the in-line arrangement. After passing through the test-section, two different patterns are recorded separately in different colors. The image separation is provided by the second beam splitter and two interference filters.

The analysis is performed in a superimposing microscope-camera system.[46] By superimposing the two color films, the amount of pattern movement can be obtained. The entire procedure including the use of two colors makes it easy to distinguish the two different patterns. Two colors and interference filters provide a simple method for high speed holography without moving any part of the instrument.

10.3 Direct analysis from far-field patterns

We have seen in Section 9.1.1 that the complex far-field pattern or the hologram itself can be studied for size and position analysis. The longitudinal distance (z_o) can be determined from the fine interference fringes in the pattern. Once z_o is known, the size can also be determined from the broad diffraction or envelope fringes.

In practice, there are situations where an object of known shape is used. For example, spherical seed particles are used for flow-field studies. Then, as proposed by Menzel and Shofner[47], complete three-dimensional position and velocity information can be obtained. The approach is to determine the object locations from double- or multiple-exposure holograms.

The situation is represented in Figure 10.5. The fixed recording source at S on the optical axis illuminates the object field. The far-field hologram is recorded at a fixed plane. The distance z_R is therefore constant. The object positions P_1 and P_2 record far-field holograms centered at the points P_1' and P_2' respectively. For opaque spherical objects, the far-field intensity pattern with respect to the effective hologram center (see Section 4.1) is

$$I = 1 - \frac{2m_o a}{r'} J_1\left(\frac{2\pi a r'}{\lambda_o |z_o|}\right) \times \sin\left(\frac{\pi r'^2}{\lambda_o m_o |z_o|}\right) + \left(\frac{m_o a}{r'}\right)^2 J_1^2\left(\frac{2\pi a r'}{\lambda_o |z_o|}\right). \tag{10.1}$$

The fine interference fringes governed by the sine function yield the longitudinal object distance z_o. The J_1 distribution is generally broad providing the envelope. A typical far-field pattern is represented by Figure 9.2.

Extrema of the fine fringes occur at the minimum or maximum of the sine function. At the extrema, the argument of the sine function is the odd multiple of $\pi/2$. The ring radii become

$$r'^2_s = \frac{(2s+1)\lambda_o m_o |z_o|}{2}, \tag{10.2}$$

where s is an integer. Zero and even values of s determine intensity minimum whereas odd values of s correspond to the maxima. Since m_o is $1 - z_o/z_R$ and z_R is a fixed quantity with the present arrangement, the object distance can be written as

$$z_o = \frac{2r'^2_s}{(2s+1)\lambda_o + 2r'^2/z_R}. \tag{10.3}$$

By measuring the extrema ratio r_s' from a microdensitometer trace of the far-field pattern, z_o is thus known.[47] From two such patterns, z_{o1} and z_{o2} can likewise be determined yielding the object's longitudinal positions. The recording source S, the object point, and the center of

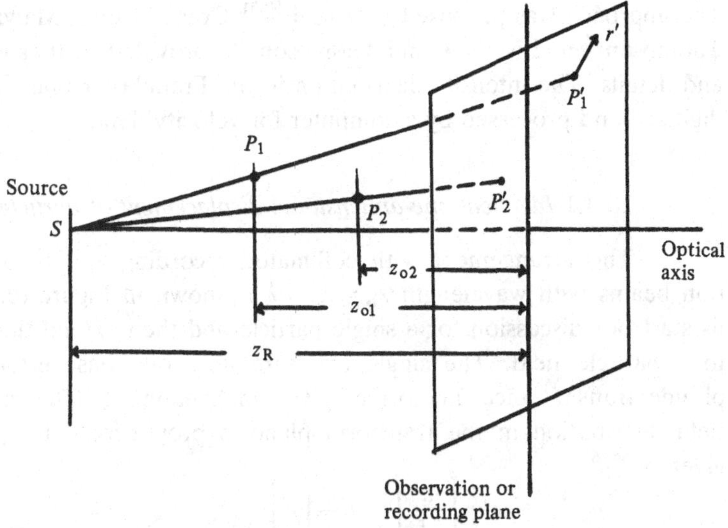

Fig. 10.5 Diagram for the study of the object's displacement from the far-field pattern. Object locations P_1 and P_2 give far-field patterns centered at P'_1 and P'_2 respectively.

the far-field pattern form a straight line, so the exact object positions in space can thus be determined.

Experimental results, accuracy and data analysis techniques have been discussed by Menzel and Shofner.[47] Generally, an accuracy of about 1% in three-dimensional velocity measurements can be obtained. Tsuno and Takahashi[48] have used the technique to study an exploding wire. The high speed phenomena were recorded using a HeNe laser and a high speed rotating prism camera. Spatial frequency analysis of the diffraction pattern for determining z_o can also be performed.[49] However, there is no significant improvement with this mode of analysis.[49]

10.4 Fraunhofer plane analysis

In Section 9.3, we saw that the Fourier transform of the hologram or the distribution in the Fraunhofer plane can yield the particle size distribution. For a doubly-exposed hologram of a moving scene, a fringe pattern is also formed within the distribution in the Fraunhofer plane. The fringe pattern contains information about the displacement of the particle field between the exposures. Thus, in velocimetry applications, the problem of the tedious and time consuming search for particle pairs in the reconstruction can be avoided.

The approach was proposed by Ewan.[50,51] Crane, Dunn, Malyak and Thompson[52] and Malyak and Thompson[53–5] provided further insights and details. The intensity distribution in the Fraunhofer plane can be digitized and processed by a computer for velocity data.

10.4.1 *Identical one-dimensional displacement of particles*

The arrangement with collimated recording and reconstruction beams with wavelength $\lambda_o = \lambda_c = \lambda$ is shown in Figure 10.6. Let us start our discussion for a single particle and then extend the study to a particle field. The single exposure hologram has certain amplitude transmittance $T(x, y)$ as given in Section 3.1. The complex field distribution in the transform plane is proportional to $\psi(\alpha, \beta)$ given by[54–6]

$$\psi(\alpha, \beta) = \frac{\exp\left[\frac{i\pi}{\lambda f}\left(1 - \frac{d}{f}\right)\gamma^2\right]}{i\lambda f} \iint_{-\infty}^{+\infty} T(x, y) \times \exp\left[-\frac{2\pi i(\alpha x + \beta y)}{\lambda f}\right] dx\, dy, \tag{10.4}$$

where $\gamma^2 = \alpha^2 + \beta^2$. Corresponding to the integrals discussed in Chapter 3, there are four terms in Equation (10.4). These correspond to the directly transmitted light, real and virtual image terms and the intermodulation terms. The directly transmitted light and the intermodulation terms can be neglected because they are confined near the origin of the transform plane. Omitting constant terms, $\psi(\alpha, \beta)$ thus becomes (except near the origin):[52,54]

$$\psi(\alpha, \beta) = \cos\left(\frac{\pi|z_o|\gamma^2}{\lambda f^2}\right) \tilde{A}\left(\frac{\alpha}{\lambda f}, \frac{\beta}{\lambda f}\right). \tag{10.5}$$

If a double-exposure (of equal amounts) hologram were recorded with a particle displaced in (ξ, η) plane by $(\Delta\xi, \Delta\eta)$ then

$$\psi(\alpha, \beta) = \frac{1}{2}\cos\left(\frac{\pi|z_o|\gamma^2}{\lambda f^2}\right)\left[\tilde{A}\left(\frac{\alpha}{\lambda f}, \frac{\beta}{\lambda f}\right) + \tilde{A}'\left(\frac{\alpha}{\lambda f}, \frac{\beta}{\lambda f}\right)\right], \tag{10.6}$$

where $\tilde{A}'(\ )$ is the Fourier transform of the displaced particle. $\tilde{A}'(\ )$ and $\tilde{A}(\ )$ are related by the shift theorem of Fourier analysis, resulting in

$$\psi(\alpha, \beta) = \tilde{A}\left(\frac{\alpha}{\lambda f}, \frac{\beta}{\lambda f}\right)\cos\left(\frac{\pi|z_o|\gamma^2}{\lambda f^2}\right)\cos\left[\frac{\pi(\alpha\Delta\xi + \beta\Delta\eta)}{\lambda f}\right] \times \exp\left[\frac{\pi i(\alpha\Delta\xi + \beta\Delta\eta)}{\lambda f}\right]. \tag{10.7}$$

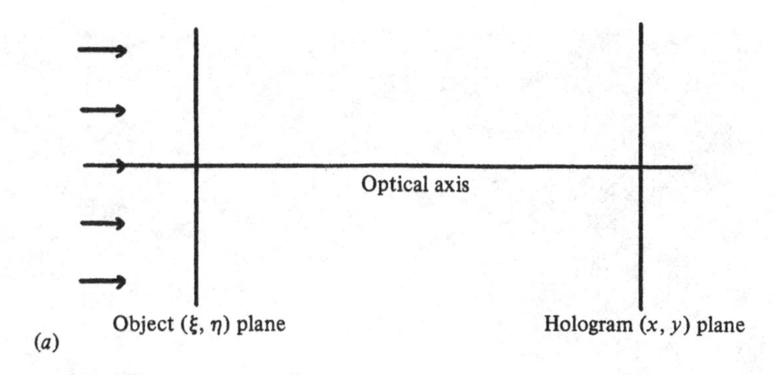

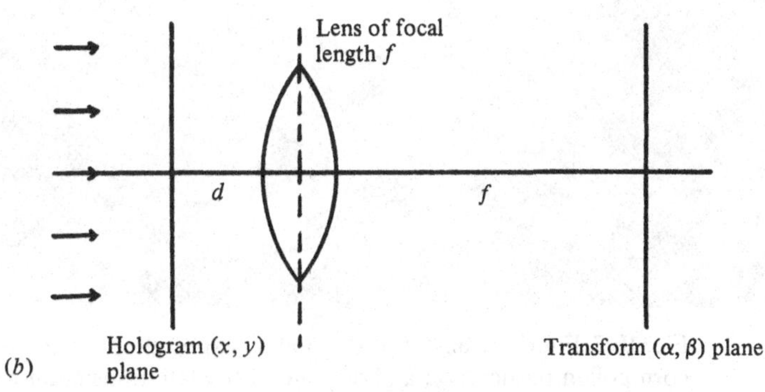

Fig. 10.6 Diagram for recording and Fraunhofer plane analysis of a doubly-exposed hologram with collimated beams of wavelength λ: (*a*) recording; (*b*) transform by a lens of focal length f.

The resulting irradiance distribution is

$$\begin{aligned} I(\alpha, \beta) &= |\psi(\alpha, \beta)|^2 \\ &= \left|\tilde{A}\left(\frac{\alpha}{\lambda f}, \frac{\beta}{\lambda f}\right)\right|^2 \\ &\quad \times \cos^2\left(\frac{\pi |z_o| \gamma^2}{\lambda f^2}\right) \cos^2\left[\frac{\pi(\alpha \Delta\xi + \beta \Delta\eta)}{\lambda f}\right]. \end{aligned} \tag{10.8}$$

The distribution of Equation (10.8) has been derived for a single object. It is still valid if all the objects are at the same distance z_o and are identically displaced. $\tilde{A}(\alpha/\lambda f, \beta/\lambda f)$ is the Fourier transform of the function describing the object shape. For spherical particles $\tilde{A}(\,)$ is $[2J_1(\,)/(\,)]$. The effect of $|\tilde{A}(\alpha/\lambda f, \beta/\lambda f)|^2$ in Equation (10.8) is a slowly varying diffraction halo. $\cos^2(\pi z_o \gamma^2/\lambda f^2)$ is a zone lens term yielding circular fringes. $\cos^2[\pi(\alpha\Delta\xi + \beta\Delta\eta)/\lambda f]$ provides straight line fringes yielding the object motion. Figure 10.7, which is a photograph of doubly-exposed 79 μm corn pollen particles sprinkled on a glass slide, shows all these patterns. The slide was translated along ξ-axis between exposures.

Fig. 10.7 Fourier transform of a doubly-exposed hologram of 79 μm corn pollen particles on a glass plate. The plate translation was horizontal. (Crane *et al.*[52].)

From Equation (10.8), the zone lens structure is present when z_o is a constant. However, this structure is present with some depth but with reduced contrast.[52,53] These fringes or rings can obscure the velocity information fringes. The zone lens term is due to the in-line approach and it can be eliminated by using the off-axis method.[53] A multiple exposure hologram can also be recorded to produce sharper fringes in the transform plane.[53]

For different recording and reconstruction wavelengths the frequency of the velocity information fringes is multiplied by a constant (see Section 9.3). For collimated recording and reconstructed beams of different wavelengths, the constant is unity and λ is replaced by λ_c.

10.4.2 *Particles moving differently*

Equation (10.8) is valid for a single particle or all particles moving identically. For different particles moving differently, the complex amplitudes due to each particle given by Equation (10.5) should be added together. Except near the origin in the transform plane, the average (expected value) of the intensity is[54]

$$\langle I(\alpha, \beta)\rangle = \left|\tilde{A}\left(\frac{\alpha}{\lambda f}, \frac{\beta}{\lambda f}\right)\right|^2 \cos^2\left(\frac{\pi|z_0|\gamma^2}{\lambda f^2}\right) \times \sum_{j=1}^{N}\left\{1 + \cos\left[\frac{2\pi}{\lambda f}(\alpha\Delta\xi_j + \beta\Delta\eta_j)\right]\right\}, \quad (10.9)$$

where N is the number of particles. If a continuous distribution of particle velocities is assumed, the summation can be converted into the integral form:

$$\langle I(\alpha, \beta)\rangle = N\left|\tilde{A}\left(\frac{\alpha}{\lambda f}, \frac{\beta}{\lambda f}\right)\right|^2 \cos^2\left(\frac{\pi|z_0|\gamma^2}{\lambda f^2}\right) \times \left[1 + \int_{-\infty}^{+\infty} p(\delta)\cos\left(\frac{2\pi\alpha\delta}{\lambda f}\right)\mathrm{d}\delta\right], \quad (10.10)$$

where one-dimensional velocity distribution of the density function $p(\delta)$ has been assumed. If the velocity distribution is uniform, say $p(\delta) = (D_2 - D_1)^{-1}$ for $D_1 < \delta < D_2$ then Equation (10.10) becomes[54]

$$\langle I(\alpha, \beta)\rangle = N\left|\tilde{A}\left(\frac{\alpha}{\lambda f}, \frac{\beta}{\lambda f}\right)\right|^2 \cos^2\left(\frac{\pi|z_0|\gamma^2}{\lambda f^2}\right) \times \left\{1 + \operatorname{sinc}\left[\frac{\pi\alpha(D_2 - D_1)}{\lambda f}\right] \times \cos\left[\frac{\pi\alpha(D_2 - D_1)}{\lambda f}\right]\right\}, \quad (10.11)$$

where sinc( ) = sin( )/( ), δ is the particle displacement between the exposures (in the ξ-direction) and D_1 and D_2 are the minimum and the maximum displacements respectively. The transform plane still contains cosine fringes modulated by a sine function. Thus, for a wide distribution $D_2 - D_1$, the fringes will not be observed. The frequency of the cosine fringes gives the mean particle displacement.

Ewan[50,51] and Malyak and Thompson[53-5] discussed the role of the Gaussian distribution $p(\delta)$. Again, the fringe frequency is proportional to the mean velocity of the particle field and the visibility decreases exponentially. The recording plate can be translated between the exposures to manipulate the velocity distribution and also to solve the sense of the direction problem.

Thus, if the functional form of the one-dimensional velocity distribution is known, the mean displacement and distribution can be measured from the frequency and contrast respectively of the cosine fringes.

The case of practical interest is when the functional form of the velocity distribution is unknown. Ewan[51] measured the one-dimensional velocity distribution (of unknown functional form). The method involves a linear photodiode array in the transform plane linked to a computer. The Fourier transform of the intensity distribution (in the transform plane) after removing the envelope function directly yields the distribution $p(\delta)$.[51,53–5] Malyak and Thompson[53–5] discussed the approach for two-dimensional velocities (still transverse, i.e. in the plane parallel to the hologram). The intensity distribution in the transform plane is digitized using a television camera and digitizer. The data are processed (to obtain a two-dimensional Fourier transform) using a computer to yield the particle displacement information. In two-dimensional cases, the marginal value of p (probability density function) can be obtained for independent orthogonal components of the displacement.[55]

Thus, from the transform plane analysis, the object displacement and hence velocity can be studied. The data can be digitized, stored and processed by a computer.

11

The off-axis approach

The in-line Fraunhofer approach generally works well due to its simplicity. Therefore this approach is very common in actual practice. However, there are practical situations where off-axis or Leith–Upatnieks holography becomes more suitable or necessary. The first, as discussed in Section 5.1, is when the density of the microobjects increases. The loss of an uninterrupted cross-section of the reference beam then ultimately causes the failure of the in-line or Gabor approach. The second limitation of the in-line approach is when the size range of the microobjects is large. The maximum object distance for the smallest microobject (diameter $d_{\rm min}$) is $z_{\rm o,max} = Nd^2_{\rm min}/\lambda_{\rm o}$ (see Section 4.3) where N is a constant. The maximum value of N is generally 100 for the collimated beam case. Now for the largest object's diameter $d_{\rm max}$, the minimum object distance is $z_{\rm o,min} = d^2_{\rm max}/\lambda_{\rm o}$ to satisfy the far-field condition. Both these requirements may not be simultaneously met for large differences between $d_{\rm min}$ and $d_{\rm max}$.

Also, there are applications of multiple reference beams (off-axis approach) in multiplexing in high speed holography (see Section 10.1). In Fraunhofer plane analysis for velocimetry applications, the off-axis approach eliminates the unwanted zone-lens fringes of the in-line method (see Section 10.4). The analysis of Chapters 6 and 7, in general, is applicable to in-line as well as off-axis techniques. In this chapter the general off-axis systems methodology applicable to particle field holography is discussed.

11.1 Basic recording and reconstruction arrangements

The off-axis holography using a pulsed laser was introduced by Brooks, Heflinger, Wuerker and Briones,[1] Wuerker, Heflinger

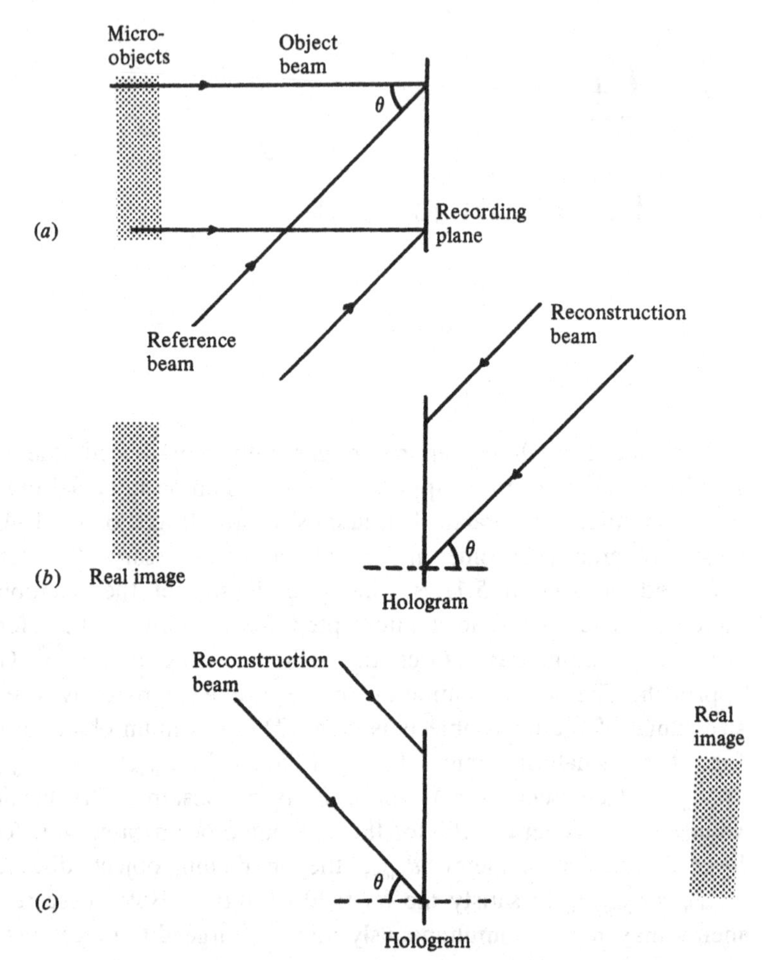

Fig. 11.1 Schematic diagram of off-axis particle-field holography with diffracted object field; (*a*) recording with collimated reference beam; (*b*) and (*c*) two approaches for reconstructing the real images.

and Zivi,[2] Buckles[3] and Ansley and Siebert[4]. Further discussions and applications are to be found in the work of Wuerker and Heflinger,[5] Withrington[6] and Matthews[7]. Descriptions and applications of the off-axis holographic arrangement for particle field holography are widely available.[1–19]

The general principle of recording and reconstruction can be described by Figure 11.1. Figure 11.1(*a*) shows the diffracted light as the information beam and a collimated reference beam. Figures 11.1(*b*) and (*c*) show two methods for obtaining real images. The

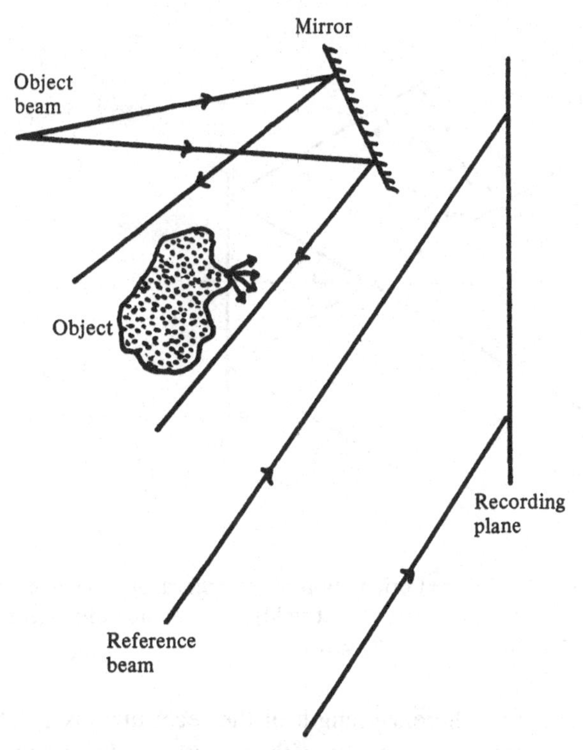

Fig. 11.2 Diagram illustrating the off-axis recording arrangement with reflected or backscattered light.

approach of Figure 11.1(*b*) is important with lens-assisted recordings to cancel out any aberrations introduced.[8–12] With this arrangement, the reconstructed field can be passed back through the original recording system. Precise alignment methods for the reference beam during the reconstruction are available.[11,12]

Recording with reflected light as the information beam is also possible.[4,7,9,17] A typical recording arrangement is illustrated by Figure 11.2.

In Figures 11.1 and 11.2, the subject beam is facing the recording plane directly. With modern holographic plates, the orientation is not of general concern. However, particularly with films, alignment errors during the reconstruction may be present. Emulsion shrinkage effects can be serious. In those situations, it is better that the film faces the bisector between the subject and reference beams. This aspect has been discussed by Trolinger.[20] The arrangement is represented in Figure 11.3 for diffracted information beam.

The path difference between the object and the reference beams

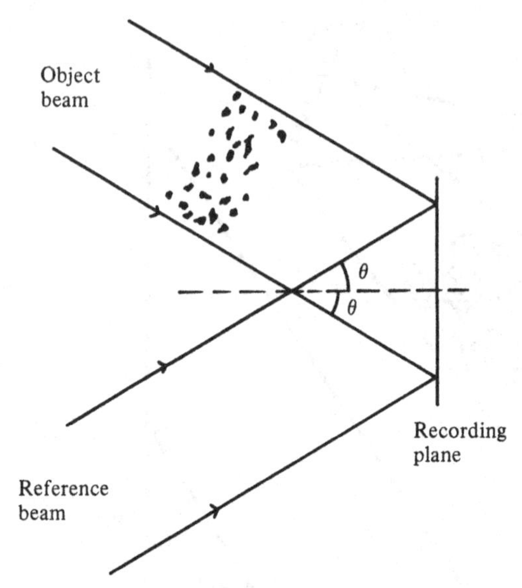

Fig. 11.3 The desired orientation of the recording arrangement (shown for the diffracted object field) for minimum alignment errors, emulsion shrinkage effects, etc.

should be within the coherence length of the recording laser. The use of single longitudinal mode lasers with an intracavity etalon or an interferometric resonator is common. The coherence length of these lasers is sufficiently large for most of the practical applications. In a particular case, the optical path lengths of reference and object beams can be matched by precise control of the paths.[21] For common applications, mirrors can be used to equalize the lengths of the object and reference paths.[14,15] Path length mismatch by two laser cavity lengths is another practical solution.[8] Optical fibers can also be used.

11.2 Relay lenses and related magnifications

The optimum reference-to-object angle and the presence of physical obstructions often require the recording medium to be placed away from the object beam. The use of relay lenses to bring the object field closer to the recording plane is common. Pre-magnification of the subject field before the recording is also very useful to achieve high resolution.[12,15] This pre-magnification and/or relay can be obtained by two-lens telescopic as well as single-lens[22] systems.

For image analysis without using relay lenses during the recording, the analysis of Chapter 6 is sufficient. For the relay lens situation, the

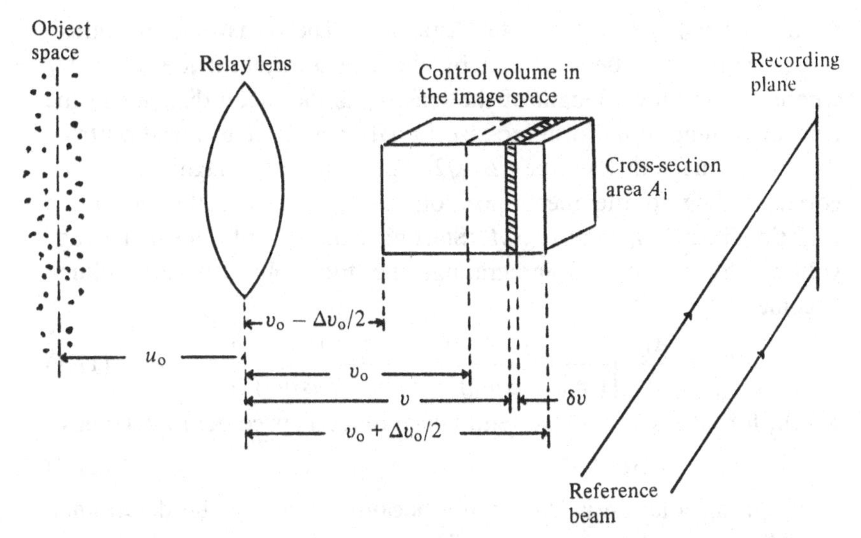

Fig. 11.4 Schematic diagram of the recording of off-axis holograms of microobjects, using a single relay lens.

relayed image is actually the object field for the hologram. A relationship between the original and the relayed fields is therefore necessary.

11.2.1 *Double lens telescopic system*

Image relay with this arrangement is discussed in Section 5.2.3. Suppose, first the analysis of Section 6.4 is incorporated if the medium is other than air. Equivalent object coordinates in air from the first lens are thus obtained. The relayed image positions with respect to the second lens (and hence with respect to the recording plane) are thus obtained from Equations (5.9) and (5.10). The constant magnification in the telescopic arrangement is very convenient.

11.2.2 *Single relay lens*

The single relay lens arrangement is simpler from the alignment point of view and it takes less space as compared to the double lens system. However, it gives non-uniform magnification across the image depth. This aspect has been discussed in detail by Vikram and McDevitt.[22] Figure 11.4 represents off-axis holography with a single relay lens. The actual object space (equivalent object distance u in air from the lens) is relayed to the distance v. The transverse magnification $m = v/u$ can be obtained by the lens formula $1/u + 1/v = 1/f$ where f is the focal length of the lens. v_o is the mean distance in the control volume of uniform cross-sectional area A_i. The variation from this mean image distance is $\pm\Delta v_o/2$. The transverse magnification m_o corresponding to the mean position is $v_o/f - 1$ and the maximum variation is $\pm\Delta m_o = \pm\Delta v_o/2f$. Starting from an infinitesimal image volume depth δV and integrating, the total object space volume becomes:[22]

$$V_o = \frac{A_i f}{3}\left[\frac{1}{(m_o - \Delta m_o)^3} - \frac{1}{(m_o + \Delta m_o)^3}\right], \tag{11.1}$$

which, for very small image (and hence object) space depth becomes

$$V_o \approx A_i \Delta v_o / m_o^4. \tag{11.2}$$

The limiting solution is easy to use because m_o has to be determined or calibrated at only one plane. The relationship of Equation (11.1) can be expanded using the binomial series method. The percentage error ρ in the volume from the relationship of Equation (11.2) is thus

$$\rho \approx \frac{1000}{3}\left(\frac{\Delta v_o}{2 m_o f}\right)^2. \tag{11.3}$$

For $m_o = 1$, $f = 10$ cm, $\Delta v_o = 1$ cm, ρ is less than 1. It is noticeable here that the off-axis approach is generally used for higher object densities and a very small control volume is thus needed. A control volume of 0.22 cm^3 was sufficient when the nuclei content in the water was as high as about 1800 cm^{-3}.[15] Thus, the non-linear magnification of the single lens arrangement is not a general problem. Equation (11.1) is anyway available. The relationship $m = v/f - 1$ can be used for the transverse magnification in general. It can also be calibrated. Suppose m_o and m_o' are the transverse magnification at image distances v_o and v_o' respectively. Then the magnification m_o'' at the distance v_o'' is

$$m_o'' = m_o + (m_o' - m_o)\left(\frac{v_o'' - v_o}{v_o' - v_o}\right), \tag{11.4}$$

which involves the position differences $v_o'' - v_o$ and $v_o' - v_o$ rather than the exact positions.

11.3 Diffused subject illumination

The use of ground glass to provide diffused subject illumination is common in off-axis particle field holography. The resulting

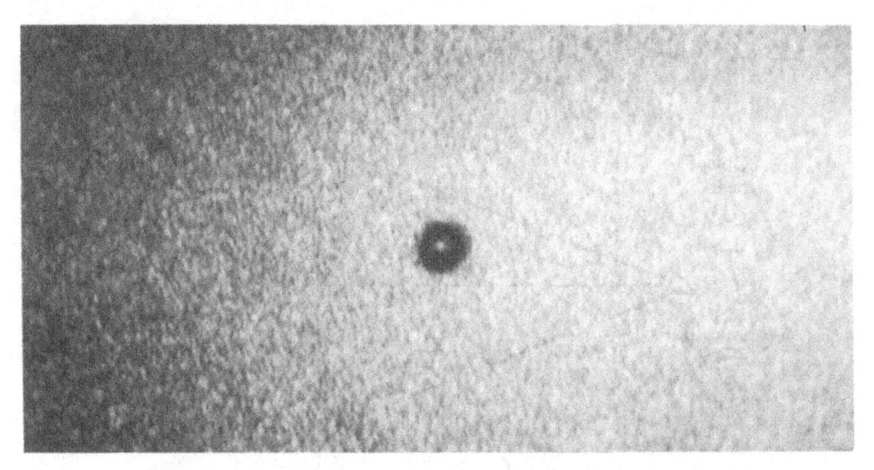

Fig. 11.5 A typical reconstructed image with diffused illumination.

reconstruction is easier to view with the naked eye.[8] The information from each object is distributed over the entire hologram resulting in a short depth of focus during the reconstruction.[14,15,23] However, the reconstruction yields a speckled background. The mean speckle separation is about twice the resolution capability of the recording system.[8,24] Therefore the resolution capability of the holography system should be several times better than that required for the smallest object to be studied. If diffraction-limited resolution is the goal, then use of the diffuser should be avoided.[21] Figure 11.5 shows a typical reconstruction of the diffused subject illumination situation. The speckled background is clearly seen. Multiple rather than the diffused illumination can also be provided[8] for different angles of view without the speckle effect.

11.4 Film resolution requirements

Film resolution requirements in different holographic processes have been discussed by DeVelis and Reynolds[25] and Smith[26] among others. In Section 4.2, the resolution requirements have been discussed for the in-line Fraunhofer case. In this section we consider the basic difference in the off-axis case. Suppose the off-axis recording of a point object is performed with a plane reference wave as shown in Figure 11.6. Suppose a point scatterer is situated on the optical axis at a distance z_o from the recording plane (x, y). Also suppose the reference beam is in the (x, y) plane making an angle ϕ

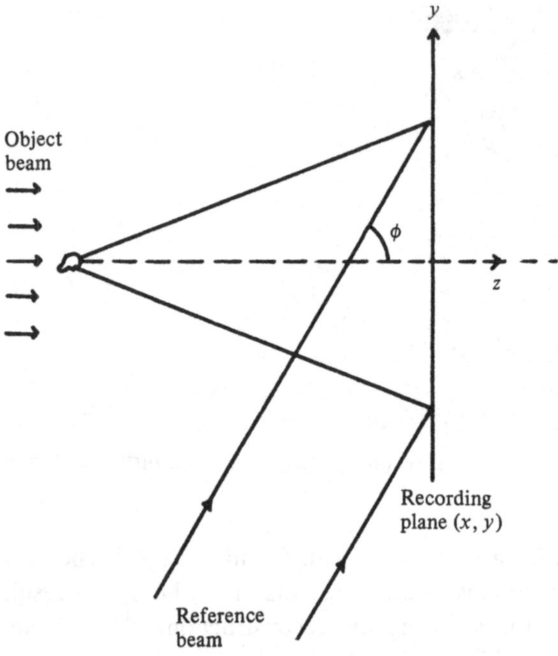

Fig. 11.6 Diagram illustrating off-axis holography of a point scatterer using a plane reference beam with an angle ϕ from the z-axis in the (y, z) plane.

with the z-axis. The object and the reference beam amplitudes at the recording plane are therefore $O \exp(i\pi r^2/\lambda_o z_o)$ and $R \exp(2\pi i y \sin\phi/\lambda_o)$ respectively where O and R are constants. The irradiance distribution at the recording plane is therefore

$$I(x, y) = O^2 + R^2 + 2OR \cos\left(\frac{\pi r^2}{\lambda_o z_o} + \frac{2\pi y \sin\phi}{\lambda_o}\right). \tag{11.5}$$

Along the y-direction, the fringe frequency is the derivative of $y^2/2\lambda_o z_o + y(\sin\phi)/\lambda_o$ with respect to y. The frequency β is given by

$$\beta = \frac{y}{\lambda_o z_o} + \frac{\sin\phi}{\lambda_o}. \tag{11.6}$$

Notice that β is only $y/\lambda_o z_o$ in the in-line case ($\phi = 0$). For a given resolution (value of y) requirement in the in-line case, the film resolution capability demand is increased by $(\sin\phi)/\lambda_o$. For $\lambda_o = 0.6943$ μm and $\phi = 20°$, the increased demand alone is about 500 lines per millimeter. Spherical wave recording again generally requires better film resolution capability.[25] A particular case of off-axis holograms – Fourier transform holograms – has relaxed film resolution requirements but with a restricted field of view.[25]

11.5 Allowable subject movement

Another factor associated with the separate reference beam is the allowable object displacement during the exposure. The scene–reference beam optical path length must not change by more than $\lambda_o/10$.[7] From geometrical considerations (Figure 11.6), the worst case is object movement parallel to the hologram. Such a movement of amount D will give a pathlength change $D \sin\phi$ and the desired condition is:[7]

$$D_{max} \leqslant \lambda_o/10 \sin\phi. \qquad (11.7)$$

A Q-switched ruby laser of 15 ns duration, $\lambda_o = 0.6943\ \mu m$ and $\phi = 20°$ yields an allowable object velocity of only about 13.5 m/s. With $\phi = 10°$ and 1°, it becomes 26.7 m/s and 265.2 m/s respectively. Consequently, a smaller reference beam angle is better. When the angle ϕ approaches zero (the in-line method), the allowable object movement is ultimately governed by the analysis of Section 4.6.

The above discussion is for a forward scatter object beam. In the back scatter case (Figure 11.2), the situation is worse for longitudinal object movement. The path change is then double the object's movement resulting in[7]

$$D_{max}(\text{longitudinal}) \leqslant \lambda_o/5. \qquad (11.8)$$

Finally, for known object motion directions, geometrical considerations in the recording arrangement can yield changes in path lengths which are highly insensitive to object displacements.[27,28]

11.6 Very small or far objects

To observe very small and/or far objects in the reconstructed image, the system noise should be as small as possible. One practical approach is to store only field diffracted or scattered by the particle field. Approaches to accomplish this are shown in Figure 11.7. The method of Figure 11.7(a) is reported by Pluta.[19] A collimated beam[19] or the divergent beam of Figure 11.7(a) is converged with a positive lens near the recording plane. The diffracted field will reach everywhere in the recording plane while the directly transmitted light is blocked with an opaque mask. Upon reconstruction, the images will appear bright in the dark background. Particles down to 1 μm in diameter can be analyzed by this technique.[19]

Another approach is to illuminate the subject field by a pencil of sheet of laser light.[7,8] The direction of propagation can be adjusted

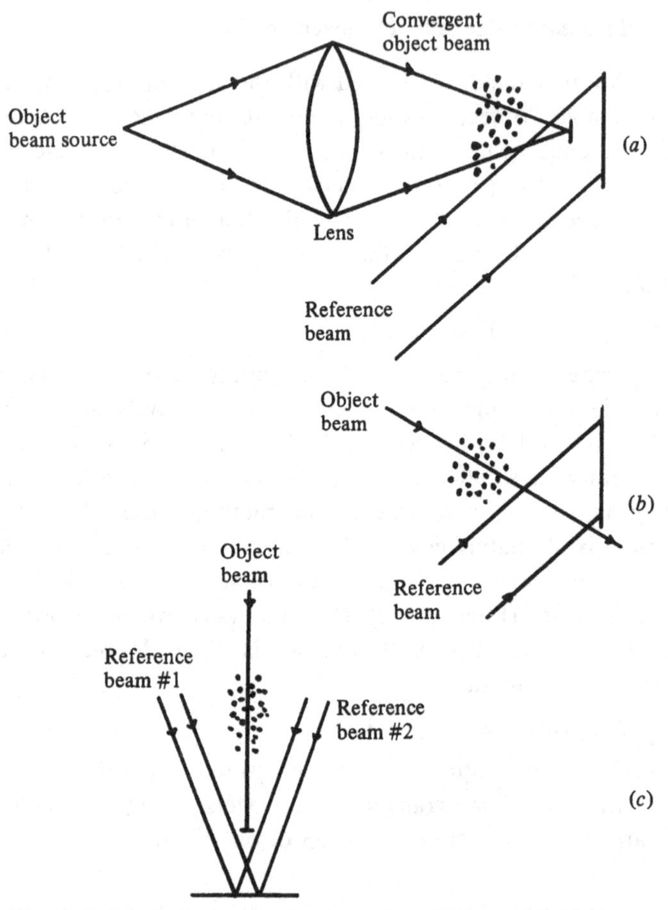

Fig. 11.7 Methods to record scattered or diffracted field alone. (*a*) Convergent scene illumination beam with an opaque stop for blocking the directly transmitted light. (*b*) Scene illumination by a pencil or sheet of light directed slightly away from the recording plane. (*c*) The two reference beam approach with an opaque stop to block the directly transmitted light of the object beam.

slightly away from the recording medium. Thus, only the scattered field will be stored in the hologram. The arrangement is represented in Figure 11.7(*b*). Again, bright images will appear in a dark background. Even if the particles are much smaller than the system resolution, they can be detected. The image brightness can be associated with the particle size. Variations in the overall intensity in the reconstructed image can also be associated with the local particle density.[7] The two reference beam arrangement of Figure 11.7(*c*) is better than the single reference approach. With weak scene beam

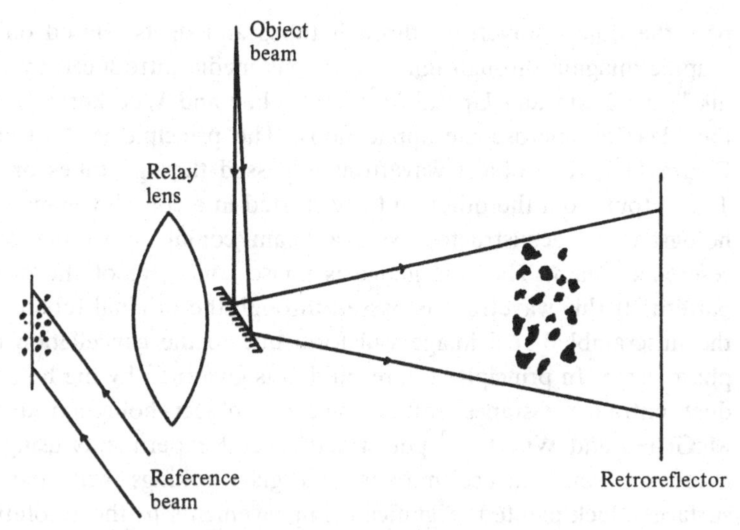

Fig. 11.8 Arrangement for off-axis holography for very far micro-objects.

signals, the desired 1:4 scene/reference beam intensity ratio is difficult to meet. The second reference beam along with the weak scene beam acts as an object beam thus solving the intensity ratio problem. The approach allows the scene beam to be attenuated by a factor of 10^{-6} as compared to the reference beam![7] Thus small, dense and/or far particles can be holographically recorded and reconstructed. The reconstructed 'second reference' intensity also acts as a reference in intensity measurements for quantitative particle density analysis.[7] The approach has been used for low angle forward (as shown in Figure 11.7(*b*) and (*c*)) as well as back scattering modes.[7] However, the forward scattering mode is better[7] because it is less sensitive to particle velocities (see Section 11.5).

Royer[29] used a modified off-axis technique to obtain 100 μm resolution for objects several meters away. The arrangement is represented in Figure 11.8. The retroreflector is used to rear illuminate the scene volume and a large aperture relay lens brings the scene closer to the recording plane. Since the total optical path of the object beam is independent of the object position, the source coherence requirement does not change over the scene depth. Akbari and Bjelkhagen[18] found an application of the arrangement in bubble chamber holography. Two reference beams from independent laser cavities can also be used for velocimetry applications.[29]

Another development in the area of imaging of remote objects is to pass the object wavefront through tubes and ducts. Based on holographic imaging through inhomogeneous media introduced by Kogelnik[30] and Leith and Upatnieks[31], McGehee and Wuerker[32] proposed the idea for microscopic applications. The principle is illustrated in Figure 11.9. The object wavefront is passed through tubes or ducts. The output from the other end is distorted in a complex manner. The hologram is reconstructed with a beam conjugate to the original reference beam. The real image is phase conjugate of the distorted pattern. If this wavefront is passed through the original tube or duct, the unscrambled real image will form due to the cancellation of the phase error. In principle, the resolution is governed by the hologram–duct entrance distance rather than the object–hologram distance. McGehee and Wuerker[32] performed several experiments using ducts made of front surface mirrors and glass tubings with the outer surfaces black-painted. Significant improvements in the resolution as compared to the unassisted case and methods to align the reconstruction beam are reported.[32]

In holography of large volumes, such as in bubble chambers, likewise large coherence length may be required. Harigel *et al.*[33] proposed a pulse stretching in a Q-switched ruby laser. The resulting coherence length of several meters is very useful in such experiments as well as in fiber optic transmission of the object field from a remote area.

11.7 Specialized techniques

There are certain less common off-axis techniques applicable to specialized applications. The use of acoustooptical switchable beam splitters is mentioned in Section 10.1.2. Basically the technique is to provide different reference beam directions for multiplexing events during a high speed process. Briones and Wuerker[9,34] electronically switched to the Pockels cell of a ruby laser so that two light pulses pass through different reference paths. During the reconstruction, one reference path at a time can be used to view the corresponding image. Thus, changing particle fields can be recorded and reconstructed with direction sense.

Marcisz and Aprahamian[35] divided the object beam (after passing through the scene volume) into two parts using a beam splitter. The undiffracted background from one part is filtered out using a lens-opaque stop system. Both these object beams are incident on the

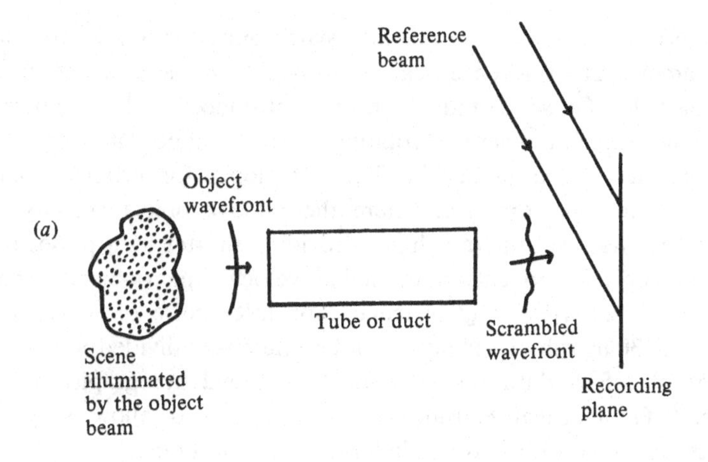

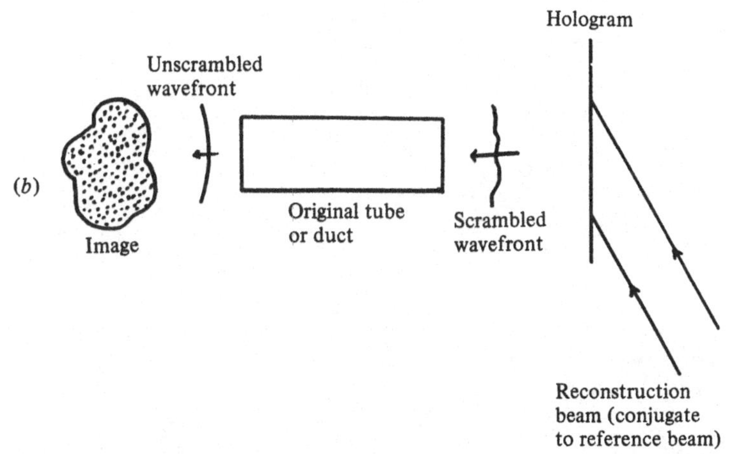

Fig. 11.9 Principle of holography through tubes and ducts: (*a*) recording; (*b*) reconstruction.

recording plane from different directions. Upon reconstruction, dark and light field images appear in corresponding directions.

Dyes and Ward[36] used a unique off-axis approach for ice crystal studies. The collimated beam, after passing through the subject volume, encounters a Ronchi ruling. The diffraction pattern is collected by a lens. The off-axis hologram between the light passing through a small hole at the zero-order and first-order diffraction is recorded.

Hypervelocity particles pose special problems in keeping the hologram fringes stationary. Klaubert and Ward[37] and Dyes, Kellen and

Klaubert[38] introduced a velocity synchronized Fourier transform holocamera. The reference beam is derived from back reflection from the particle. Gassend and Boerner[39] introduced a local reference beam holocamera where a rotating mirror compensates the scene motion effects. The method has been developed for airborne applications, rocket injectors, etc. where the relative object velocity with respect to the holocamera is high. Providing shorter light pulses is the natural approach to encounter higher velocity microobjects. Special techniques can yield a Q-switched ruby laser pulse of 3–5 ns duration.[35] A 30 ns ruby laser pulse can be effectively divided into several pulses of 1.5 ns duration each for high speed holography applications.[40] Geometrical considerations[27,28] can also allow very high object velocities with reasonably large pulse durations.

On the front of image contrast enhancement, holographic subtraction[41] is useful in double-exposure off-axis particle field holography.

References

Chapter 1

1. R. A. Dobbins, L. Crocco and I. Glassman, Measurement of mean particle sizes of sprays from diffractively scattered light, *AIIA J.*, **1**, 1882–6 (1963).
2. J. H. Ward, Determination of particle-size distribution from composite diffraction-pattern data, *J. Opt. Soc. Am.*, **59**, 1566 (1968).
3. W. L. Anderson, Particle counting and sizing by Fourier-plane measurements of scattered light, *J. Opt. Soc. Am.*, **59**, 1566–7 (1968).
4. W. L. Anderson and R. E. Beissner, Counting and classifying small objects by far-field light scattering, *Appl. Opt.*, **10**, 1503–8 (1971).
5. A. L. Wertheimer and W. L. Wilcock, Light scattering measurements of particle distributions, *Appl. Opt.*, **15**, 1616–20 (1976).
6. J. Swithenbank, J. M. Beer, D. S. Taylor, D. Abbot and G. C. McCreath, A laser diagnostic technique for the measurement of droplet and particle size distribution, *Prog. Astronaut. Aeronaut.*, **53**, 421–47 (1977).
7. A. L. Wertheimer, H. N. Frock and E. C. Muly, Light scattering instrumentation for particulate measurement in processes, in *Effective Utilization of Optics in Quality Assurance, Proc. S.P.I.E.*, **129**, 49–58 (1978).
8. C. Gorecki, Particle analyser using Fourier techniques, in *Optics and the Information Age, Proc. S.P.I.E.*, **813**, 295–6 (1987).
9. H. N. Frock and D. L. Walton, Particle size instrumentation for the ceramic industry, *Ceramic Bull.*, **59**, 650–1 (1980).
10. B. J. Thompson, Fraunhofer diffraction patterns of opaque objects with coherent background, *J. Opt. Soc. Am.*, **53**, 1350 (1963).
11. B. J. Thompson, Diffraction by opaque and transparent particles, *J. SPIE*, **2**, 43–6 (1964).
12. G. B. Parrent, Jr. and B. J. Thompson, On the Fraunhofer (far-field) diffraction patterns of opaque and transparent objects with coherent background, *Opt. Acta.*, **11**, 183–93 (1964).
13. B. A. Silverman, B. J. Thompson and J. H. Ward, A laser fog disdrometer, *J. Appl. Meteorology*, **3**, 792–801 (1964).

14. B. J. Thompson, A new method of measuring particle size by diffraction techniques, *Japan. J. Appl. Phys.*, **4** (Supplement 1), 302–7 (1965).
15. D. Gabor, Microscopy by reconstructed wavefronts, *Proc. Roy. Soc.*, **A197**, 454–87 (1949).
16. J. B. DeVelis, G. B. Parrent, Jr and B. J. Thompson, Image reconstruction with Fraunhofer holograms, *J. Opt. Soc. Am.*, **56**, 423–7 (1966).
17. B. J. Thompson, G. B. Parrent, J. H. Ward and B. Justh, A readout technique for laser fog disdrometer, *J. Appl. Meteorology*, **5**, 343–8 (1966).
18. B. J. Thompson and J. H. Ward, Particle sizing – the first direct use of holography, *Sci Res.*, **1**, 37–40 (1966).
19. C. Knox, Holographic microscopy as a technique for recording dynamic microscopic subjects, *Science*, **153**, 989–90 (1966).
20. J. B. DeVelis and G. O. Reynolds, *Theory and Applications of Holography*, Addison-Wesley Publishing Company, Reading, Massachusetts (1967).
21. G. O. Reynolds and J. B. DeVelis, Hologram coherence effects, *IEEE Trans. Antennas propagation*, **AP-15**, 41–8 (1967).
22. H. Fujiwara and K. Murata, Some effects of temporal coherence on in-line Fraunhofer holography, *Opt. Acta.*, **19**, 85–104 (1972).
23. G. A. Tyler and B. J. Thompson, Fraunhofer holography applied to particle size analysis: a reassessment, *Opt. Acta.*, **23**, 685–700 (1976).
24. T. Asakura, Y. Matsushita and H. Mishina, In-line Fraunhofer hologram field under illumination of partially coherent light, *Optik*, **47**, 185–93 (1977).
25. W. K. Witherow, A high resolution holographic particle sizing system, *Opt. Eng.*, **18**, 249–55 (1979).
26. C. S. Vikram and M. L. Billet, Some salient features of in-line Fraunhofer holography with divergent beams, *Optik*, **78**, 80–3 (1988).
27. E. N. Leith and J. Upatnieks, Reconstructed wavefronts and communication theory, *J. Opt. Soc. Am.*, **52**, 1123–30 (1962).
28. E. N. Leith and J. Upatnieks, Wavefront reconstruction with continuous-tone objects, *J. Opt. Soc. Am.*, **53**, 1377–81 (1963).
29. E. N. Leith and J. Upatnieks, Wavefront reconstruction with diffused illumination and three-dimensional objects, *J. Opt. Soc. Am.*, **54**, 1295–301 (1964).
30. R. E. Brooks, L. O. Heflinger, R. F. Wuerker and R. A. Briones, Holographic photography of high-speed phenomena with conventional and *Q*-switched ruby lasers, *Appl. Phys. Lett.*, **7**, 92–4 (1965).
31. B. J. Thompson, J. H. Ward and W. R. Zinky. Application of hologram techniques for particle size analysis, *Appl. Opt.*, **6**, 519–26 (1967).
32. B. J. Thompson, Fraunhofer holography, in *Holography, Proc. SPIE*, **15**, 25–9 (1968).

33. W. R. Zinky, Hologram techniques for particle-size analysis, *Annals New York Acad. Sci.*, **158**, 741–52 (1969).
34. B. J. Thompson, Particle size examination, in *The Engineering Uses of Holography*, E. R. Robertson and J. M. Harvey (eds.), Cambridge University Press, Cambridge (1970) pp. 249–59.
35. D. I. Staselko and V. A. Kosnikovskii, Holographic recording of three-dimensional ensembles of fast-moving particles, *Opt. Spectrosc.*, **34**, 206–10 (1973).
36. B. J. Thompson, Holographic particle sizing techniques, *J. Phys. E.; Sci. Instrum.*, **7**, 781–8 (1974).
37. J. D. Trolinger, Particle field holography, *Opt. Eng.*, **14**, 383–92 (1975).
38. R. A. Belz and R. W. Menzel, Particle field holography at Arnold Engineering Development Center, *Opt. Eng.*, **18**, 256–65 (1979).
39. B. J. Thompson and P. Dunn, Advances in far-field holography – Theory and applications, in *Recent advances in Holography, Proc. SPIE*, **215**, 102–11 (1980).
40. S. L. Cartwright, P. Dunn and B. J. Thompson, Particle sizing using far-field holography: new developments, *Opt. Eng.*, **19**, 727–33 (1980).
41. W. Grabowski, Measurement of the size and position of aerosol droplets using holography, *Opt. Laser Technol.*, **15**, 199–205 (1983).
42. B. J. Thompson, Holographic methods of dynamic particulate measurements – current status, in *High Speed Photography and Photonics, Proc. SPIE*, **348**, 626–33 (1983).
43. Y. J. Lee and J. H. Kim, A review of holography applications in multiphase flow visualization study, in *Physical and Numerical Flow Visualization*, M. L. Billet, J. H. Kim and T. R. Heidrick (Eds.), The American Society of Mechanical Engineers, New York (1985) pp. 47–58.
44. J. D. Trolinger, Particle and flow field holography, in *Holography: Critical Review of Technology, Proc. SPIE*, **532**, 40–62 (1985).
45. B. J. Thompson, Holographic methods for particle size and velocity measurement – recent advances, in *Holographic Optics II: Principles and Applications, Proc. SPIE*, **1136**, 308–26 (1989).
46. C. S. Vikram, Trends in far-field holography, *Optics and Lasers in Engineering*, **13**, 27–38 (1990).
47. C. S. Vikram (Ed), *Selected Papers in Holographic Particle Diagnostics, SPIE* Milestone Series Volume MS 21, The International Society for Optical Engineering, Bellingham, Washington (1990).

Chapter 2

1. R. J. Collier, C. B. Burckhardt and L. H. Lin, *Optical Holography*. Academic Press, New York (1971).
2. H. M. Smith, *Principles of Holography*, Second Edition, John Wiley & Sons, New York (1975).

3. H. J. Caulfield, *Handbook of Optical Holography*, Academic Press, New York (1979).
4. P. Hariharan, *Optical Holography*, Cambridge University Press, Cambridge (1984).
5. K. Patorski, Fraunhofer diffraction patterns of tilted planar objects, *Opt. Acta.*, **30**, 673–9 (1983).
6. W. D. Furlan, E. E. Sicre and M. Garavaglia, Quasi-geometrical paraxial approach to Fraunhofer diffraction by thick planar objects using phase-space signal representations, *J. Mod. Opt,* **35**, 735–41 (1988).
7. R. W. Meier, Magnification and third-order aberrations in holography, *J. Opt. Soc. Am.*, **55**, 987–92 (1965).

Chapter 3

1. G. A. Tyler and B. J. Thompson, Fraunhofer holography applied to particle size analysis; a reassessment, *Optica Acta*, **23**, 685–700 (1976).
2. M. Born and E. Wolf, *Principles of Optics*, Sixth (corrected) edition, Pergamon Press, Oxford (1980).
3. J. D. Trolinger, R. A. Belz and W. M. Farmer, Holographic techniques for the study of dynamic particle fields, *Appl. Opt.*, **8**, 957–61 (1969).
4. C. S. Vikram and M. L. Billet, Optimizing image-to-background irradiance ratio in far-field in-line holography, *Appl. Opt.*, **23**, 1995–98 (1984).
5. Ref. 2, p. 753.
6. D. Gabor, A new microscopic principle, *Nature*, **161**, 777–8 (1948).
7. D. Gabor, Microscopy by reconstructed wavefronts, *Proc. Roy. Soc.*, **A197**, 454–87 (1949).
8. D. Gabor, Microscopy by reconstructed wavefronts, *Proc. Phys. Soc*, **B64**, 449–69 (1951).
9. E. N. Leith and J. Upatnieks, Reconstructed wavefronts and communication theory, *J. Opt. Soc. Am.*, **52**, 1123–30 (1962).
10. E. N. Leith and J. Upatnieks, Wavefront reconstruction with diffuse illumination and three-dimensional objects, *J. Opt. Soc. Am.*, **54**, 1295–301 (1964).
11. J. B. DeVelis, G. B. Parrent, Jr., and B. J. Thompson, Image reconstruction with Fraunhofer holograms, *J. Opt. Soc. Am.*, **56**, 423–7 (1966).
12. C. Özkul, Nonlinearities of an aperture-limited in-line far-field hologram. *Appl. Opt.*, **25**, 3924–6 (1986).
13. W. L. Bragg and G. L. Rogers, Elimination of the unwanted image in diffraction microscopy, *Nature*, **167**, 190–1 (1951).
14. J. D. Trolinger, Particle field holography, *Opt. Eng.*, **14**, 383–92 (1975).
15. K. Murata, H. Fujiwara and T. Asakura, Several problems in in-line Fraunhofer holography for bubble-chamber track decoding, in *The*

Engineering Uses of Holography, E. R. Robertson and J. M. Harvey (Eds), Cambridge University Press, Cambridge (1970) pp. 289–305.
16. J. T. Bartlett and R. J. Adams, Development of a holographic technique for sampling particles in moving aerosols, *Microscope*, **20**, 375–84 (1972).
17. M. Howells, Fundamental limits in x-ray holography, in *X-Ray Microscopy* II, D. Sayre, M. Howells, J. Kirz and H. Rarback (Eds.), Springer-Verlag, Berlin (1988) pp. 263–71.
18. B. J. Thompson, G. P. Parrent, B. Justh and J. Ward, A readout technique for the laser fog disdrometer. *J. Appl. Meterol.*, **5**, 343–8 (1966).
19. B. J. Thompson, J. Ward and W. Zinky, Application of hologram techniques for particle size analysis, *J. Opt. Soc. Am.*, **55**, 1566A (1965); *Appl. Opt.*, **6**, 519–26 (1967).
20. J. B. DeVelis and G. O. Reynolds, *Theory and Applications of Holography*, Addison-Wesley Publishing Company, Reading, Massachusetts (1967) Chap. 5.
21. L. Onural and P. D. Scott, Digital decoding of in-line holograms for imaging fractal aggregates, *Electron. Lett.*, **22**, 1118–19 (1986).
22. L. Onural and P. D. Scott, Digital decoding of in-line holograms, *Opt. Eng.*, **26**, 1124–32 (1987).
23. G. Liu and P. D. Scott, Phase retrieval and twin-image elimination for in-line Fresnel holograms, *J. Opt. Soc. Am. A*, **4**, 159–65 (1987).
24. G. Liu, Object reconstruction from noisy holograms, *Opt. Eng.*, **29**, 19–24 (1990).
25. K. A. Nugent, Twin-image elimination in Gabor holography, *Opt. Commun.*, **78**, 293–9 (1990).
26. L. Hua, G. Xie, D. T. Shaw and P. D. Scott, Resolution enhancement in digital in-line holography, in *Optics, Illumination, and Image Sensing for Machine Vision V Proc. SPIE*, **1385**, 142–51 (1991).

Chapter 4

1. G. A. Tyler and B. J. Thompson, Fraunhofer holography applied to particle size analysis: a reassessment, *Optica Acta*, **23**, 685–700 (1976).
2. G. B. Parrent, Jr. and B. J. Thompson, On the Fraunhofer (far field) diffraction patterns of opaque and transparent objects with coherent background, *Optica Acta*, **11**, 183–93 (1964).
3. S. L. Cartwright, P. Dunn and B. J. Thompson, Particle sizing using far-field holography: new developments, *Opt. Eng.*, **19**, 727–33 (1980).
4. P. Dunn and B. J. Thompson, Object shape, fringe visibility, and resolution in far-field holography, *Opt. Eng.*, **21**, 327–32 (1982).
5. F. Slimani, G. Grehan, G. Gouesbet and D. Allano, Near-field Lorenz–Mie Theory and its application to microholography, *Appl. Opt.*, **23**, 4140–8 (1984).

6. J. B. DeVelis and G. O. Reynolds, *Theory and Application of Holography*, Addison-Wesley Publishing Company, Reading, Massachusetts (1967) Chap. 7.
7. W. Grabowski, Measurement of the size and position of aerosol droplets using holography, *Optics and Laser Technology*, **15**, 199–205 (1983).
8. B. J. Thompson, Holographic particle sizing techniques, *J. Phys. E: Sci Instrum*, **7**, 781–8 (1974).
9. B. J. Thompson, J. H. Ward, and W. R. Zinky, Application of hologram techniques for particle size analysis, *Appl. Opt.*, **6**, 519–26 (1967).
10. C. S. Vikram and M. L. Billet, Some salient features of in-line Fraunhofer holography with divergent beams, *Optik*, **78**, 80–3 (1988).
11. B. J. Thompson, G. B. Parrent, J. H. Ward and B. Justh, A readout technique for the laser fog disdrometer, *J. Appl. Meterology*, **5**, 343–8 (1966).
12. J. B. DeVelis and G. O. Reynolds, *Theory and Applications of Holography*, Addison-Wesley Publishing Company, Reading, Massachusetts (1967) Chap. 8.
13. W. K. Witherow, A high resolution particle sizing system, *Opt. Eng.*, **18**, 249–55 (1979).
14. G. O. Reynolds and J. B. DeVelis, Hologram coherence effect, *IEEE Trans. Antennas Propag.*, **AP-15**, 41–8 (1967).
15. H. Fujiwara and K. Murata, Some effects of spatial and temporal coherence on in-line Fraunhofer holography, *Optica Acta*, **19**, 85–104 (1972).
16. T. Asakura, Y. Matsushita and H. Mishina, In-line Fraunhofer hologram field under illumination of partially coherent light, *Optik*, **47**, 185–93 (1977).
17. M. Born and E. Wolf, *Principles of Optics*, Sixth edition, Pergamon Press, Oxford (1980) Chap. 10.
18. R. J. Collier, C. B. Burckhardt and L. H. Lin, *Optical Holography*, Academic Press, New York (1971) Chap. 7.
19. Ref. 18, Chap. 11.
20. B. B. Brenden, Miniature multiple-phase *Q*-switched ruby laser holocamera for aerosol analysis, *Opt. Eng.*, **20**, 907–11 (1981).
21. C. S. Vikram and M. L. Billet, On allowable object velocity in Fraunhofer holography, *Optik*, **80**, 155–60 (1987).
22. J. Crane, P. Dunn, P. H. Malyak and B. J. Thompson, Particulate velocity and size measurements using holographic and optical processing methods, in *High Speed Photography and Photonics, Proc. SPIE*, **348**, 634–42 (1983).
23. Ref. 17, Chap. 8.
24. Lord Rayleigh, *Collected Papers*, Cambridge University Press, Cambridge (1902) Vol. 3, p. 84. See also Dover Publications, New York (1964).
25. D. N. Grimes and B. J. Thompson, Two-point resolution with partially coherent light, *J. Opt. Soc. Am.*, **57**, 1330–4 (1967).

26. D. Gabor, Laser speckles and its elimination, *IBM Jour. Res. Dev.*, **14**, 509–14 (1970).
27. K. W. Pavitt, M. C. Jackson, R. J. Adams and T. J. Bartlett, Holography of fast moving cloud droplets, *J. Phys. E: Sci Instrum.*, **3**, 971–5 (1970).
28. J. T. Bartlett and R. J. Adams, Development of a holographic technique for sampling particles in moving aerosols, *Microscope*, **20**, 375–84 (1972).
29. J. W. Goodman, Some fundamental properties of speckle, *J. Opt. Soc. Am.*, **66**, 1145–50 (1976).
30. R. A. Briones, L. O. Heflinger and R. F. Wuerker, Holographic microscopy, *Appl. Opt.*, **17**, 944–50 (1978).
31. L. O. Heflinger, G. L. Stewart and C. R. Booth, Holographic motion pictures of microscopic plankton, *Appl. Opt.*, **17**, 951–4 (1978).

Chapter 5

1. J. D. Trolinger, Particle field holography, *Opt. Eng.*, **14**, 383–92 (1975).
2. W. K. Witherow, A high resolution holographic particle sizing system, *Opt. Eng.*, **18**, 249–55 (1979).
3. P. Hariharan, *Optical Holography*, Cambridge University Press, Cambridge (1984) p. 175.
4. J. B. DeVelis and G. O. Reynolds, *Theory and Applications of Holography*, Addison-Wesley Publishing Company, Reading, Massachusetts (1967) Chap. 8.
5. B. J. Thompson, J. H. Ward and W. R. Zinky, Application of hologram techniques for particle size analysis, *Appl. Opt.*, **6**, 519–26 (1967).
6. B. J. Thompson, Fraunhofer holography, in *Holography, Proc. SPIE*, **15**, 25–39 (1968).
7. M. E. Fourney, J. H. Matkin and A. P. Waggoner, Aerosol Size determination via holography, *Rev. Sci. Instrum.*, **40**, 205–13 (1969).
8. D. M. Robinson, Interim tests on a holographic technique for photographing high-speed mil-size particles, in *Holographic Instrumentation Applications*, B. Ragent and R. M. Brown (Eds.), NASA SP-248 (1970) pp. 221–35.
9. B. J. Thompson and P. Dunn, Advances in far-field holography – theory and applications, in *Recent Advances of Holography III, Proc. SPIE*, **215**, 102–11 (1980).
10. B. J. Thompson, Droplet characteristics with conventional and holographic imaging techniques, in *Liquid Particle Size Measurement Techniques*, J. M. Tishkoff, R. D. Ingebo and J. B. Kennedy (Eds.), ASTM, STP-848 (1984) pp. 111–22.
11. B. A. Silverman, B. J. Thompson and J. H. Ward, A laser fog disdrometer, *J. Appl. Meteorology*, **3**, 792–801 (1964).

12. J. H. Ward and B. J. Thompson, In-line hologram systems for bubble chamber recording, *J. Opt. Soc. Am.*, **57**, 275–6 (1967).
13. R. A. Belz and N. S. Dougherty, Jr., In-line holography for recording liquid sprays, in *Proc. Symp. Engineering Applications of Holography, SPIE, Redondo Beach, California* (1972) pp. 209–18.
14. B. J. Thompson and J. H. Ward, Particle sizing – the first direct use of holography, *Sci. Res.*, **1**, 37–40 (1966).
15. B. J. Thompson, Application of Fraunhofer holograms, in *Record of the IEEE 9th Annual Symposium on Electron, Ion and Laser Beam Technology*, R. F. W. Pease (Ed.), IEEE (1967) pp. 295–303.
16. J. H. Ward, Holographic particle sizing, in *Applications of Lasers to Photography and Information Handling*, R. D. Murry (Ed.), SPSE, Washington (1968) pp. 217–33.
17. W. R. Zinky, Hologram techniques for particle-size analysis, *Annals New York Acad. Sci.*, **158**, 741–52 (1969).
18. B. J. Thompson, Particle size examination, in *The Engineering Uses of Holography*, E. R. Robertson and J. M. Harvey (Eds.), Cambridge University Press, Cambridge (1970) pp. 249–59.
19. B. J. Thompson, Applications of holography, in *Laser Applications*, Volume 1, M. Ross (Ed.), Academic Press, New York (1971) pp. 1–60.
20. D. I. Staselko and V. A. Kosnikovskii, Holographic recording of three-dimensional ensembles of fast-moving particles, *Opt. Spectrosc.*, **34**, 206–10 (1973).
21. B. J. Thompson, Holographic particle sizing techniques, *J. Phys. E.; Sci. Instrum.*, **7**, 781–8 (1974).
22. B. J. Thompson, Particle size measurements, in *Handbook of Optical Holography*, H. J. Caulfield (Ed.), Academic Press, New York (1979) pp. 609–10.
23. W. Grabowski, Measurement of the size and position of aerosol droplets using holography, *Opt. Laser Technol.*, **15**, 199–205 (1983).
24. B. J. Thompson and W. R. Zinky, Holographic detection of submicron particles, *Appl. Opt.*, **7**, 2426–8 (1968).
25. H. Royer, Holographic velocity of submicron particles, *Opt. Commun.*, **20**, 73–5 (1977).
26. C. Knox and R. E. Brooks, Holographic motion picture microscopy, *Proc. Roy. Soc. Lond.*, **B-174**, 115–21 (1969).
27. E. A. Boettner and B. J. Thompson, *Opt. Eng.*, **12**, 56–9 (1973).
28. H. Rieck, Holographic small particle analysis with ultraviolet ruby laser light, in *Engineering Uses of Holography*, E. R. Robertson and J. M. Harvey (Eds.), Cambridge University Press, Cambridge (1970) pp. 261–6.
29. T. Murakami and M. Ishikawa, Holographic measurements of velocity distribution of particles accelerated by a shock wave, in *Thirteenth International Congress on High Speed Photography and Photonics, Proc. SPIE*, **189**, 326–9 (1979).
30. R. Hickling, Scattering of light by spherical liquid droplets using computer-synthesized holograms, *J. Opt. Soc. Am.*, **58**, 455–60 (1968).

31. R. Hickling, Holography of liquid droplets, *J. Opt. Soc. Am.*, **59**, 1334–9 (1969).
32. A. Tonomura, A. Fukuhara, H. Watanabe and T. Komoda, Optical reconstruction of image from Fraunhofer electron-hologram, *Japan J. Appl. Phys.*, **7**, 295 (1968).
33. A. Tonomura, Applications of electron holography, *Rev. Mod. Physics,* **59**, 639–69 (1987).
34. H. M. A. El-Sum and P. Kirpatrick, Microscopy by reconstructed wavefronts, *Phys. Rev.*, **85**, 763 (1952).
35. H. M. A. El-Sum, Reconstructed wavefront microscopy, Ph.D. Thesis, Standford University (1952). Available from University Microfilm Inc., Ann Arbor, Michigan.
36. A. V. Baez, A study in diffraction microscopy with special reference to x-rays, *J. Opt. Soc. Am.*, **42**, 756–62 (1952).
37. M. R. Howells, M. A. Iarocci and J. Kirz, Experiments in x-ray holographic microscopy using synchrotron radiation, *J. Opt. Soc. Am. A*, **3**, 2171–8 (1986).
38. J. C. Solem and G. F. Chapline, X-ray biomicroholography, *Opt. Eng.*, **23**, 193–203 (1984).
39. D. Sayre, M. Howells, J. Kirz and H. Rarback (Eds.), *X-Ray Microscopy II*, Springer-Verlag, Berlin (1988).
40. R. A. London, M. D. Rosen and J. E. Trebes, Wavelength choice for soft x-ray laser holography of biological samples, *Appl. Opt.*, **28**, 3397–404 (1989).
41. C. Jacobson, M. Howells, J. Kirz and S. Rothman, X-ray holographic microscopy using photoresists, *J. Opt. Soc. Am. A*, **7**, 1847–61 (1990).
42. C. M. Vest, *Holographic Interferometry*, John Wiley & Sons, New York (1979) p. 49.
43. M. J. Landry and G. S. Phipps, Holographic characteristics of 10E75 plates for single and multiple-exposure holograms, *Appl. Opt.*, **14**, 2260–6 (1975).
44. G. S. Phipps, C. E. Robertson and F. M. Tamashiro, Reprocessing of nonoptimally exposed holograms, *Appl. Opt.*, **19**, 802–11 (1980).
45. J. B. DeVelis, G. B. Parrent, Jr. and B. J. Thompson, Image reconstruction with Fraunhofer holograms, *J. Opt. Soc. Am.,* **56**, 423–7 (1966).
46. C. S. Vikram and M. L. Billet, Optimizing image-to-background irradiance ratio in far-field in-line holography, *Appl. Opt.*, **23**, 1995–8 (1984).
47. P. Dunn and J. M. Walls, Absorption and phase in-line holograms: a comparison, *Appl. Opt.*, **18**, 2171–4 (1979).
48. H. Royer, P. Lecoq and E. Ramsmeyer, Application of holography to bubble chamber visualization, *Opt. Commun.*, **37**, 84–6 (1981).
49. R. Bexon, J. Gibbs and G. D. Bishop, Automatic assessment of aerosol holograms, *J. Aerosol Sci.*, **7**, 397–407 (1976).
50. R. Bexon, M. G. Dalzell and M. C. Stainer, In-line holography and the assessment of aerosols, *Opt. Laser Technol.*, **8**, 161–5 (1976).
51. R. Bexon, Magnification in aerosol sizing by holography, *J. Phys.*

E: Sci. Instrum., **6**, 245–8 (1973).
52. C. Özkul, Nonlinearities of an aperture-limited in-line far-field hologram, *Appl. Opt.*, **25**, 3924–6 (1986).
53. K. W. Pavitt, M. C. Jackson, R. J. Adams and T. J. Bartlett, Holography of fast-moving cloud droplets, *J. Phys. E: Sci. Instrum.*, **3**, 971–5 (1970).
54. P. Dunn and J. M. Walls, Improved microimages from in-line absorption holograms, *Appl. Opt.*, **18**, 263–4 (1979).
55. Ref. 3, Chap. 4.
56. J. T. Bartlett and R. J. Adams, Development of a holographic technique for sampling particles in moving aerosols, *Microscope*, **20**, 375–84 (1972).
57. S. Cartwright, Far-field holography using a liquid-crystal light value, *J. Opt. Soc. Am.*, **72**, 410 (1982).
58. S. L. Cartwright, Aspects of far-field holography, Ph.D. Thesis, University of Rochester, 1982.
59. J. W. Goodman, *Introduction to Fourier Optics*, McGraw-Hill, New York (1968).
60. R. J. Collier, C. B. Burckhardt and L. H. Lin, *Optical Holography*, Academic Press, New York (1971) Chap. 6.
61. H. Lipson, *Optical Transforms*, Academic Press, London (1972).
62. J. D. Gaskill, *Linear Systems, Fourier Transforms, and Optics*, John Wiley & Sons, New York (1978).
63. P. M. Duffieux, *The Fourier Transform and its Application to Optics*, Second Edition, John Wiley & Sons, New York (1983).
64. F. T. S. Yu, *Optical Information Processing*, John Wiley & Sons, New York, (1983).
65. D. G. Falconer, Optical processing of bubble chamber photographs, *Appl. Opt.*, **5**, 1365–9 (1966).
66. K. Murata, H. Fujiwara and T. Asakura, Several problems in in-line Fraunhofer holography for bubble-chamber track recording, in *The Engineering Uses of Holography*, E. R. Robertson and J. M. Harvey (Eds.), Cambridge University Press, Cambridge (1970) pp. 289–305.
67. Ref. 61, Chap. 8.
68. Ref. 62, Chap. 11.
69. Ref. 63, pp. 187–90.
70. K. G. Birch, A spatial frequency filter to remove zero frequency, *Optica Acta*, **15**, 113–27 (1968).
71. M. Young, Spatial filtering microscope for linewidth measurements, *Appl. Opt.*, **28**, 1467–73 (1989).
72. J. D. Trolinger, W. M. Farmer and R. A. Belz, Multiple exposure holography of time varying three-dimensional fields, *Appl. Opt.*, **7**, 1640–1 (1968).
73. J. D. Trolinger and T. H. Gee, Resolution factors in edgeline holography, *Appl. Opt.*, **10**, 1319–23 (1971).
74. J. D. Trolinger, R. A. Belz and W. M. Farmer, Holographic techniques for the study of dynamic particle fields, *Appl. Opt.*, **8**, 957–61 (1969).

75. R. Jones and C. Wykes, *Holographic and Speckle Interferometry*, Cambridge University Press, London (1983) Chap. 4.
76. J. A. Liburdy, Holocinematographic velocimetry: resolution limitation for flow measurement, *Appl. Opt.*, **26**, 4250–5 (1987).
77. D. Gabor, G. W. Stroke, D. Brumn, A. Funkhouser and A. Labeyrie, Reconstruction of phase objects by holography, *Nature*, **208**, 1159–62 (1965).
78. L. H. Tanner, Some applications of holography in fluid mechanics, *J. Sci. Instrum.*, **43**, 81–3 (1966).
79. L. H. Tanner, On holography of phase objects, *J. Sci. Instrum.*, **43**, 346, (1966).
80. L. H. Tanner, The application of lasers to time-resolved flow visualization, *J. Sci. Instrum.*, **43**, 353–8 (1966).
81. Ref. 42, Chaps. 5 and 6.
82. H. J. Raterink and J. H. J. Van der Meulen, Application of holography in cavitation and flow research, in *Applications of Holography and Optical Data Processing*, E Morom, A. A. Friesem and E. Wiener-Avnear (Eds.), Pergamon Press, Oxford (1977) pp. 643–52.
83. J. Katz, Cavitation phenomena within regions of flow separation, *J. Fluid Mech.*, **140**, 397–436 (1984).
84. Y. J. Lee and J. H. Kim, A review of holography applications in multiphase flow visualization study, in *Physical and Numerical Flow Visualization*, M. L. Billet, J. H. Kim and T. R. Heidrick (Eds.), The American Society of Mechanical Engineers, New York, pp. 47–58 (1985).
85. G. M. Lomas, Interference phenomena in Fraunhofer holograms and in their reconstructions of tapered glass fibers, *Appl. Opt.*, **8**, 2037–41 (1969).
86. S. L. Cartwright, P. Dunn and B. J. Thompson, Particle sizing using far-field holography: new developments, *Opt. Eng.*, **19**, 727–33 (1980).
87. I. Prikryl, Holographic imaging of nonopaque droplets or particles, *J. Opt. Soc. Am.*, **71**, 1569 (1981).
88. I. Prikryl and C. M. Vest, Holographic imaging of semitransparent droplets or particles, *Appl. Opt.*, **21**, 2541–7 (1982).
89. Z. K. Lu and H. Meng, Holography of semi-transparent spherical droplets, in *Optics in Engineering Measurement, Proc SPIE*, **599**, 19–25 (1986).
90. F. B. Paterson, Hydrodynamic cavitation and some considerations of the influence of free gas content. *Proceedings 9th Symposium on Naval Hydrodynamics, Paris* (1972) pp. 1131–86.
91. G. E. Davis, Scattering of light by an air bubble in water. *J. Opt. Soc. Am.*, **45**, 572–81 (1955).
92. E. M. Gates and J. Bacon, A note on the determination of cavitation nuclei distribution by holography, *J. Ship. Res.*, **22**, 29–31 (1978).
93. S. Gowing, C. S. Vikram and S. Burton, Comparison of holography,

light scattering and venturi techniques for bubble measurements in a water tunnel, in *Cavitation and Multiphase Flow Forum, 1988*, O. Furuya (Ed.), The American Society of Mechanical Engineers, New York (1988) p. 25–8.

94. D. M. Robinson, A calculation of edge smear in far-field holography using a short-cut edge trace technique, *Appl. Opt.*, **9**, 496–7 (1970).
95. R. A. Belz, An investigation of the real image reconstructed by an in-line Fraunhofer hologram aperture-limited by film effects, Ph.D. dissertation, University of Tennessee, August (1971).
96. R. A. Belz, An investigation of the real image of an aperture-limited, in-line Fraunhofer hologram, in *Holography and Optical Filtering*, NASA SP-299 (1973) pp. 193–202.
97. R. A. Belz and F. M. Shofner, Characteristics and measurements of an aperture-limited in-line hologram image, *Appl. Opt.*, **11**, 2215–22 (1972).
98. W. P. Dotson, Jr., A theoretical calculation of edge smear in far-field holography, in *Proceedings Symp. Engineering Applications of Holography, SPIE, Redondo Beach, California* (1972) pp. 113–22.
99. G. A. Tyler and B. J. Thompson, Fraunhofer holography applied to particle size analysis: A reassessment, *Opt. Acta*, **23**, 685–700 (1976).
100. C. Ozkul, Effect of finite aperture on linewidth measurement using in-line Fraunhofer holography, *Opt. Laser Technol.*, **18**, 36–8 (1986).
101. C. S. Vikram, Accurate linewidth measurements in aperture-limited in-line Fraunhofer holography, *J. Mod. Opt.*, **37**, 2047–54 (1990).
102. G. Haussmann and W. Lauterborn, Determination of size and position of fast moving gas bubbles in liquids by digital 3-D processing of hologram reconstructions, *Appl. Opt.*, **19**, 3529–35 (1980).
103. J. Katz, T. J. O'Hern and A. J. Acosta, An underwater holographic camera system for the detection of microparticles, in *Cavitation and Multiphase Flow Forum, 1984*, J. W. Hoyt (Ed.), The American Society of Mechanical Engineers, New York (1984) pp. 22–25.
104. T. J. O'Hern, J. Katz and A. J. Acosta, Holographic measurements of cavitation nuclei in the sea, in *Cavitation and Multiphase Flow Forum, 1985*, J. W. Hoyt and O. Furuya (Eds.), The American Society of Mechanical Engineers, New York (1985) pp. 39–42.
105. C. Knox, Holographic microscopy as a technique for recording dynamic subjects, *Science*, **153**, 989–90 (1966).
106. K. Murata, H. Fujiwara and T. Asakura, Use of diffused illumination on in-line Fraunhofer holography, *Japan, J. Appl. Phys.*, **7**, 301–2 (1968).
107. R. R. Roberts and T. D. Black, Infrared holograms recorded at 10.6 μm and reconstructed at 0.6328 μm, *Appl. Opt.*, **15**, 2018–19 (1976).
108. G. H. Tyler, Image coding with far-field holograms, *Appl. Opt.*, **17**, 2768–72 (1978).

109. H. Heidt and R. Furchert, Holographic droplet sizing with automatic evaluation in pest control, *Microscopia Acta, Suppl.* **1**, 51–61 (1977).
110. J. Lidl, Image analysis of droplet holograms by TV camera and table calculator via IEC-BUS, *Proceedings 3rd Int. Conf. Lasers 77, Opto-Electronics, Munich* (1977) pp. 767–72.
111. P. Payne, K. L. Carder and R. G. Steward, Image analysis techniques for holograms of dynamic oceanic particles, *Appl. Opt.*, **23**, 204–10 (1984).
112. S. P. Feinstein and M. A. Girard, Automated holographic drop-size analyzer, *Appl. Opt.*, **20**, A52 (1981).
113. L. M. Weinstein, C. B. Beeler and A. M. Lindemann, High-speed holocinematographic velocimeter for studying turbulent flow control physics, AIIA-85-0526, presented at the AIAA Shear Flow Control Conference, 12–14 March 1985, Boulder, Colorado.
114. A. C. Stanton, H. J. Caulfield and G. W. Stewart, An approach for automated analysis of particle holograms, *Opt. Eng.*, **23**, 577–82 (1984).
115. H. J. Caulfield, Automatic analysis of particle holograms, *Opt. Eng.*, **24**, 462–3 (1985).
116. C. S. Vikram and M. L. Billet, Test section window sidewall effects and their elimination in particle field holography, *Opt. Lasers Eng.*, **12**, 67–73 (1990).
117. C. S. Vikram, Optimizing image quality in Fraunhofer holography with variable intensity reconstruction beam, *Optik*, **86**, 58–60 (1990).
118. J. D. Trolinger, H. Tan and R. Lal, A space flight holography system for flow diagnostics in micro-gravity, in *Focus on Electro-Optic, Sensing and Measurement*, G. Kychakoff (Ed.), Laser Institute of America (1988) pp. 139–143.
119. J. D. Trolinger, R. B. Lal and A. K. Batra, Holographic instrumentation for monitoring crystal growth in space, in *Optical Testing and Metrology III: Recent Advances in Industrial Optical Inspection, Proc. SPIE*, **1332**, 151–65 (1990).

Chapter 6

1. H. M. A. El-Sum, Reconstructed wavefront microscopy, Ph.D. Thesis, Stanford University (1952). Available from University Microfilm Inc., Ann Arbor, Michigan.
2. G. L. Rogers, Experiments in diffraction microscopy, *Proc. Royal Soc. (Edinburgh)*, **63A**, 193–221 (1952).
3. P. Kirpatrick and H. M. A. El-Sum, Image formation by reconstructed wave fronts. I. Physical principles, *J. Opt. Soc. Am.*, **46**, 825–31 (1956).
4. G. L. Rogers, Two hologram methods in diffraction microscopy, *Proc. Royal Soc. (Edinburgh)*, **64A**, 209–21 (1956).

5. E. N. Leith, J. Upatnieks and K. A. Hains, Microscopy by wavefront reconstruction, *J. Opt. Soc. Am.*, **55**, 981–6 (1965).
6. R. W. Meier, Magnification and third-order aberrations in holography, *J. Opt. Soc. Am.*, **55**, 987–92 (1965).
7. J. B. DeVelis and G. O. Reynolds, *Theory and Applications of Holography*, Addison-Wesley Publishing Company, Reading, Massachusetts, 1967, Chap. 3.
8. M. E. Fourney, J. H. Matkin and A. P. Waggoner, Aerosol size and velocity determination via holography, *Rev. Sci. Instrum.*, **40**, 205–13 (1969).
9. G. Haussmann and W. Lauterborn, Determination of size and position of fast moving gas bubbles in liquids by digital 3-D image processing of hologram reconstructions, *Appl. Opt.*, **19**, 3529–35 (1980).
10. C. S. Vikram and M. L. Billet, Volume magnification in particle field holography, *Appl. Opt.*, **26**, 1147–50 (1987).
11. R. Bexon, Magnificaton in aerosol sizing by holography, *J. Phys. E: Sci. Instrum.*, **6**, 245–8 (1973).
12. R. Bexon, J. Gibbs and G. D. Bishop, Automatic assessment of aerosol holograms, *J. Aerosol Sci.*, **7**, 397–407 (1976).
13. R. Bexon, M. G. Dalzell and M. C. Stainer, In-line holography and the assessment of aerosols, *Opt. Laser Technology*, **8**, 161–5 (1976).
14. W. K. Witherow, A high resolution holographic particle sizing system, *Opt. Eng.*, **18**, 249–55 (1979).
15. C. S. Vikram and M. L. Billet, Analysis of particle field holograms on moveable stage during recontruction, *J. Mod. Opt.*, **36**, 405–11 (1989).
16. C. S. Vikram and T. E. McDevitt, Simple determination of magnification due to recording configuration in particle field holography, *Appl. Opt.*, **28**, 208–9 (1989).
17. I. S. Gradshteyn and I. M. Ryzhik, *Tables of Integrals, Series and Products*, Academic Press, Orlando (1980).
18. R. Kingslake, *Lens Design Fundamentals*, Academic Press, New York (1978) Chap. 3.
19. W. J. Smith, Image formation: geometrical and physical optics, in *Handbook of Optics*, W. G. Driscoll and W. Vaughan (Eds.), McGraw-Hill, New York (1978) Section 2.
20. C. S. Vikram and M. L. Billet, Holographic image formation of objects inside a chamber, *Optik*, **61**, 427–32 (1982).
21. C. S. Vikram and M. L. Billet, Magnification with divergent beams in Fraunhofer holography of object inside a chamber, *Optik*, **63**, 109–14 (1983).

Chapter 7

1. D. I. Staselko and V. A. Kosnikovskii, Holographic recording of three-dimensional ensembles of fast-moving particles, *Opt. Spectrosc.*, **34**, 206–10 (1973).

2. C. Özkul, Contribution to the study of far-field diffraction: application to the linewidth measurement by microholography and by the electronic scanning system for diffraction pattern analysis (in French), D. Sc. Thesis, Rouen University, France (1987).
3. R. W. Meier, Magnification and third-order aberrations in holography, *J. Opt. Soc. Am.*, **55**, 987–92 (1965).
4. E. N. Leith, J. Upatnieks and K. A. Haines, Microscopy by wavefront reconstruction, *J. Opt. Soc. Am.*, **55**, 981–6 (1965).
5. M. Born and E. Wolf, *Principles of Optics*, Sixth (corrected) edition, Pergamon Press, Oxford (1980) Chap. 9.
6. C. S. Vikram and M. L. Billet, Aberration limited resolution in Fraunhofer holography with collimated beams, *Opt. Laser. Technology*, **21**, 185–7 (1989).
7. H. Royer, P. Lecoq and E. Ramsmeyer, Application of holography to bubble chamber visualization, *Opt. Commun.*, **37**, 84–6 (1981).
8. H. Royer, The use of microholography in bubble chambers (in French), *J. Opt. (Paris)*, **12**, 347–50 (1981).
9. G. W. Stroke, Lensless Fourier-transform method for optical holography, *Appl. Phys. Lett.*, **6**, 210–3 (1965).
10. S. Kikuta, S. Aoki, S. Kosaki and K. Kohra, X-ray holography of lensless Fourier-transform type, *Opt. Commun.*, **5**, 86–9 (1972).
11. B. Reuter and H. Mahr, Experiments with Fourier transform holograms using 4.48 nm x-rays, *J. Phys. E: Sci. Instrum.*, **9**, 746–51 (1976).
12. R. A. Briones, Particle holography at extended distances and μm resolutions, in *Recent Advances in Holography, Proc. SPIE*, **215**, 112–15 (1980).
13. H. Akbari and H. I. Bjelkhagen, Holography in 15-foot bubble chamber at Fermilab, in *Practical Holography, Proc. SPIE.*, **615**, 7–12 (1986).
14. H. Akbari and H. Bjelkhagen, Pulsed holography for particle detection in bubble chambers, *Opt. Laser Technology*, **19**, 249–55 (1987).
15. W. Grabowski, Measurement of the size and position of aerosol droplets using holography, *Opt. Laser Technology*, **15** 199–205 (1983).
16. C. S. Vikram and M. L. Billet, Aberration-free Fraunhofer holography of micro-objects, in *Optical Testing and Metrology, Proc. SPIE*, **954**, 54–7 (1989).
17. J. C. Zeiss, J. Z. Knapp, J. S. Crane, P. Dunn and B. J. Thompson, Holographic measurement of particulate contamination in sealed sterile containers, in *Industrial and Commercial Applications of Holography, Proc. SPIE*, **353**, 106–13 (1982).
18. J. S. Crane, P. Dunn, B. J. Thompson, J. Z. Knapp and J. Zeiss, Far-field holography of ampule contaminants, *Appl. Opt.*, **21**, 2548–53 (1982).
19. W. J. Smith, Image formation: geometrical and physical optics, in *Handbook of Optics*, W. G. Driscoll and W. Vaughn (Eds.), McGraw-Hill, New York (1978) Section 2.

20. B. R. Parkin, Hydrodynamics and fluid mechanics, *Naval Research Reviews*, **XXXV**, 20–34 (1983).
21. C. S. Vikram and M. L. Billet, Fraunhofer holography in cylindrical tunnels: neutralizing window curvature effects, *Opt. Eng.*, **25**, 189–91 (1986).
22. C. S. Vikram and M. L. Billet, Modifying tunnel test sections for optical applications, *Opt. Eng.*, **25**, 1324–6 (1986).
23. C. S. Vikram and M. L. Billet, Window curvature effects and tolerances in Fraunhofer holography in cylindrical tunnels, *Appl. Phys.*, **B40**, 99–102 (1986).
24. L. Levi, *Applied Optics – A Guide to Optical System Design*, Vol. 2., Wiley, New York (1980) Chap. 10.
25. I. Banyasz, G. Kiss and P. Varga, Holographic image of a point source in the presence of misalignment, *Appl. Opt.*, **27**, 1293–97 (1988).
26. R. Naon, H. Bjelkhagen, R. Burnstein and L. Voyvodic, A system for viewing holograms, *Nucl. Instrum. Meth. Phys. Res.*, **A283**, 24–36 (1989).
27. C. Özkul and N. Anthore, Residual aberration corrections in far field in-line holography using an auxiliary off-axis hologram, *Appl. Opt.*, **30**, 372–3 (1991).
28. K. Goto and M. Kitaoka, Aberrations in nonparaxial holography, *J. Opt. Soc. Am. A*, **5**, 397–402 (1988).

Chapter 8

1. B. J. Thompson, G. B. Parrent, J. H. Ward and B. Justh, A readout technique for the laser fog disdrometer, *J. Appl. Meteorology*, **5**, 343–8 (1966).
2. J. B. DeVelis and G. O. Reynolds, *Theory and Applications of Holography*, Addison-Wesley Publishing Company, Reading, Massachusetts (1967) Chap. 8.
3. J. H. Ward, Holographic particle sizing, in *Applications of Lasers to Photography and Information Handling*, R. D. Murray (Ed). SPSE, Washington (1968) pp. 217–33.
4. W. R. Zinky, Hologram techniques for particle-size analysis, *Annals New York Acad. Sci.*, **158**, 741–52 (1969).
5. H. Rieck, Holographic small particle analysis with ultraviolet ruby laser light, in *Engineering Uses of Holography*, E. R. Robertson and J. M. Harvey (Eds.), Cambridge University Press, Cambridge (1970) pp. 261–6.
6. F. Slimani, G. Grehan, G. Gouesbet and D. Allano, Near-field Lorenz–Mie theory and its application to microholography, *Appl. Opt.*, **23**, 41440–8 (1984).
7. P. Dunn and B. J. Thompson, Object shape and resolution in far-field holography, *J. Opt. Soc. Am.*, **69**, 1402 (1979).
8. S. L. Cartwright, P. Dunn and B. J. Thompson, Particle sizing using

far-field holography: new developments, *Opt. Eng.*, **19**, 727–33 (1980).
9. P. Dunn and B. J. Thompson, Object shape, fringe visibility, and resolution in far-field holography, *Opt. Eng.*, **21**, 327–32 (1982).
10. C. S. Vikram and M. L. Billet, Some salient features of in-line Fraunhofer holography with divergent beams, *Optik*, **78**, 80–3 (1988).
11. W. K. Witherow, A high resolution holographic particle sizing system, *Opt. Eng.*, **18**, 249–55 (1979).
12. C. S. Vikram and M. L. Billet, Gaussian beam effects in far-field in-line holography, *Appl. Opt.*, **22**, 2830–5 (1983).
13. S. Johansson and K. Biedermann, Multiple-sine-slit microdensitometer and MTF evaluation of high resolution emulsions. 2: MTF data and other recording parameters of high resolution emulsions for holography, *Appl. Opt.*, **13**, 2288–91 (1974).
14. G. Molesini, D. Bertani and M. Cetica, In-line holography with interference filters as Fourier processors, *Optica Acta*, **29**, 479–84 (1982).
15. C. Özkul, D. Allano and M. Trinité, Filtering effects in far-field in-line holography, *Opt. Eng.* **25**, 1142–48 (1986).
16. C. Özkul, Imagery by an objective with a central semi-transparent obstruction and application to the microholography (in French), *J. Optics (Paris)*, **16**, 29–35 (1985).
17. C. Özkul, D. Allano, M. Trinité and N. Anthore, In-line far-field holography and diffraction pattern analysis: new developments, in *Progress in Holographic Applications, Proc. SPIE*, **600**, 151–8 (1986).

Chapter 9

1. B. J. Thompson, Diffraction by opaque and transparent particles, *J. Soc. Photog. Instr. Eng.*, **2**, 43–6 (1963).
2. B. J. Thompson, Fraunhofer diffraction patterns of opaque objects with coherent background, *J. Opt. Soc. Am.*, **53**, 1350 (1963).
3. G. B. Parrent, Jr. and B. J. Thompson, On the Fraunhofer (far field) diffraction patterns of opaque and transparent objects with coherent background, *Optica Acta*, **11**, 183–93 (1964).
4. B. A. Silverman, B. J. Thompson and J. H. Ward, A laser fog disdrometer, *J. Appl. Meteorology*, **3**, 792–801 (1964).
5. B. J. Thompson, A new method of measuring particle size by diffraction techniques, *Japanese J. Appl. Phys.*, **4**, (Supplement 1), 302–7 (1965).
6. C. Özkul, Traitement optique des figures de diffraction de Fraunhofer pour une analyse avec une ligne de microphotodiodes (in French), *Optica Acta*, **28**, 1543–9 (1981).
7. C. Özkul, D. Allano, M. Trinité and N. Anthore, In-line far-field holography and diffraction pattern analysis: new developments, in

Progress in Holographic Applications, Proc. SPIE, **600**, 151–8 (1986).

8. C. Özkul, A. Leduc, D. Allano and M. Abdelghani-Idrissi, Processing of diffraction patterns scanned with a photodiode array: influence of the optical transfer function of diodes on line width measurements, in *Image Processing III, Proc. SPIE*, **1135**, 131–40 (1989).
9. T. Tsuno and R. Takahashi, Measurement of change in a cross section and position of small particles by diffraction techniques, *Proceedings 10th Int. Congr. High Speed Photography, Nice* (1972) pp. 222–226
10. M. Yonemura, Phase visualization of diffraction patterns and its application, in *Optics and the Information Age, Proc. SPIE*, **813**, 409–10 (1987).
11. C. Özkul, Fresnel diffraction of a gaussian beam by one-dimensional opaque objects, *Optica Acta*, **30**, 505–10 (1983).
12. F. Slimani, G. Grehan, G. Gouesbet and D. Allano, Near-field Lorenz–Mie theory and its application to microholography, *Appl. Opt.*, **23**, 4140–8 (1984).
13. B. J. Thompson and P. H. Malyak, Accuracy of measurement in coherent imaging of particulates in a three-dimensional sample, in *Particle Sizing and Spray Analysis, Proc. SPIE*, **573**, 12–20 (1985).
14. S. Ganci, Fourier diffraction through a tilted slit, *Eur. J. Phys.*, **2** 158–60 (1981).
15. K. Patorski, Fraunhofer diffraction patterns of tilted planer objects, *Optica Acta*, **30**, 673–9 (1983).
16. Yu. V. Chugui and B. E. Krivenkov, Fraunhofer diffraction by volumetric bodies of constant thickness, *J. Opt. Soc. Am. A*, **6**, 617–26 (1989).
17. C. S. Vikram and M. L. Billet, On the problem of automated analysis of particle field holograms: proposal for direct diffraction measurements without holography, *Optik*, **73**, 160–2 (1986).
18. C. S. Vikram and M. L. Billet, Far-field holography at non-image planes for size analysis of small particles, *Appl. Phys. B*, **33**, 149–53 (1984).
19. C. S. Vikram and M. L. Billet, Generalized far-field holography at non-image planes, in *Optical Testing and Metrology, Proc. SPIE*, **661**, 44–49 (1986).
20. A. C. Stanton, H. J. Caulfield and G. W. Stewart, An approach for automated analysis of particle holograms, *Opt. Eng.*, **23**, 577–82 (1984).
21. H. J. Caulfield, Automated analysis of particle holograms, *Opt. Eng.*, **24**, 462–3 (1985).
22. R. A. Dobbins, L. Crocco and I. Glassman, Measurement of mean particle sizes of sprays from diffractively scattered light, *AIIA J.*, **1**, 1882–6 (1963).
23. J. H. Ward, Determination of particle-size distributions from composite diffraction-pattern data, *J. Opt. Soc. Am.*, **59**, 1566 (1968).

24. W. L. Anderson, Particle counting and sizing by Fourier-plane measurements of scattered light, *J. Opt. Soc. Am.*, **59**, 1566 (1968).
25. W. L. Anderson and R. E. Beissner, Counting and classifying small objects by far-field light scattering, *Appl. Opt.*, **10**, 1503–8 (1971).
26. J. Swithenbank, J. M. Beer, D. S. Taylor, D. Abbot and G. C. McCreath, A laser diagnostic technique for the measurement of droplet and particle size distribution, *Prog. Astronaut. Aeronaut.*, **53**, 421–47 (1977).
27. C. Gorecki, Optical/digital analyzer of Fe_2O_3, FeO substrate by Fourier techniques, *Opt. Eng.*, **27**, 466–70 (1988).
28. C. Gorecki, Optical sizing by Fourier transformation *J. Opt. (Paris)*, **20**, 25–9 (1989).
29. B. C. R. Ewan, Fraunhofer plane analysis of particle field holograms, *Appl. Opt.*, **19**, 1368–72 (1980).
30. B. C. R. Ewan, J. Swithenbank and C. Sorusbay, Measurement of transient spray size distributions, *Opt. Eng.*, **23**, 620–5 (1984).
31. C. F. Hess, J. D. Trolinger and T. R. Wilmot, Particle field holography data reduction by Fourier transform analysis, in *Applications of Holography, Proc. SPIE*, **523**, 102–9 (1985).
32. C. F. Hess and J. D. Trolinger, Particle field holography data reduction by Fourier transform analysis, *Opt. Eng.*, **24**, 470–4 (1985).
33. G. Gouesbet and G. Gréhan (Eds.), *Optical Particle Sizing: Theory and Practice*, Plenum Press, New York (1988).
34. E. D. Hirleman and P. A. Dellenback, Adaptive Fraunhofer diffraction particle sizing instrument using a spatial light modulator, *Appl. Opt.*, **28**, 4870–8 (1989).
35. B. H. Miles, P. E. Sojka and G. B. King, Malvern particle size measurements in media with time varying index of refraction gradients, *Appl. Opt.*, **29**, 4563–73 (1990).
36. T. Wang and S. F. Clifford, Use of rainfall-induced optical scintillations to measure path-averaged rain parameters, *J. Opt. Soc. Am.*, **65**, 927–7 (1975).
37. T. Wang, G. Lerfald, R. S. Lawrence and S. F. Clifford, Measurement of rain parameters by optical scintillation, *Appl. Opt.*, **16**, 2236–41 (1977).
38. T. Wang, K. B. Earnshaw and R. S. Lawrence, Simplified optical path-averaged rain gauge, *Appl. Opt.*, **17**, 384–90 (1978).
39. P. E. Klingsporn, Determination of the diameter of an isolated surface defect based on Fraunhofer diffraction, *Appl. Opt.*, **19**, 1435–8 (1980).
40. P. Latimer, Determination of diffraction size and shape from diffracted light, *Appl. Opt.*, **17**, 2162–70 (1978).
41. W. D. Furlan, E. E. Sicre and M. Garavaglia, Quasi-geometrical paraxial approach to Fraunhofer diffraction by thick planar objects using phase-space signal representations, *J. Mod. Opt.*, **35**, 735–41 (1988).
42. J. Uozumi, H. Kimura and T. Asakura, Fraunhofer diffraction by

Koch fractals, *J. Mod. Opt.*, **37**, 1011–31 (1990).
43. Z. Xihua and Z. Yafen, Spectral functions of self-similar structures, *J. Mod. Opt.*, **37**, 1617–28 (1990).
44. S. Nakadate and H. Saito, Particle-size-distribution measurement using a Hankel transform of a Fraunhofer diffraction spectrum. *Opt. Lett.*, **8**, 578–80 (1983).
45. G. Roy and P. Tessier, Particle size distributions obtained from diffraction patterns: contribution of refraction and reflection, *J. Aerosol Sci.*, **21**, 515–26 (1990).
46. G. E. Sommargren and H. J. Weaver, Diffraction of light by an opaque sphere. 1: Description and properties of the diffraction pattern, *Appl. Opt.*, **29**, 4446–657 (1990).
47. R. Nyga, E. Schmitz and W. Lauterborn, In-line holography with a frequency doubled Nd:YAG laser for particle size analysis, *Appl. Opt.*, **29**, 3365–8 (1990).

Chapter 10

1. E. A. Boettner and B. J. Thompson, Multiple exposure holography of moving fibrous particulate matter in the respiratory range, *Opt. Eng.*, **12**, 56–9 (1973).
2. J. D. Trolinger, W. M. Farmer, and R. A. Belz, Multiple exposure holography of time varying three-dimensional fields, *Appl. Opt.*, **7**, 1640–1 (1968).
3. J. D. Trolinger, R. A. Belz, and W. M. Farmer, Holographic techniques for the study of dynamic particle fields, *Appl. Opt.*, **8**, 957–61 (1969).
4. M. E. Fourney, J. H. Matkin, and A. P. Waggoner, *Rev. Sci. Instrum.*, **40**, 205–13 (1969).
5. J. M. Webster, The application of holography as a technique for size and velocity analysis of high velocity droplets and particles, *J. Photo. Sci*, **19**, 38–44 (1971).
6. E. I. LeBaron and E. A. Boettner, Fiber motion analysis by two-pulse holography, *Appl. Opt.*, **19**, 891–4 (1980).
7. R. A. Belz and R. W. Menzel, Particle field holography at Arnold Engineering Development Center, *Opt. Eng.*, **18**, 256–65 (1979).
8. B. B. Brenden, Miniature multiple-pulsed *Q*-switched ruby laser holocamera for aerosol analysis, *Opt. Eng.* **20**, 907–11 (1981).
9. H. Royer and F. Albe, Holographic investigation and measurements in a cloud of moving microparticles, (in French), *Proceedings 10th Int. Cong. High Speed Photography, Nice* (1972) pp. 391–4.
10. H. Royer, An application of high-speed microholography: the metrology of fogs (in French), *Nouv. Rev. Opt.*, **5**, 87–93 (1974).
11. H. Royer, Particle velocity measurements via high-speed holography (in French), *Proceedings 11th Cong. High Speed Photography, London* (1974) pp. 259–64.
12. H. Royer, Holographic velocimetry of submicron particles, *Opt. Commun.*, **20**, 73–5 (1977).

13. T. Uyemura, Y. Yamamoto and K. Tenjinbayashi, Application of pulse laser holography, *Proceedings 13th Int. Cong. High Speed Photography and Photonics, Tokyo* (1978) pp. 346–9.
14. J. A. Liburdy, Holocinematographic velocimetry: resolution limitation for flow measurements, *Appl. Opt.*, **26**, 4250–4 (1987).
15. J. Trolinger, H. Tan and R. Lal, A space flight holography system for flow diagnostics in micro-gravity, in *Focus on Electro-Optic, Sensing and Measurement*, G. Kychakoff (Ed.), Laser Institute of America (1988) pp. 139–43.
16. P. R. Schuster and J. W. Wagner, Holographic velocimetry for flow diagnostics, *Exp. Mech.*, **28**, 402–8 (1988).
17. J. Crane, P. Dunn, P. H. Malyak, and B. J. Thompson, Particle velocity and size measurements using holographic and optical processing methods, in *High Speed Photography and Photonics, Proc. SPIE*, **348**, 634–42 (1982).
18. R. A. Briones and R. F. Wuerker, Holography of solid propellant combustion, *Prog. Astronaut. Aeronaut.*, **63**, 251–76 (1977).
19. R. A. Briones and R. F. Wuerker, Holography of solid propellant combustion, *Proc. SPIE*, **125**, 90–104 (1977).
20. C. Knox and R. E. Brooks, Holographic motion picture microscopy, *Proc. Roy. Soc. Lond B.*, **174**, 115–21 (1969).
21. G. L. Stewart, J. R. Beers, and C. Knox, in *Developments in Laser Technology, Proc. SPIE*, **41**, 183–8 (1974).
22. L. O. Heflinger, G. L. Stewart and C. R. Booth, Holographic motion pictures of microscopic plankton, *Appl. Opt.*, **17**, 951–4 (1978).
23. H. J. Raterink and J. H. J. Van der Meulen, Application of holography in cavitation and flow research, in *Applications of Holography and Optical Data Processing*, E. Marom, A. A. Friesem and E. Wiener-Avnear (Eds.), Pergamon Press, Oxford (1977) pp. 643–52.
24. M. J. Landry and A. E. McCarthy, use of the multiple cavity laser holographic system for FBW analysis, *Opt. Eng.*, **14**, 69–72 (1975).
25. K. Hinsch and F. Bader, Acoustooptic modulators as switchable beam-splitters in high speed holography, *Opt. Commun.*, **12**, 51–4 (1974).
26. W. Lauterborn and K. J. Ebeling, High-speed holocinematography of cavitation bubbles, in *High-Speed Photography (Photonics), Proc. SPIE*, **97**, 96–103 (1977).
27. W. Lauterborn and K. J. Ebeling, High-speed holography of laser-induced breakdown in liquids, *Appl. Phys. Lett.*, **31**, 663–4 (1977).
28. W. Lauterborn, High-speed photography and holography of laser induced breakdown in liquids, *Proceedings 13th Int. Cong. High Speed Photography and Photonics, Tokyo* (1978) pp. 330–3.
29. K. J. Ebeling and W. Lauterborn, High speed holocinematography using spatial multiplexing for image separation, *Opt. Commun.*, **21**, 67–71 (1977).
30. K. J. Ebeling and W. Lauterborn, Acoustooptic beam deflection for

spatial frequency multiplexing in high speed holocinematography, *Appl. Opt.*, **17**, 2071–6 (1978).
31. W. Hentschel and W. Lauterborn, High speed holographic movie camera, *Opt. Eng.*, **24**, 687–91 (1985).
32. W. Lauterborn and W. Hentschel, Cavitation bubble dynamics studied by high speed photography and holography: part one, *Ultrasonics*, **23**, 260–8 (1985).
33. W. Lauterborn and W. Hentschel, Cavitation bubble dynamics studied by high speed photography and holography: part two, *Ultrasonics*, **24**, 59–65 (1986).
34. M. Novaro, High speed holographic camera, *Proceedings 10th Int. Cong. High Speed Photography, Nice* (in French), (1972) pp. 205–6.
35. J. D. Trolinger and M. P. Heap, Coal particle combustion studied by holography, *Appl. Opt.*, **18**, 1757–62 (1979).
36. R. A. Belz and N. S. Dougherty, Jr., In-line holography of reacting liquid sprays, *Proceedings Symp. Engineering Applications of Holography, SPIE, Redondo Beach* (1972) pp. 209–18.
37. B. J. Thompson and P. Dunn, Advances in far-field holography-theory and applications, in *Recent Advances in Holography, Proc. SPIE*, **215**, 102–11 (1980).
38. G. M. Afeti, Bubble break-up at low viscosity liquid surface, in *Cavitation and Multiphase Flow Forum, 1987*, O. Furuya (Ed.), The American Society of Mechanical Engineers, New York (1987) pp. 26–9.
39. J. H. Ward and B. J. Thompson, In-line hologram system for bubble-chamber recording, *J. Opt. Soc. Am.*, **57**, 275–6 (1967).
40. H. Royer, P. Lecoq and E. Ramsmeyer, Application of holography to bubble chamber visualization, *Opt. Commun.*, **37**, 84–6 (1981).
41. H. Royer, The use of microholography in bubble chambers (in French), *J. Opt (Paris)*, **12**, 347–50 (1981).
42. P. Haridas, E. Hafen, I. Pless, J. Harton, S. Dixit, G. Goloskie and S. Benton, in *Applications of Holography, Proc. SPIE*, **523** 110–18 (1985).
43. H. Akbari and H. Bjelkhagen, Pulsed holography for particle detection in bubble chambers, *Opt. Laser Technol.*, **19**, 249–55 (1987).
44. A. M. Bekker, N. I. Bukhtoyarova and M. V. Stabnikov, Holography and image processing in track detectors, in *Holography '89, Proc. SPIE.* **1183**, 506–16 (1990).
45. G. A. Ruff, L. P. Bernal and G. M. Faeth, High speed in-line holocinematography for dispersed-phase dynamics, *Appl. Opt.*, **29**, 4544–6 (1990).
46. T. Murakami and M. Ishikawa, Holographic measurements of velocity distribution of particles accelerated by a shock wave, *Proceedings 13th Int. Cong. High Speed Photography and Photonics, Tokyo* (1978) pp. 326–9.
47. R. Menzel and F. M. Shofner, An investigation of Fraunhofer holography for velocimetry applications, *Appl. Opt.*, **9**, 2073–9 (1970).

48. T. Tsuno and R. Takahashi, Measurement of change in a cross section and position of small particles by diffraction techniques, *Proceedings 10th Int. Cong. High Speed Photography, Nice* (in French) (1972) pp. 222–6.
49. F. M. Shofner, T. G. Russell and R. Menzel, Two-dimensional spatial frequency analysis of the Fraunhofer hologram of a small opaque sphere, *Appl. Opt.*, **8**, 2043–6 (1969).
50. B. C. R. Ewan, Holographic particle velocity measurement in the Fraunhofer plane, *Appl. Opt.*, **18**, 623–6 (1979).
51. B. C. R. Ewan, Particle velocity distribution measurement by holography, *Appl. Opt.*, **18**, 3156–60 (1979).
52. J. Crane, P. Dunn, P. H. Malyak and B. J. Thompson, Particulate velocity and size measurements using holographic and optical processing methods, in *High Speed Photography and Photonics, Proc. SPIE*, **348**, 634–42 (1983).
53. P. H. Malyak and B. J. Thompson, Particle velocity measurements using holographic methods, in *High Speed Photography, Videography, and Photonics, Proc. SPIE*, **427**, 172–79 (1983).
54. P. H. Malyak and B. J. Thompson, Particle displacement and velocity measurement using holography, *Opt. Eng.*, **23**, 567–76 (1984).
55. P. H. Malyak and B. J. Thompson, Obtaining particle velocity and displacement distributions from double-exposure holograms using optical and digital processing, in *Particle Sizing and Spray Analysis, Proc. SPIE*, **573**, 2–11 (1985).
56. J. W. Goodman, *Introduction to Fourier Optics*, McGraw Hill, New York (1968) p. 86.

Chapter 11

1. R. E. Brooks, L. O. Heflinger, R. F. Wuerker and R. A. Briones, Holographic photography of high-speed phenomena with conventional and *Q*-switched ruby lasers, *Appl. Phys. Lett.*, **7**, 92–4 (1965).
2. R. F. Wuerker, L. O. Heflinger and S. M. Zivi, Holographic interferometry and pulsed laser holography, in *Holography, Proc. SPIE*, **15**, 97–104 (1968).
3. R. G. Buckles, Recording of sequential events in a translucent biological objects, in *Holography, Proc. SPIE*, **15**, 111–15 (1968).
4. D. A. Ansley and L. D. Siebert. Coherent pulse laser holography, in *Holography, Proc. SPIE*, **15**, 127–31 (1968).
5. R. F. Wuerker and L. O. Heflinger, Pulsed laser holography, in *The Engineering Uses of Holography*, E. R. Robertson and J. M. Harvey (Eds.), Cambridge University Press, Cambridge (1970) pp. 99–114.
6. R. J. Withrington, Bubble chamber holography, in *The Engineering Uses of Holography*, E. R. Robertson and J. M. Harvey (Eds.), Cambridge University Press, Cambridge (1970) pp. 267–77 (1970).
7. B. J. Matthews, Measurement of fine particulate in pollution con-

trol, in *Developments in Holography, Proc. SPIE*, **25**, 157–68 (1971).
8. J. D. Trolinger, Particle field holography, *Opt. Eng.*, **14**, 383–92 (1975).
9. R. A. Briones and R. F. Wuerker, Holography of solid propellant combustion, in *Advances in Laser Technology for the Atmospheric Sciences, Proc. SPIE*, **125**, 90–104 (1977).
10. R. A. Briones, L. O. Heflinger and R. F. Wuerker, Holographic microscopy, *Appl. Opt.*, **17**, 944–50 (1978).
11. L. O. Heflinger, G. L. Stewart and C. R. Booth, Holographic motion pictures of microscopic plankton, *Appl. Opt.*, **17**, 951–4 (1978).
12. J. D. Trolinger and M. P. Heap, Coal particle combustion studied by holography, *Appl. Opt.*, **18**, 1757–62 (1979).
13. R. A. Belz and R. W. Menzel, Particle field holography at Arnold Engineering Development Center, *Opt. Eng.*, **18**, 256–65 (1979).
14. G. Haussmann and W. Lauterborn, Determination of size and position of fast moving gas bubbles in liquids by digital 3-D image processing of hologram reconstructions, *Appl. Opt.*, **19**, 3529–35 (1980).
15. W. Hentschel, H. Zarschizky and W. Lauterborn, Recording and automatic analysis of pulsed off-axis holograms for determination of cavitation nuclei size spectra, *Opt. Commun.*, **53**, 69–73 (1985).
16. P. Haridas, E. Hafen, I. Pless, J. Harton, S. Dixit, D. Goloskie and S. Benton, Detection of short-lived particles using holography, in *Applications of Holography, Proc. SPIE*, **523**, 110–18 (1985).
17. Y. J. Lee and J. H. Kim, A review of holography applications in multiphase flow visualization, in *Physical and Numerical Flow Visualization*, M. L. Billet, J. H. Kim and T. R. Heidrick (Eds.), The American Society of Mechanical Engineers, New York (1985) pp. 47–58.
18. H. Akbari and H. Bjelkhagen, Pulsed holography for particle detection in bubble chambers, *Opt. Laser Technol.*, **19**, 249–55 (1987).
19. M. Pluta, Holographic microscopy, in *Advances in Optical and Electron Microscopy*, Vol. 10, R. Barer and V. E. Cosslett (Eds.), Academic Press, London (1977) pp. 99–213.
20. J. D. Trolinger, Analysis of holographic diagnostics systems, *Opt. Eng.*, **19**, 722–6 (1980).
21. R. A. Briones, Particle holography at extended distances and μm resolutions, in *Recent Advances in Holography, Proc. SPIE*, **215**, 112–15 (1980).
22. C. S. Vikram and T. E. McDevitt, Single relay lens magnification effects in off-axis particle field holography, *Opt. Eng.*, **27**, 570–3 (1988).
23. K. Hinsch and F. Bader, Acoustooptic modulators as switchable beam-splitters in high-speed holography, *Opt. Commun.*, **12**, 51–5 (1974).

24. D. Gabor, Laser speckle and its elimination, *IBM J. Res. and Dev.*, **14**, 509–14 (1970).
25. J. B. DeVelis and G. O. Reynolds, *The Theory and Applications of Holography*, Addison-Wesley Publishing Company, Reading, Massachusetts (1967) Chap. 5.
26. H. M. Smith, *Principles of Holography*, Second Edition, John Wiley & Sons, New York (1975) Chap. 9.
27. C. S. Vikram, Holographic recording of moving objects, *Nouv. Rev. Opt. Appl.*, **3**, 305–8 (1972).
28. R. L. Kurtz and L. M. Perry, Real-time holographic motion picture camera capable of recording front surface detail from a random velocity vector, *Appl. Opt.*, **12**, 2815–21 (1973).
29. H. Royer, Holography of very far objects, in *Progress in Holographic Applications, Proc. SPIE*, **600**, 127–30 (1986).
30. H. Kogelnik, Holographic image projection through inhomogeneous media, *Bell Sys. Techn. J.*, **44**, 2451–5 (1965).
31. E. N. Leith and J. Upatnieks, Holographic imagery through diffusing media, *J. Opt. Soc. Am.*, **56**, 523 (1966).
32. J. H. McGehee and R. F. Wuerker, Holographic microscopy through tubes and ducts, in *Applications of Holography, Proc. SPIE*, **523**, 70–4 (1985).
33. G. Harigel, C. Baltay, M. Bregman, M. Hibbs, A. Schaffer, H. Bjelkhagen, J. Hawkins, W. Williams, P. Nailor, R. Michaels and H. Akbari, Pulse stretching in a *Q*-switched ruby laser for bubble chamber holography, *Appl. Opt.*, **25**, 4102–10 (1986).
34. R. A. Briones and R. F. Wuerker, Holography of solid propellant combustion, *Prog. Astronaut. Aeronaut.*, **63**, 251–76 (1977).
35. T. J. Marcisz and R. Aprahamian, The holographic investigation of a hypersonic target material subject to impact with a water droplet, in *High Speed Photography, Proc. SPIE*, **97**, 113–22 (1977).
36. W. A. Dyes and J. H. Ward, An ice crystal hologram camera, in *Holography, Proc. SPIE*, **15**, 133–6 (1968).
37. E. C. Klaubert and J. C. Ward, Fourier transform hologram camera with automatic velocity synchronization, in *Evaluation of Motion-Degraded Images*, NASA SP-193, National Aeronautics and Space Administration, Washington, DC (1969) pp. 119–22.
38. W. A. Dyes, P. F. Kellen and E. C. Klaubert, Velocity synchronized Fourier transform hologram camera system, *Appl. Opt.*, **9**, 1105–12 (1970).
39. M. L. A. Gassend and W. M. Boerner, Hypervelocity particle measurement by a holographic method, in *High Speed Photography, Proc. SPIE*, **97**, 104–12 (1977).
40. M. Novaro, High speed holographic camera, in *Proceedings 10th Int. Cong. High Speed Photography, Nice*, (in French) (1972) pp. 205–6.
41. K. M. Hagenbuch and S. P. Sarraf, Holographic image subtraction of high-velocity dust using a Pockels cell, *Opt. Lett.*, **7**, 325–7 (1982).

Index